D0554212

SOCIAL COMMUNICATION AMONG PRIMATES

Social Communication among Primates

EDITED BY STUART A. ALTMANN

THE UNIVERSITY OF CHICAGO PRESS
CHICAGO & LONDON

Library of Congress Catalog Card Number: 65–25120

The University of Chicago Press, Chicago & London
The University of Toronto Press, Toronto 5, Canada

Published 1967
Printed in the United States of America

To the memory of

Professor K. R. L. Hall

Fundamentally, the social sciences are the study of the means of communication between man and man or, more generally, in a community of any sort of beings.

NORBERT WIENER, 1948

Preface

STUART A. ALTMANN

An international symposium on Communication and Social Interactions in Primates was held in Montreal, Canada, December 27 through 31, 1964, during the annual meeting of the American Association for the Advancement of Science. Twenty-six speakers from seven countries discussed various aspects of social interactions in primates—the group of animals that includes not only man, but also the prosimians, monkeys, and apes. The chapters in this book are based on talks that were given at that symposium; several other contributions to the symposium will be published separately.

A surprising diversity of scientific disciplines were represented at the symposium: anthropology (Itani, Ripley, Sade, Simonds, Sugiyama), psychology (Bernstein, Carpenter, Gartlan, Hall, Miller, Rosenblum, Warren), psychiatry and neurology (Jensen, Ploog, Rioch, Robinson), mathematics (J. Altmann), linguistics (Sebeok), and zoology (S. Altmann, Jolly, Kaufmann, Kummer, Moynihan, Rowell, Struhsaker, Vandenbergh). Clearly, the study of primate social behavior has not become the private domain of any one group of scientists. In view of the complexity of the subject matter, this is fortunate.

It was not surprising, then, that the discussions ranged widely. This was true not only of the aspects of social behavior that were investigated, but also of the species of primates that these scientists worked on. Bernstein presented a comparison of social responses in nine species of primates. Two species of prosimian primates, the Aye-aye and the ring-tailed lemur, were discussed by Petter and Jolly, respectively. The New World monkeys were represented by material on howlers (S. Altmann, Carpenter) and titi monkeys (Moynihan); Ploog's report on studies of squirrel monkeys, which is included in this volume, was presented by title at the symposium.

Not unexpectedly, the Old World monkeys were the most frequent subjects of research, but no longer is this work devoted solely to rhesus monkeys.

Although this species continued to get its share of attention (in studies presented by Kaufmann, Koford, Miller, Robinson, Rowell, Sade, and Vandenbergh), there were also reports on Japanese macaques (Itani, Tsumori), pigtail macaques (Jensen, Rosenblum), bonnet macaques (Rosenblum, Simonds), hamadryas baboons (Kummer), sub-Saharan baboons (J. Altmann, Hall, Rowell), vervets (Gartlan, Struhsaker), patas (Hall), and langurs (Ripley, Sugiyama).

The great apes were represented by George Schaller's film on the mountain gorilla and Jane Goodall's film on chimpanzees. The latter was shown by the National Geographic Society as their annual film presentation.

Many aspects of the behavior of these primates were discussed—the neurological and physiological concomitants of behavior (Miller, Ploog, Robinson); ontogeny (Jensen, Rosenblum); dominance, aggression, and subordination (J. Altmann, Gartlan, Kaufmann, Kummer, Sade); social signaling (S. Altmann, Hall, Itani, Moynihan, Struhsaker); reproductive behavior (Jolly, Rowell); social dynamics (Bernstein, Petter, Ripley, Sugiyama, Tsumori, Vandenbergh); and general research strategy (Carpenter). This profusion of species and of aspects under study was an indication of two healthy research trends in this area: primate behavior is being studied both in its many guises and along its many frontiers. In both respects, the study of primate behavior is one of the most rapidly developing areas of biological research (Fig. 1).

Despite this great diversity of subject matter and approach, certain major themes developed as the symposium progressed. One is that primate societies are both varied and variable. Changes in behavior may come as a result of an annual breeding season (Jolly), the female's estrus cycle (Rowell), the formation of new groups (Vandenbergh), or the take-over of an existing group by a new adult male (Sugiyama). Relatively small changes in the social and physical environment of infant primates may significantly affect their later behavior (Jensen, Rosenblum), and thereby considerably alter the social structure.

Differences between species of primates may be even greater than such intraspecific variations. For example, the contrasts are striking between the nocturnal, one-family groups of Aye-ayes (described by Petter), the one-male, multifemale groups of patas monkeys (described by Hall), and the multimale, multifemale groups of baboons and macaques (described by several participants).

Second, the symposium revealed a strong trend toward naturalistic or biological approaches and these were, in many cases, explicitly evolutionary. When one considers the diversity of the speakers, this trend is quite remarkable. It was evident in the discussions of the problem of natural units of social behavior. Beyond that, this trend toward a naturalistic perspective was revealed in the fact that even those who had carried out the most minute analyses of behavior in laboratory settings were both aware of and interested in the extent to which their results were relevant to an understanding of how these animals solve their own, naturally occurring problems, and the extent to which their experimental setup provided an adequate paradigm of processes of adaptation.

A third research trend that was revealed by the symposium was the tendency toward quantitative studies. For several species of primates, the initial, descriptive phase in the research cycle has advanced sufficiently far, in both field and laboratory, that more exact analysis can now be carried out. Such analysis is particularly important and fruitful when dealing with primates, in part because their behavior is so multiply contingent and hence highly

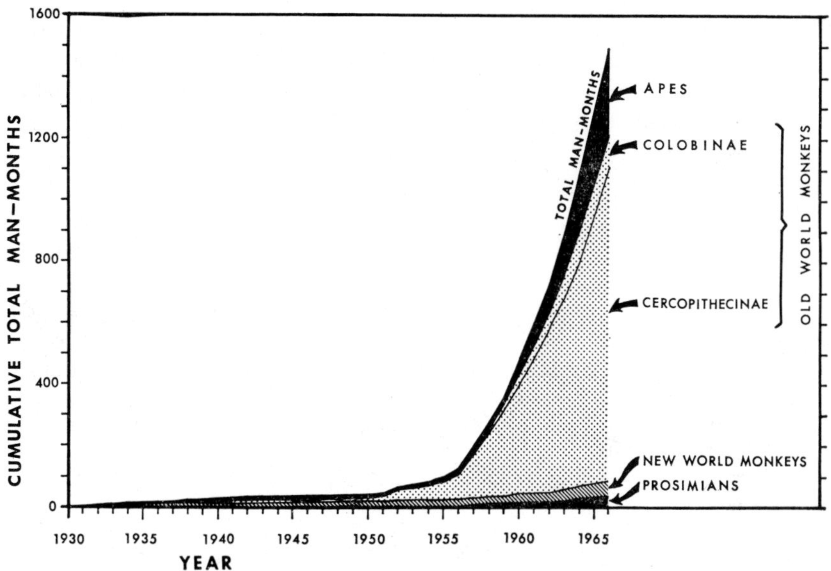

FIG. 1.—Cumulative man-months devoted to field studies of ecology and behavior of nonhuman primates (after Altmann 1966). For the purpose of this tabulation, field studies include studies in natural habitats, in artificial habitats to which animals have voluntarily adapted but to which they are not confined by barriers (for example, Indian temple monkeys), and studies on quasinatural island colonies (for example, Cayo Santiago). It does not include studies in zoos, indoor or outdoor fenced enclosures, or laboratories. It is based on results of an inquiry sent to all known primate fieldworkers early in 1965 and on reports in the literature. The results for 1965 are incomplete, hence are an underestimate.

The data indicate that the growth rate of research in this area has been fairly steady since 1955, with a doubling of actual (not cumulative) research activity every 5 years; the incomplete data for 1965 were not counted in making this calculation. Thus, the growth rate in this research area is now much faster than in scientific research as a whole, which doubles about every 15 years (Price 1963). The amount of research in this field during the period 1962 through 1965 alone was greater than all the research that preceded it!

variable, and in part because of the complex, multi-individual interactions that occur within primate social groups.

Another research trend was part of the *raison d'être* for the organization of the symposium and is epitomized by Norbert Wiener's quotation on the epigraph page of this book. "Fundamentally, the social sciences are the study of the means of communication between man and man or, more generally, in a community of any sort of beings." The concept of a society as a communication system has far-reaching research implications that we are just beginning to appreciate.

One last research trend in research on primate behavior that was revealed by the symposium was so pervasive as to pass almost unnoticed. No one, either among the participants or the audience, was sufficiently impressed to comment on such seeming incongruities as a psychiatrist and a zoologist discussing infant development, a linguist and a zoologist discussing communication in monkeys, or a mathematician and an anthropologist discussing the control of aggression. It would seem that, in studies of animals of this group at least, the old barriers to communication between members of various scientific disciplines are, if not dead, at least moribund.

STUART A. ALTMANN

REFERENCES

Altmann, Stuart A. 1966. Primate field studies: development of a research area. In preparation.

Price, Derek J. deS. 1963. *Little science, big science.* New York and London: Columbia University Press.

Wiener, Norbert. 1948. Time, communication and the nervous system. *Ann. N.Y. Acad. Sci.* 50:197–220.

Contents

PART III CAUSAL MECHANISMS

PART IV SOCIAL DYNAMICS

PART V COMMUNICATION PROCESSES

Part I

REPRODUCTIVE BEHAVIOR

1

BREEDING SYNCHRONY IN WILD *Lemur catta*

ALISON JOLLY

First of all, I should like to briefly introduce *Lemur catta,* its looks, and a few of its gestures. Next, I will give a detailed account of the behavior of one troop in the wild during March and April, 1964, which included the 1964 breeding season. Third, I will summarize the evidence for polyestry and for breeding synchrony in captive and wild prosimians. Finally, I will speculate on the possible mechanisms of breeding synchrony.

The location, hours, and general method of study are described in Jolly (in press).

Lemur catta looks rather like a Paris-styled raccoon (Plate 1.1). It is the size of a cat; it has light, bright gray fur, a black-and-white mask, and a black-and-white ringed tail. It is usually quadrupedal, with striped tail dangling as it leaps about the branches, or raised in a jaunty question mark as it strolls on the ground. A troop of twenty *L. catta* give the impression more of squirrels than of primates.

Their gestures and vocalizations are also different from most primates, although they may be easily homologized with those of other Lemuroidea (Andrew 1963, Jolly in press). Grooming, for instance, is only licking, or scraping with the "tooth-comb" (the procumbent lower canines and incisors). Lemuroidea hold on to each other's fur or limbs while grooming but do not use their hands to part the fur or to pick up particles.

Scent marking includes a whole range of gestures and is associated with much visual display. The most dramatic confrontation between *L. catta* is the male "tail waving." A male first rubs his tail several times between his antebrachial

Alison Jolly, New York Zoological Society, Bronx Park, New York.

glands—the spurs on his forearms—then stands on all fours, brings his tail forward over his head, and quivers it, pointed toward another animal, which usually spats and runs away. Occasionally, during a stink fight, two males tail-wave at once or in succession, facing each other (at a safe distance) like the two halves of a heraldic design.

More serious fighting occurred in jump fights, when animals leaped at each other on their hind legs, slashing downward with the canines. By the end of the breeding season, most males had 4- or 5-cm. gashes on limbs or flank.

In this chapter, several kinds of interactions are grouped as *agonistic:* spats, chasing, and jump fights. The group of *friendly* interactions includes contact, grooming, and play. *Scent marking* includes genital marking of branches, palmar marking of branches, marking the tail, and tail-waving. (These are fully described in Jolly in press.) The placement of the interactions in these categories could be disputed: in particular, tail-waving was not included among agonistic interactions because it occasionally seemed to indicate sexual approach toward a female, although it usually was an obvious high-intensity threat. Scent marking I generally thought of as "self-advertisement" to avoid oversimplifying it to its frequent aggressive content.

BEHAVIOR OF TROOP 1 IN MARCH–APRIL, 1964

Troop 1 of *L. catta* was observed for about 150 hours from March 17 to May 1, 1964. It consisted of five males, nine females, one subadult of each sex, and seven juveniles, a total of twenty-three animals. (It also had an *L. macaco collaris* male, an escaped captive which had joined the troop. The *L. macaco's* attempts to approach *L. catta* females precipitated much chasing and fighting but are not included in the data here.)

Male genitalia apparently change little during the year, nor are the testes retracted (Hill 1953). However, males were only seen to have erections beginning with week before mating. It seems very probable that the males are sexually seasonal.

Three associated changes took place in the external genitalia of females. The genitalia as a whole grew larger, from about 1.5 cm in length in resting condition to 3 cm in length in full estrus. Second, the unpigmented center enlarged by comparison with the black rim—at first invisible—then smaller in diameter than the rim, finally large with only a thin, black outline. Third, the center flushed from pale caucasian skin color to bright reddish-pink. These three changes took place at different rates in different animals.

Figure 1 shows two of these changes in the genitalia of individual females; that is, the relative diameter of center and color of center, which were easy to judge by eye. All the females flushed pink at some time, but some never reached large center size.

Unfortunately, at first, individual females could not be recognized reliably;

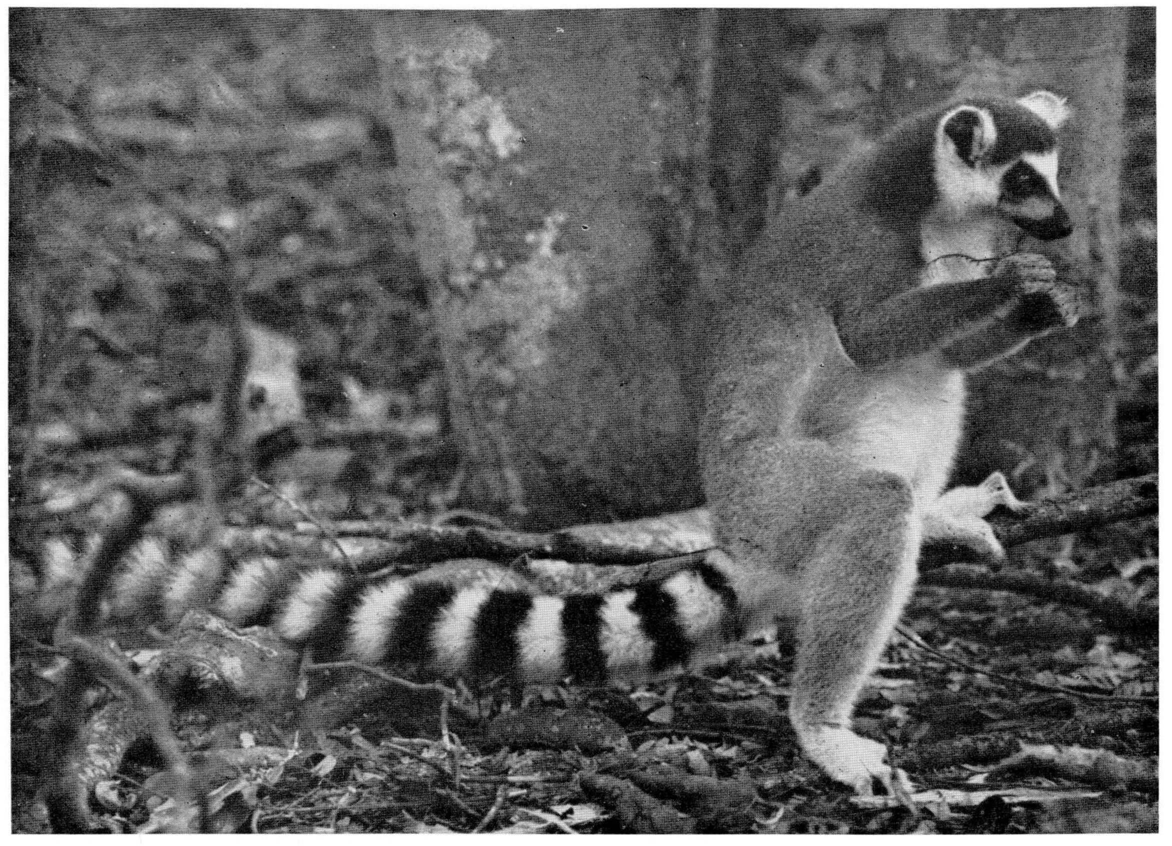

PLATE 1.1.—*Lemur catta.* (Photographed in Madagascar by Dr. C. H. Fraser Rowell.)

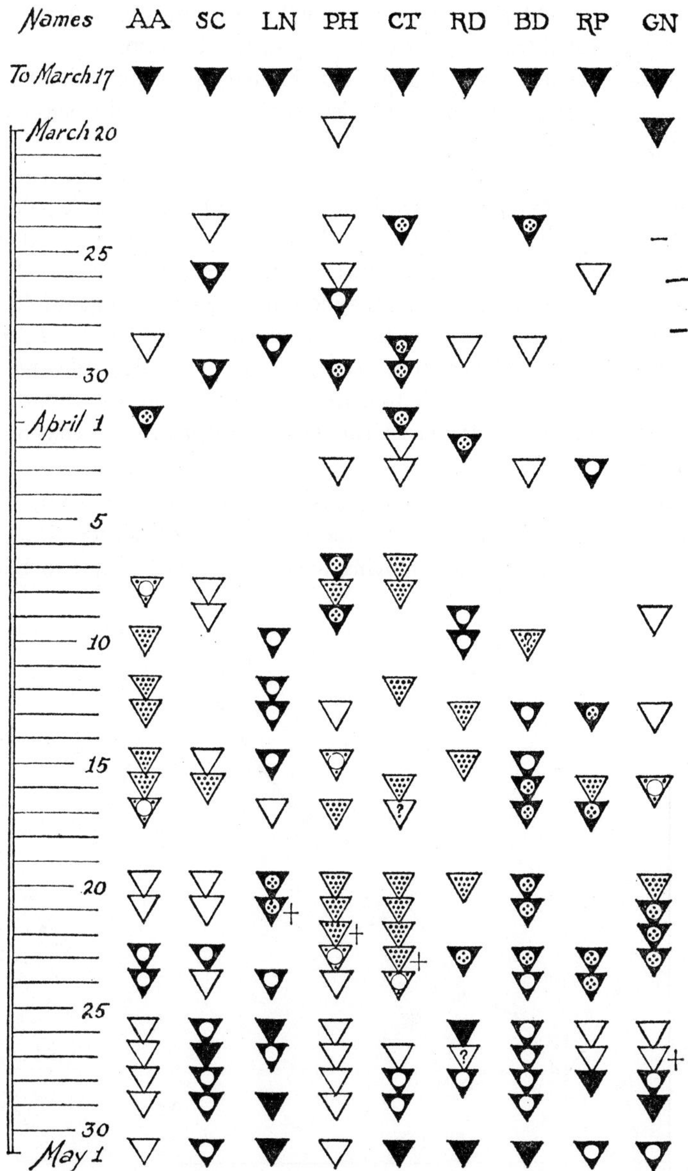

FIG. 1.—Genitalia of individual females in Troop 1. ▼ entirely black; light center, diameter smaller than black rim; ▽ light center, diameter larger than black rim; pink center, diameter smaller than black rim; pink center, diameter larger than black rim; large center, intermediate color; + observed mating.

so the data are incomplete. In the first week, however, troop counts indicate that all females were black, or with very small light centers, as indicated by the row of black inverted triangles across the top of the diagram.

Five of the nine females had a short preliminary pinkening about 3 weeks before the main estrus. Then they faded again. (This is also supported by troop counts.)

All nine of the females were in full pink estrus within 10 days of each other. Observed mating took place on April 21, 22, 23, and 27. Those females which were seen the day after they had mated had already begun to lose their pink color so it is possible that they are either normally receptive for only 1 day, or that, as soon as fertilization occurs, they lose both color and receptivity. (The last female seen mating had been pursued for some days by the *L. macaco*, who may have prevented her from mating when still pink.) Judging by the genitalia, then, the earliest female to mate would have been Aunt Agatha (AA) very probably on April 16, the last Grin (GN), on April 27, the mating period thus covering 12 days.

The troop worked up to their 12-day period of breeding through a crescendo of social interactions. Figure 2 shows the total observed friendly interactions (contact, grooming, play) in a moving average summed over two blocks of

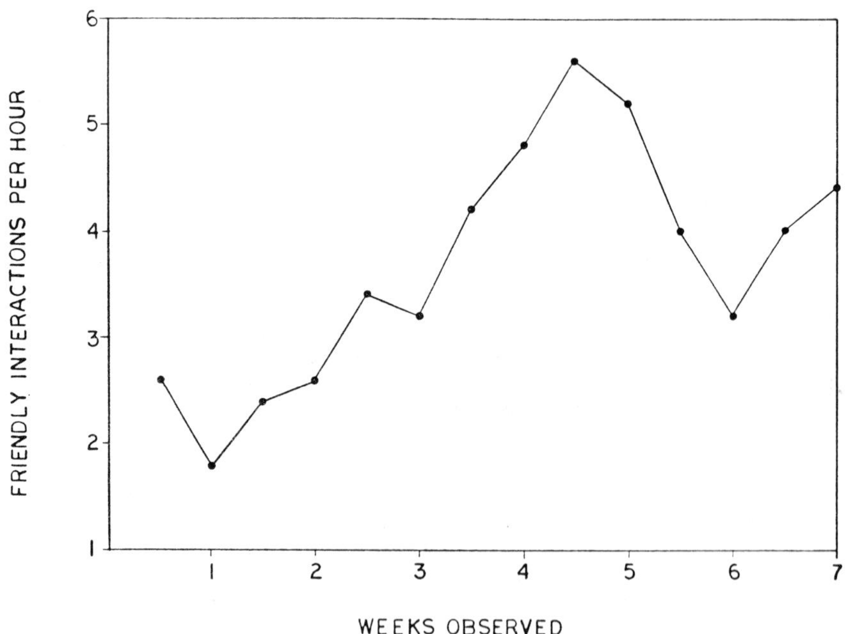

FIG. 2.—Troop 1: Observed friendly interactions. Moving average of two successive ½-week blocks, standardized for day length and hour of observation. All observed mating took place in week 6.

one-half week (3 days) each. All observed mating took place in the two blocks summed as week 6. Observations in 1963 gave a base rate of about two friendly interactions observed per hour out of the breeding season. This climbed rapidly to 5.5 per hour in week 4, then dropped as contact-grooming behavior was replaced by sexual behavior.

Figure 3 breaks down the proportions of friendly interactions among adults. The curves are adjusted for number of adults in the troop. If they all interacted at random, the expected proportions of male-male, male-female, and female-female interactions are each set equal to 1. At first females tended to sit by and groom other females, while males groomed males—particularly the two subordinate males. At week 3½, this changed abruptly, and males took to grooming females. During weeks 5½ and 6½ the proportions began to return to their original values.

The total number of agonistic interactions (Fig. 4) likewise climbed steeply in the month before mating, from 5.5 per hour to 21 per hour. These could be counted by ear, from the number of spat calls, even when the animals were not observed and so these counts represent the total for the troop. Again, the number decreased during the period of mating, as minor spats were replaced by stink fights and jump fights. Figure 5 shows the proportions of observed

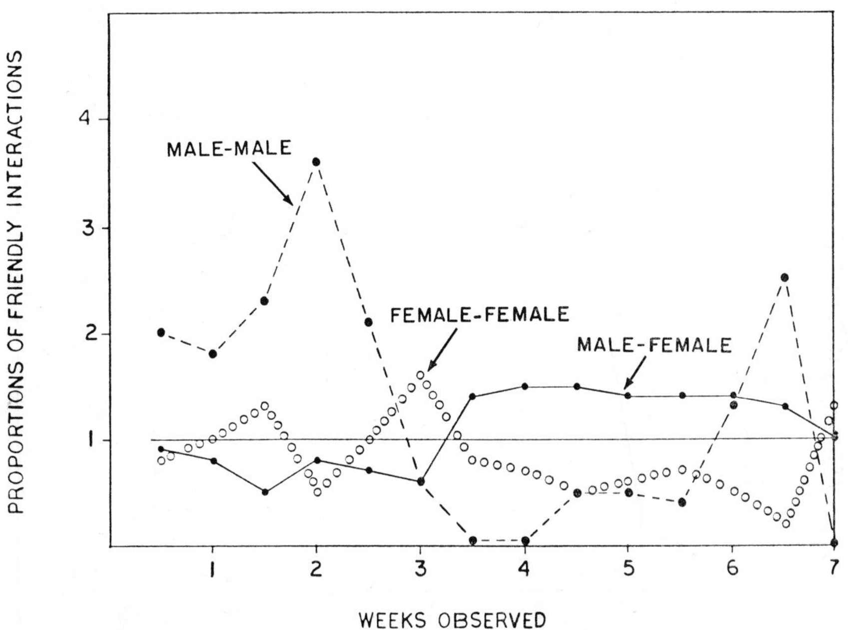

FIG. 3.—Troop 1: Proportions of friendly interactions. (Moving average of two successive ½-week blocks.) On the hypothesis of random interactions between all individuals, the expected proportions of male-male, male-female, and female-female interactions are each set equal to 1.

agonistic encounters between adults, with the enormous relative increase in male-male aggression. (The initial small peak in weeks 2 and 3 corresponds in time to the initial slight flushing of the female genitalia.)

Finally, Figure 6 gives the increase in observed scent marking, which climbed slowly at first, then reached a sharp peak with the stink fights between males during the period of mating.

All of these graphs taken together give a picture of increasing excitement in the troop: an increasing number of social interactions of all kinds, expressed, of course, in different ways by the males and females. This rise began roughly 4 weeks before the week of observed mating, starting at or near the low year-round levels for *L. catta*. After the brief breeding season, total aggression and scent marking again began to fall, although the number of contact-grooming interactions rebounded a little. Observations in 1963 indicate that 3 to 4 weeks

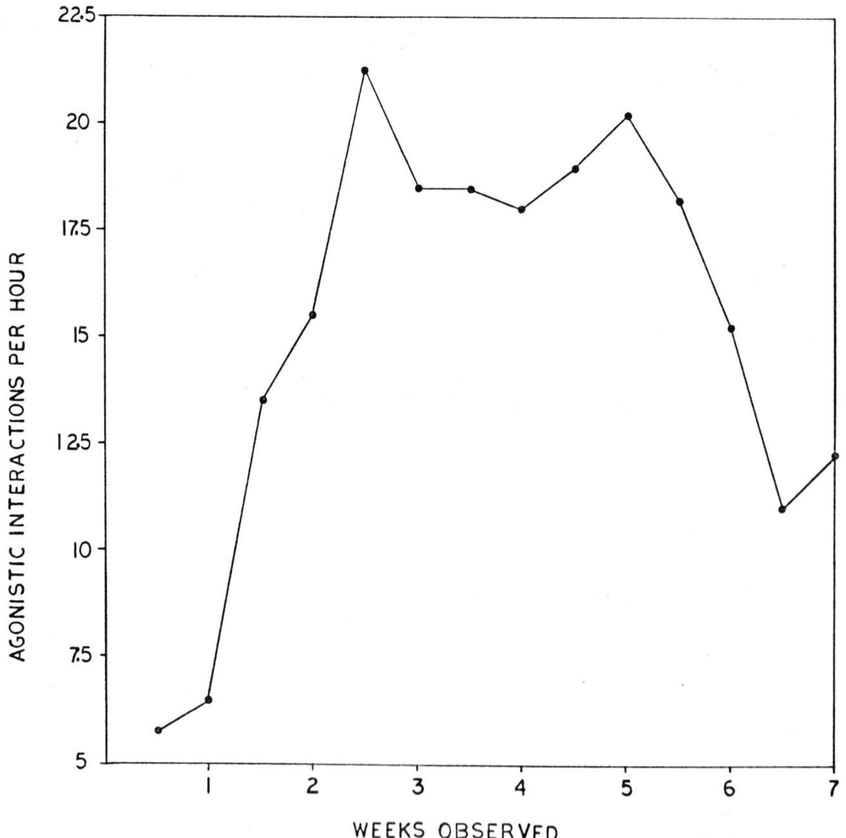

FIG. 4.—Troop 1: Total agonistic interactions. Moving average of two successive ½-week blocks, standardized for day length and hour of observation.

after mating the troop experiences its lowest level of social interaction—in fact, it acts exhausted. It therefore seems unlikely that there is any marked post-seasonal cycle after the main period of estrus.

Mating itself occurred with the maximum participation, or rather interference, by the troop. Two to four males gathered around the receptive female, alternately approaching her and chasing each other. The eventual mate either chased, stink fought, or jump fought with the other males for 5 to 25 minutes before mounting the female. As mentioned above, the males may slash each other deeply in these fights. The female might "present" with tail and hindquarters raised, but then she would crouch in a flat "doormat" position on a branch to be mounted. Presenting and mounting rarely occurred in *L. catta* except as a prelude to copulation, or at least in periods of great sexual excitement. After only 2 to 3 seconds, the male leaped off to chase rival males. Three to twelve such mountings, with or without intromission, took place, the male

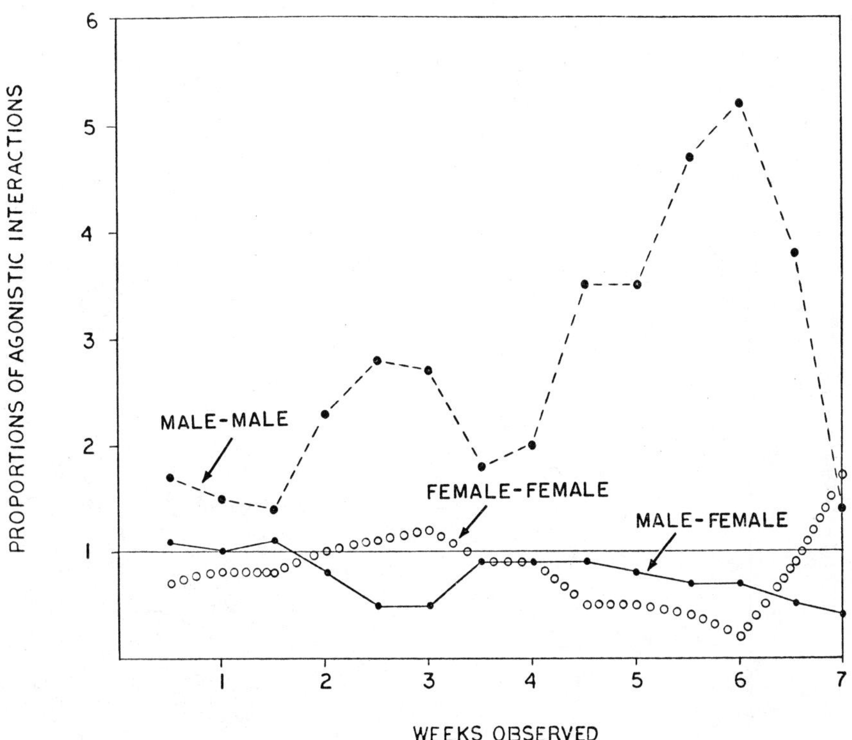

FIG. 5.—Troop 1: Proportions of agonistic interactions. (Moving average of two successive $\frac{1}{2}$-week blocks.) On the hypothesis of random interaction between all individuals, the expected proportions of male-male, male-female, and female-female interactions are each set equal to 1.

tearing away to chase his rivals up to twenty times between approaches to the female.

Twice juveniles, once a subadult female, and once an adult female with pink genitalia approached a mating pair during this period and flung themselves on top of the male, cuffing and clawing. The male perforce dismounted to chase his attacker, but on two occasions when the male cuffed a juvenile his female turned to cuff him.

At each interruption, the female moved away by one branch or so (1–2 m). Finally, when the pair thus withdrew some 10 m from the edge of the troop, copulation took place in two to four periods of about 3 minutes each. Between periods, the male and female sat without touching each other, each licking its own genitalia, or else they gently groomed each other's fur.

After this they might leave together, presumably forming a consort pair for several hours, or else they return at once to the troop. On one occasion, a female copulated with three different males in succession, but other females were with only one male at a time, in either the morning or afternoon.

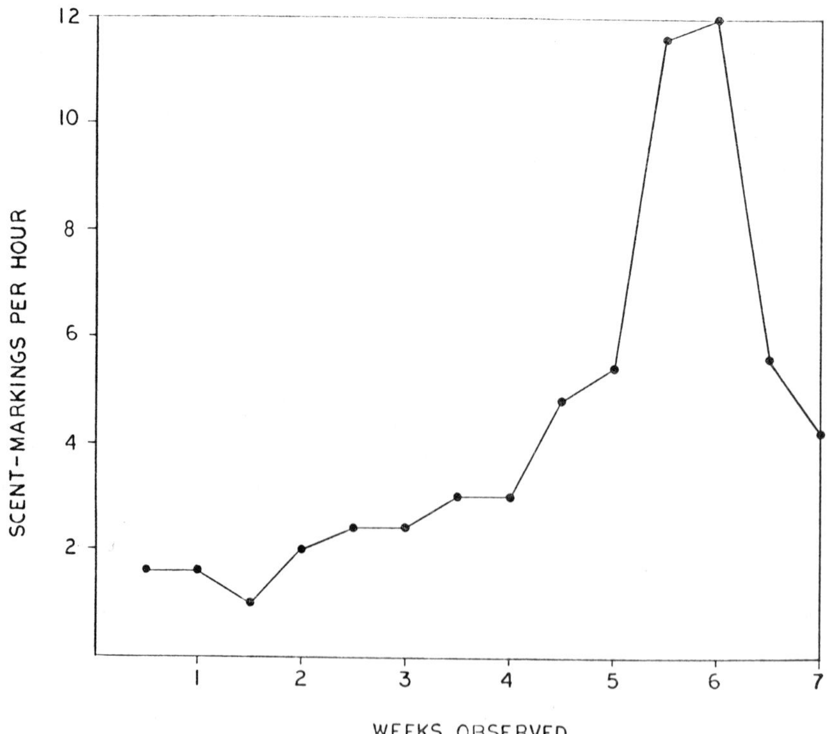

FIG. 6.—Troop 1: Observed scent marking. Moving average of two successive $\frac{1}{2}$-week blocks, standardized for day length and hour of observation.

Subordinate male no. 5 accomplished three of six observed matings. This may mean either that (1) the dominant males had previously mated with other, unobserved females; (2) the dominant males had mated, perhaps during the night, with the same female; or (3) the dominance hierarchy does not hold for access to receptive females. I favor the third possibility, because, during periods of observed mating, subordinate males successfully challenged and chased dominant ones. The adult male hierarchy in all other *L. catta* disputes is rigidly linear; so it seems likely that there is here a real breakdown of dominance.

POLYESTRY AND SYNCHRONY IN PROSIMIANS

Since the work of Zuckerman (1932, 1933), it has been generally assumed that all prosimians are seasonally polyestrous. There are many data to sunpport this assumption: in captivity, 3- to 6-week cycles appear in *Tarsius* (Catchpole and Fulton 1943); in almost all the lorisoids;[1] in the small, nocturnal lemuroids (Petter-Rousseaux 1962, 1964); and in *Lemur* itself (Cowgill *et al.* 1962). Members of all three branches of prosimians then seem to be seasonally polyestrous under laboratory conditions.

The only prosimians which have been studied in the wild are *L. macaco* (Petter 1962), *L. catta*, and *Propithecus verreauxi*. These are all diurnal and highly social Lemuroidea, although *Propithecus* belongs to the family Indriidae, not Lemuridae. Of these species, the only one seen mating in the wild is *L. catta*. (*Galago senegalensis* is known to breed seasonally [Butler 1957]).

In *L. catta* itself, breeding synchrony seems definite. Females of adjacent troops 2, 3, and 4 had bright pink genitalia in the same week as the observed mating in Troop 1. On April 23, in Troop 2, all females had pink genitalia, and mating took place.[2] The flushing lasted a few days longer in Troop 2. All Troop 1 females had faded by April 25, while two of six Troop 2 females remained pink on May 1.

L. catta births in 1963 reinforce the evidence for widespread synchrony. Four of seven infants in Troop 1 were born in one week, between August 20 and 27. In the same week infants were first born in troops 2 and 3, and also in another troop some 30 km away.

P. verreauxi shows a similar birth peak. Five infants whose births could be dated within 3 to 4 days were born within 10 days of one another, between July 5 and July 15, 1963. A sixth, born sometime between July 13 and July 26, was noticeably smaller than the first five. This gives a size criterion for supposing that the seven other infants seen in the region, including four in distant, isolated forests, were also born in about the same 10 days, July 5 to July 15. It

[1] Manley, personal communication.

[2] T. E. Rowell and C. H. F. Rowell, personal communication.

seems likely then that *Propithecus,* like *L. catta,* has synchronized breeding in the wild.

Petter (1962) saw the same crescendo of scent marking and chasing in wild *L. macaco* as in *L. catta,* but did not see mating, possibly leaving the area before mating actually occurred. Cowgill *et al.* (1962) report some synchrony of estrus in captive *L. macaco* (*L. fulvus*).

This means that breeding synchrony is possible, or probable, in all three of these advanced, social lemuroids. Furthermore, *L. macaco* is seasonally polyestrous in captivity but may well breed synchronously, once a year, in the wild. *L. catta* very probably does the same, since even in the wild some females' genitalia flushed slightly and then faded 3 weeks before full estrus and mating.

A reasonable hypothesis then is that prosimians are fundamentally seasonally polyestrous but that the social lemuroids have superimposed breeding synchrony on the primitive condition.

POSSIBLE MECHANISMS

There are two kinds of mechanisms, social and environmental, which might synchronize estrus in *Lemur.* I have tried to show the rise of social interaction before the breeding season, the spiraling excitement which must be communicated from animal to animal. *L. catta* has a wide variety of olfactory signals, from the genital marking of branches to quivering their scented tails in another animal's face. They sniff each other's genitalia and especially the marked branches, often superimposing their own marks. Since olfactory stimuli alone can induce estrus in mice (Parkes and Bruce 1961) and play a major role in reproduction of other mammals, it seems likely that the troop's synchrony depends on olfactory and other social communication within the troop.

However, there must also be an external trigger to account for the similar 1963 birth peaks in widely distant troops. Day length is the obvious one; sunset advanced by almost an hour during the March–April study period in this forest, which lay at about 25° S. Lat. All other climatic factors fluctuate drastically in this region from year to year and seem unlikely triggers.

Cowgill *et al.* (1962) suggested that *L. macaco* (*L. fulvus*) breeds at full moon. Of fifteen matings observed in their laboratory, thirteen fell within the 5 days before or after a full moon. Estral reddening of the two females observed lasted two to six days. Their lemurs' first three estral periods, after they were moved to the Northern Hemisphere, did not coincide with the moon, but after this, six of seven estral periods overlapped a day of full moon.

Evidence from the wild agrees with the moon hypothesis, but hardly proves it. *L. catta* Troop 1 females were in estrus in the week of full moon of April 9, 1963, and during the 2 weeks preceding the full moon of April 26, 1964. In Troop 2, in 1964 females were in estrus during the week before and at least 4 days after the moon. Observed mating in Troop 1 took place from 5 days before to 2 days after the moon. This fits the hypothesis adequately, but if the breeding season lasted about two weeks a year, there is 50 per cent chance in any year that the season overlaps a full moon! One confirming point is that Troop 1 females first grew pink around the preceding full moon of March 28.

Other laboratory data, such as the series given by Petter-Rousseaux (1962) for Cheirogaleinae, by Lowther (1940) for *Galago,* and by Catchpole and Fulton (1945) for *Tarsius* are randomly distributed. This is a case where negative laboratory data prove little, except that the lighting or social conditions were unnatural, while positive laboratory results are highly suggestive. In the same way, positive results in the wild are statistically inadequate, but one season's field observation of synchronous breeding which did not overlap full moon would disprove the whole idea.

CONCLUSION

There are three obvious implications of these data.

The first is that Zuckerman's (1932, 1933) original hypothesis, that primate societies originated with the cohesive force of year-round breeding, is no longer tenable. *L. catta* have the usual primate form of year-round, lifelong society; yet they have always bred seasonally and now apparently for no more than 2 weeks a year.

The second implication is that there is a challenge to elucidate the mechanism of synchronous breeding, whether social, by day length, or by the moon, as well as the evolutionary function of synchronous breeding. I have not talked at all about its function, for I could think of no explanation which fully satisfied me.

The last is that lemurs and other prosimians are rewarding animals to study. They not only occupy a crucial place in the primate evolutionary tree, they are not only unique animals threatened with extinction, they are not only furry and fubsy and appealing—they are a mine of unforeseen behavior, and of insights for the primatologist. Since I will not be doing more field work in the near future I hope that other primatologists will do it instead, for the prosimians clearly offer us many more discoveries.

ACKNOWLEDGMENTS

I should like to thank particularly Dr. T. E. Rowell, both for her observations of *L. catta* and for her helpful criticisms of this manuscript, and Dr. C. H. Fraser Rowell for his photography of wild lemurs. The people and institutions who made this work possible include the de Heaulme family, the New York Zoological Society, the American Lutheran Mission, the Malagasy Government, and many others. Research was financed by U.S. National Science Foundation grant GB-169.

REFERENCES

Andrew, R. J. 1963. The origin and expression of the calls and facial expressions of the primates. *Behaviour* 20:1–109.

Buettner-Janusch, J. 1964. The breeding of galagos in captivity and some notes on their behaviour. *Folia Primat.* 2:93–110.

Butler, H. 1957. The breeding cycle of the Senegal galago, *Galago senegalensis senegalensis*. *Proc. Zool. Soc. London* 129:147–49.

———. 1960. Some notes on the breeding cycle of the Senegal galago, *Galago senegalensis senegalensis*. *Ibid.* 135:423–30.

Catchpole, H. R., and Fulton, J. F. 1943. The oestrous cycle in tarsiers. *J. Mammal.* 24:90–93.

Cowgill, U. M., Bishop, A., Andrew, R. J., and Hutchinson, G. E. 1962. An apparent lunar periodicity in the sexual cycle of certain prosimians. *Proc. Natl. Acad. Sci. U.S.* 48:238–41.

Hill, W. C. O. 1953. *Primates: comparative anatomy and taxonomy.* V. I. Strepsirhini. Edinburgh: Edinburgh University Press.

Jolly, A. in press. *Lemur behavior: a Madagascar field study.* Chicago: Univ. of Chicago Press.

Lowther, F. de L. 1940. A study of the activities of a pair of *Galago senegalensis moholi* in captivity, including the birth and post-natal development of twins. *Zoologica* 25:433–62.

Parkes, A. S., and Bruce, N. M. 1961. Olfactory stimuli in mammalian reproduction. *Science* 1934:1049–54.

Petter, J. J. 1962. Recherches sur l'écologie et l'éthologie des Lémuriens malgaches. *Mem. Museum Natl. Hist. Nat. (Paris) Ser. A.* 27:1–146.

Petter-Rousseaux, A. 1962. Recherches sur la biologie de la réproduction des primates inférieurs. *Mammalia* 26 (Suppl. 1): 1–88.

———. 1964. Reproductive physiology and behavior of the Lemuroidea. In *Evolutionary and genetic biology of the primates,* Vol. 2. Ed. J. Buettner-Janusch. New York and London: Academic Press.

Zuckerman, S. 1932. *The social life of monkeys and apes.* London: K. Paul.

———. 1933. *Functional affinities of man, monkeys and apes.* London: K. Paul.

2

FEMALE REPRODUCTIVE CYCLES AND THE BEHAVIOR OF BABOONS AND RHESUS MACAQUES

THELMA E. ROWELL

INTRODUCTION

The menstrual cycle in monkeys is described by Zuckerman (1930, *et seq.*), and its physiology has since been intensively studied, mainly in the rhesus macaque but to some extent in other species. The effect of this cycle on the intraspecific behavior of these highly social animals has not received comparable attention, although a general account of macaques was given by Carpenter (1942). The two types of study are nearly incompatible because natural social interactions can only be studied in established, undisturbed groups, and the emotional effects of catching and handling females from such groups can be sufficient to disrupt their cycles. Nonetheless, the presence of sexually active females has been suggested as the main cohesive (Zuckerman) or disruptive (Washburn and DeVore 1961) factor in the social life of monkeys and apes.

It is common knowledge that baboons mate when the females are swollen. The object of the present study is to find out why—to make a quantitative description of the behavior changes in small undisturbed groups of monkeys during the cycles of their females. Interest has focused particularly on the "sexual swelling" which occurs rather unsystematically throughout the Catarrhina, on the function of the swelling, and on how the behavior of species whose females swell differs from the behavior of others. The effect of environment on cycling female primates must also be considered.

In wild baboons, menstrual cycles were relatively rare in the life of the adult female. Noncycling periods of a 6-month pregnancy plus a 5- to 6-month

Thelma E. Rowell, Department of Zoology, Makerere University College of the University of East Africa, Kampala, Uganda.

lactation interval were separated by periods of cycling—one to three cycles, each about 5 weeks long. Baboons, kept for long periods in captivity, undergoing repeated cycles without breeding, seem to develop relatively enormous swellings (Plate 2.1)—far larger than would seem practicable in the wild. There is the possibility here of an environmental effect, and if there is such a physiological effect, there is the possibility of a behavioral one also. In rhesus too it appears that there may be differences between sexual behavior in cages and in free-range environments (Kaufmann 1965).

MATERIALS

The data on captive rhesus monkeys are mainly those already published on the colony at Madingley, Cambridge (Rowell 1963), supplemented by observations on two adolescent females living with a male in Kampala.

Field observations were made on three adjacent troops of baboons in the Queen Elizabeth National Park, Uganda, during a period of 20 months. Figures given refer to total interactions observed in about 250 hours of observation. There were about one hundred thirty baboons in the population, roughly divided into classes as follows: one-third were adults in a 1:1 sex ratio; one-third were large juveniles in a sex ratio of 2 males: 1 female (because males take longer to mature); and one-third were small juveniles and babies in about equal sex ratios. There was no breeding season apparent in this population: pregnant, lactating, and cycling females were always present. A modal annual birth distribution is not ruled out—huge figures would be required to do so. (Rowell 1965, 1966.)

A caged group of baboons kept at Makerere College Zoology Department included an adult and a subadult (large juvenile) male, five adult females, and six others. Figures for these are based on two hundred fifty daily, 60-minute watches. Caged rhesus and baboon colonies were kept in comparable cages, relative to their size. The baboon cage was 12 × 50 × 8 feet. The rhesus cages were 8 × 18 × 8 feet. Animals in both colonies were handled as little as possible.

For the physiology of rhesus cycles I have referred to van Wagenen (1945) and Zuckerman, van Wagenen, and Gardener (1938).

For physiological data on baboon cycles I have relied on Gilman and Gilbert (1946). The present small sample of Ugandan (*Papio "doguera"*) and Kenyan (*P. "cynocephalus"*) baboons behaves as if drawn from the same population as the colossal numbers of chacma (*P. ursinus*) females in that study (with the reservation about nonbreeding females in captivity recorded above).

Baboons, as Gilman and Gilbert pointed out, are unique in the number of external signs they show of internal changes. In this study the cycle has been divided into stages based on external appearance. In the normal cycle the perineal skin begins to swell during or just after menstruation. It increases rapidly in size for about a week (state, Inflating), then remains relatively con-

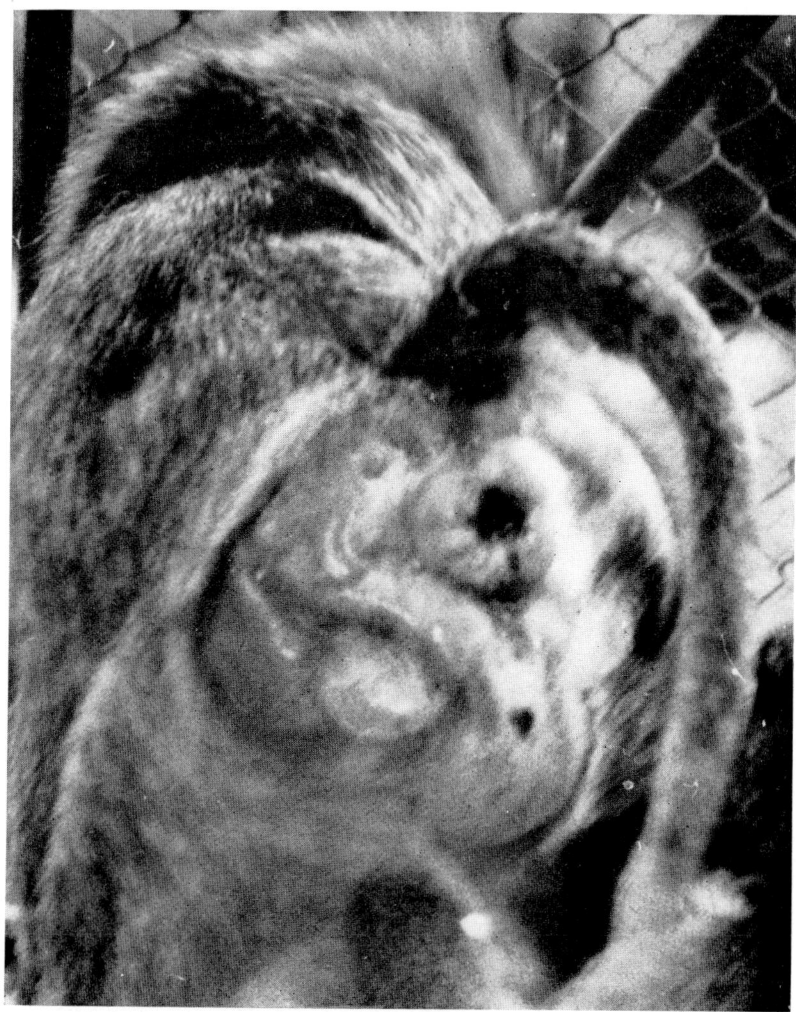

Plate 2.1

A. The largest swelling seen in a Ugandan baboon. Nulliparous animal, about 7 years old, had been caged alone since maturity. The swellings in the wild population were smaller than this.

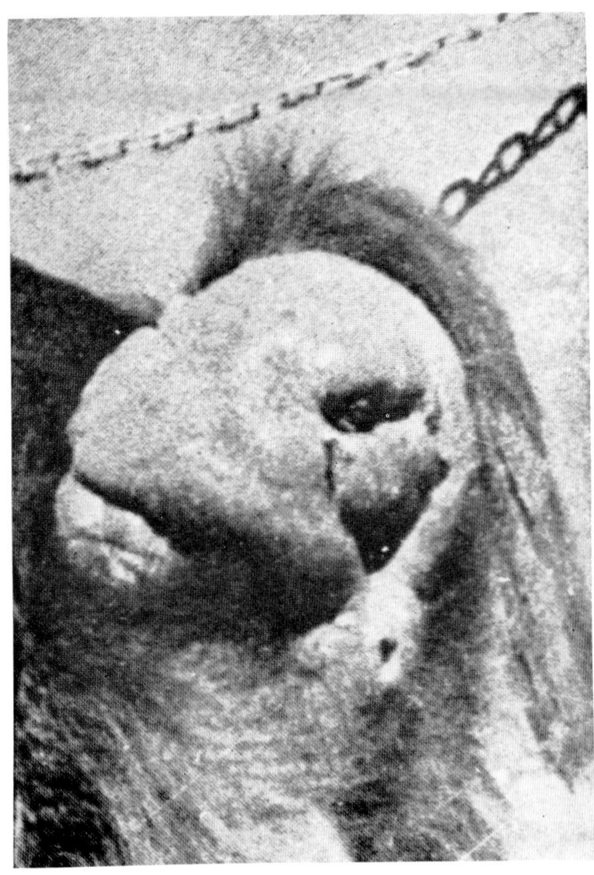

B. Photo from Gilman and Gilbert (1945). "(Chacma) female undergoing normal menstrual cycle."

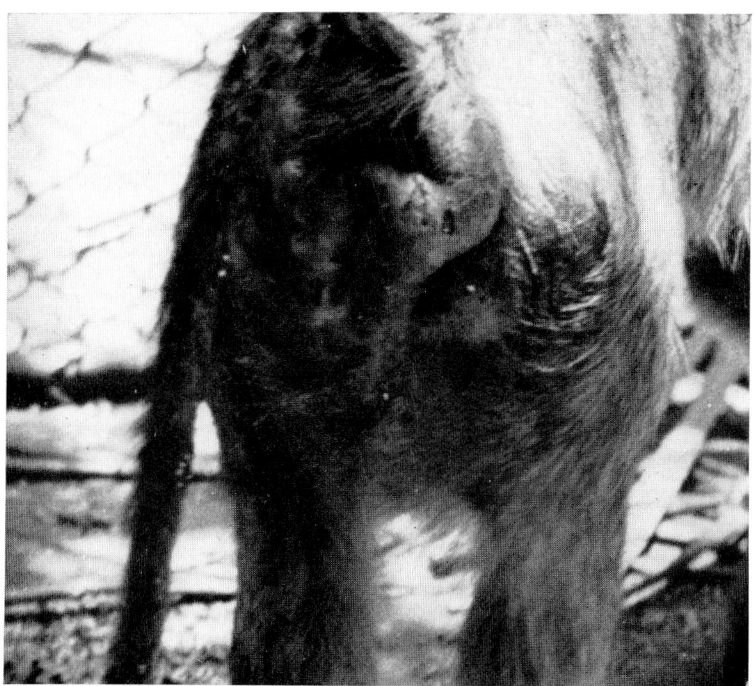

C. A Kenyan lowland baboon with full swelling.

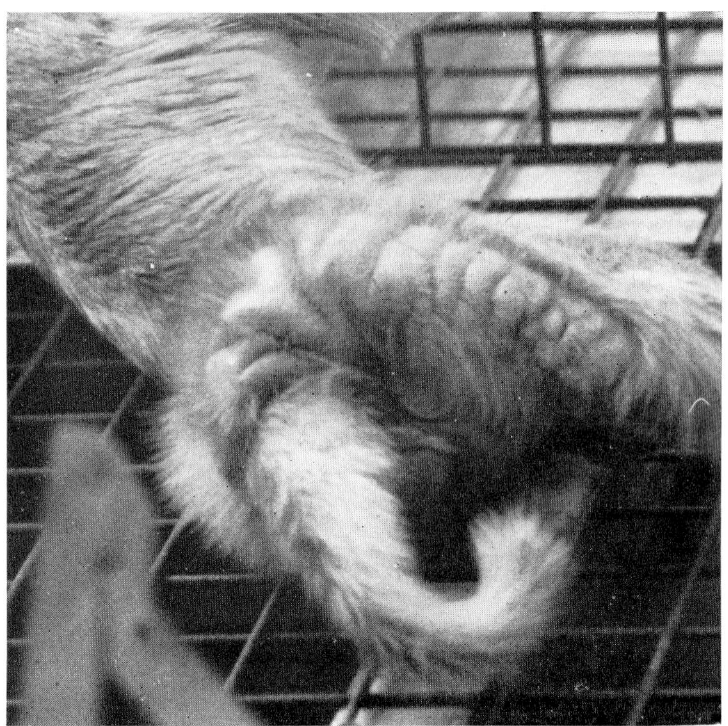

D. An adolescent rhesus monkey, about 3 years, fully swollen.

stant (enlarging slightly) for about 10 days (state, Swollen, Plates 2.1*A* and 2.1*C*). It then decreases rapidly over about 5 days (Deflating), becoming flabby, then wrinkled, and finally flat. It remains flat for about a week (Flat); then menstruation is followed by the next swelling. If the female becomes pregnant, the black, naked skin over the hips, which never normally swells, starts to lose the black color and turns bright red by the second half of pregnancy. The beginning of this change is noticeable 2 or 3 weeks after menstruation is expected (Pregnant). All states are easily recognized in the field, except that young cycling females in Flat may be confused with large immature females, and older cycling females in Flat may be confused with females late in lactation. These may all be included as Flat in wild data. Lactating caged females have been excluded, because of the complications of "aunt" behavior with their babies. Breeding in the caged group has been, for the purposes of this study, regrettably successful, and only twenty-five cycles have so far been recorded. This can, therefore, only be an interim report.

METHOD

All social interactions were recorded. Most involved only two animals, but more complex ones were broken down, if necessary, to two-baboon interactions, and these were recorded on punch cards together with the cycle states of all females in the group at that time. For this chapter, interactions have been put into a simple—and therefore sometimes arbitrary—classification:

Agonistic: includes all attacking and fleeing, threat, and obvious avoidance of one by another.

Friendly: includes grooming, lip-smacking, embracing, touching, presenting (except to potent males).

Sexual: includes all male-female behavior that could possibly be included here—any interaction with copulation, mounting, intention mounting, touching genitals, or following by males, and presenting or touching genitals by females.

Only interactions with adult females have been used, and all interactions with babies, and with own-child juveniles, have been excluded. Individual differences have been ignored, since the data are so far insufficient to analyze them in detail. (Female to female homosexual behavior was rare in baboons compared with rhesus; mounting of female by female was never seen in wild or caged animals.)

Sexual behavior in caged baboons was further divided into five categories for males: mount, intention mount, handling genitals, contact other part of female body or lip-smack, and associate (following, sitting with female). Of female behavior in sexual interactions, only presenting and avoiding are discussed in detail here.

Table 1 gives the number of days of observation for each state for each female in the cage. Numbers used in later tables are based on total instances per total days observation on each state, taking all females together.

Comparison of the frequency of interactions in different cycle states in the wild population was difficult because the vegetation was so thick that it was rarely possible to see a whole troop at once. The figures are "adjusted" on the assumption that in the adult female population there were three inflating or deflating females to four swollen to sixteen pregnant to thirty flat or lactating. These figures were calculated on the basis of the long-term cycle described on pages 15–17, and fitted reasonably well with censuses when they could be taken.

TABLE 1

DAYS OF OBSERVATION AT DIFFERENT CYCLE STATES (CAGED FEMALES)

Female	Inflating	Swollen	Deflating	Flat	Pregnant
E.	38	21	19	7	93
J.	22	48	8	56	69
H.	26	39	21	102	34
W.	26	51	31	71	39
R.	11	35	19	15	49
Total	123	194	98	251	284

RESULTS OF BABOON OBSERVATIONS

Nonsexual Behavior of Cycling Females

In a wild group of baboons, a swollen female can easily be picked out before her rump is visible. She walks about more than any other class and sits only for brief intervals, whereas other females (in this habitat) walk briefly from one sitting place to the next. The swollen female maintains a position near the males even when the group is foraging or moving and there is no sexual behavior. Pregnant females, on the other hand, typically wander farthest of all classes of animals from the main group. In the caged group the restlessness of the swollen females was noticed, but was less obvious, and the positioning was not oberved because the total space was inadequate. Pregnant females in the cage were remarkable for their inactivity and sleepiness, in contrast to the wild animals. In the cage, the number of interactions between females and all animals other than potent males or their own young remained about the same, whatever the cycle states of the females, but the type of interaction does change. Figure 1 gives the proportion of agonistic to friendly interactions between females. The data were considered in relation to the cycle state of females who started such interactions (first histogram) and females who were attacked (second histogram). Pregnant females were involved mainly in friendly interactions. The difference between the other states does not reach the 5 per cent level of significance, but I noticed it, and thus presumably so do the baboons. Swollen females were involved in more fights, both as initiators and as receivers. Deflating females rarely started a fight, but were attacked fairly often. These data are complicated by individual relationships, which will be the subject of a future analysis.

Interactions with Males

Caged.—Interactions with both of the potent males in the caged group increased while the female was inflating, and again, more sharply, while she was swollen (Fig. 2). The increase was nearly all in sexual behavior as defined above. Friendly interactions with swollen females also increased; this may merely reflect the imprecision of the classification. Agonistic interactions, which were in any case very few, did not change. Sexual behavior by and with the adult male (with full dentition and mane, about 7 to 8 years old) was much more restricted to females in the swollen state than was that by and with the big juvenile male (3 to 4 years old, canine teeth just appearing in gums, no mane, shortish nose).

Interactions in which either animal could be regarded as behaving sexually

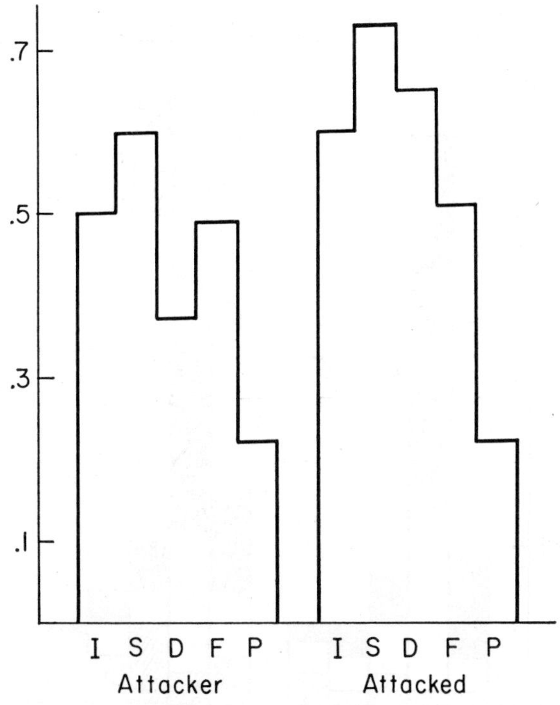

Fig. 1.—Caged baboons. Interactions between females: the proportion of agonistic to friendly interactions in each cycle state. The first histogram shows interactions started by the females in question; the second shows those begun by another animal.

I = inflating, S = swollen, D = deflating, F = flat, P = pregnant were analyzed further. Figure 3*a–e* and Fig. 4*a–e* show how five classes of behavior by the two males varied with females' cycle state. The histograms show the percentage of interactions which included each class of behavior.

1. *Adult male:* All types of behavior increased in interactions with swollen females. That is, these interactions were longer and more complex. By contrast, interactions with deflating females were simple, as well as infrequent, and there was a high proportion of "contact"—usually a friendly response to presenting, which is a very common pattern between all classes of baboons. Other patterns were rare, and deflating females and males rarely stayed close together (*e*). On the whole, positive sexual behavior (*a–c*) occurred in a relatively constant proportion of interactions—for instance, the perineum of a swollen female was touched or noted very little more than the nonswollen rump (*c*). Interactions with swollen females were prolonged by the animals remaining in close proximity (*e*). "Associate" was also high for pregnant females; here the most frequent pattern was presenting by the female, followed by her grooming. The male made no response other than the grooming posture. This is hardly sexual behavior, but all interactions which included a female presenting to a male were included in this category.

2. *The juvenile male,* compared with the adult male, showed a higher pro-

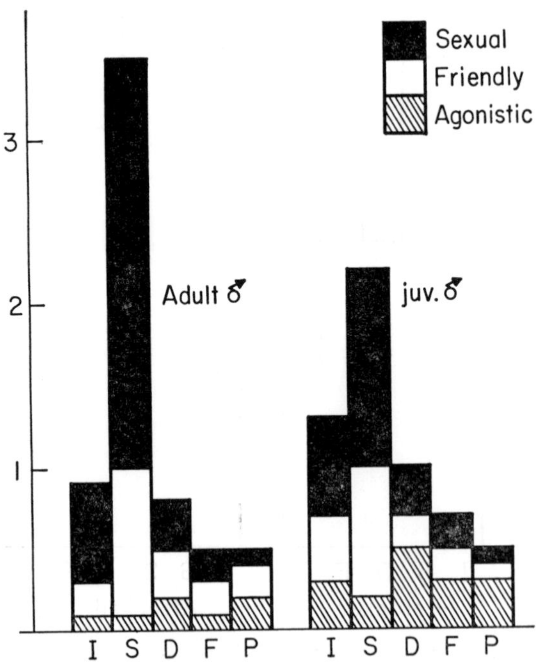

FIG. 2.—Caged baboons. Interactions between five adult females and the two potent males. Histograms show the total interactions per female day, and the proportion of three different types of interaction—agonistic, friendly, and sexual. Symbols as in Figure 1.

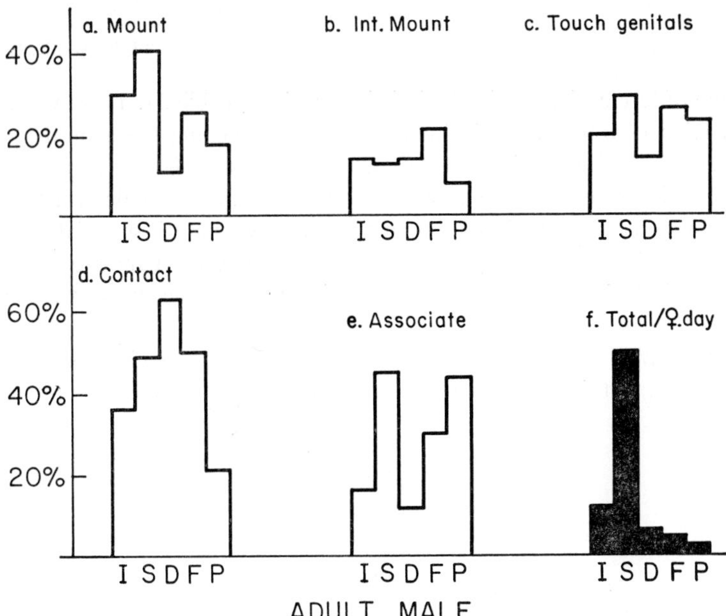

FIG. 3.—Caged baboons. Analysis of behavior of one adult male in "sexual" interactions, by cycle state. Each histogram shows the proportion of interactions which included that type of behavior with females of each cycle state. The solid histogram shows the total "sexual" interactions per female day for each cycle state. Symbols as in Figure 1.

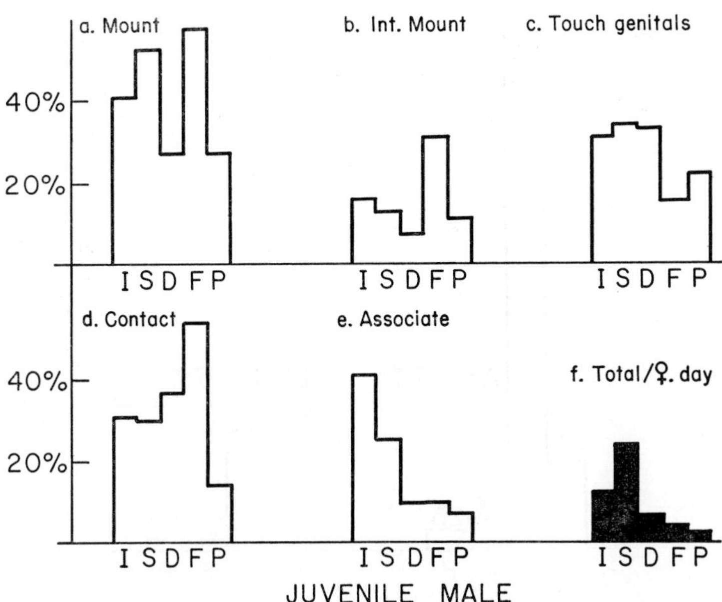

FIG. 4.—Caged baboons. Analysis of behavior of one juvenile male in "sexual" interactions, by cycle state. Each histogram shows the proportion of interactions which included that type of behavior with females of each cycle state. The solid histogram shows the total "sexual" interactions per female day for each cycle state. Symbols as in Figure 1.

portion of interactions which included mounting (Table 2). There was a larger proportion of interactions including nosing or touching the genital area with females of *any* degree of swelling than with nonswollen females. Interactions of the juvenile male (like the adult male) with deflating females least often included mounting or intention mounting. A comparison of "associating" by the two males shows that the young male was close to the female while she was swelling, but his place was taken by the big male as the swelling reached full size.

Wild.—Wild baboons show the same changes of behavior frequency with cycle state, but in more extreme form (Fig. 5). Sexual behavior was most

TABLE 2

PROPORTION OF SEXUAL INTERACTIONS WHICH INCLUDED COPULATION

MALES	WILD		CAGED	
	Total Interactions	Copulation (%)	Total Interactions	Copulation (%)
Adult	128	28	369	44
Large juvenile	38	74	211	58

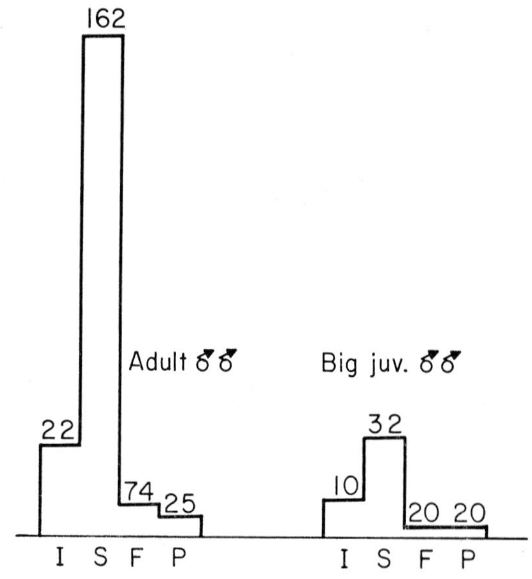

FIG. 5.—Wild baboons. Sexual interaction between adult females and potent males. Numbers are of interactions seen. The histograms show the frequencies per state, based on population estimates explained on page 18. Symbols as in Figure 1.

frequent with inflating and swollen females; the adult males were more restricted to swollen females than the big juvenile males. In both habitats, large juvenile males were involved in fewer sexual interactions than adult males. However, the proportion of interactions which included successful mounting was much higher for the young males (Table 2). Again both these differences were much more pronounced in the wild.

Mounting and Presenting

Two behavior patterns, and the responses to them, will now be discussed in more detail. Mounting and presenting are both especially interesting because

TABLE 3

MOUNTING OBSERVED IN WILD TROOPS, COMPARED WITH VALUES EXPECTED IF MATING WAS RANDOM WITHIN EACH CLASS OF MALE MOUNTER

MOUNTEES	MOUNTERS					
	Adult Males		Big Juvenile Males		Small Juvenile Males	
	Observed	Expected	Observed	Expected	Observed	Expected
Adult male	9	12.5	0	15.6	0	4.2
Big juvenile male	7	15.6	6	19.6	1	5.3
Small juvenile male	1	7.8	3	9.8	3	2.2
Small juvenile female	2	7.8	6	9.8	5	2.2
Big juvenile female	3	7.8	9	9.8	0	2.2
Adult females						
Flat	8	7.1	17	8.9	3	2.4
Partly swollen	4	0.7	9	0.9	1	0.2
Swollen	36	0.9	28	1.2	4	0.3
Pregnant	1	3.8	11	4.7	7	1.3
Totals	71		89		24	

they have been considered by many workers to be primarily sexual patterns secondarily used to express hierarchical relationships. (Bopp [1953] discusses this in relation to baboons and stresses the latter function, referring to presenting as an "Inferioritätsreaktion.")

Mounting.—Table 3 shows the number of mountings seen in the wild troops. The expected numbers on the assumption that mounting was random for each class of mounter are also given. These figures include all mounts: it was not possible in the field to distinguish between "sexual" and "dominance" mounting; the motor patterns were identical, and mounting between all partners showed a complete range between brief covering and ejaculation.

Only males mounted. All males mounted other males less, and adult females more, than would be expected on the basis of random mounting. Selection of inflating and swollen females for mounting increases steadily as the males get larger.

Males take a keen interest in the sexual activities of others. They often dance around a copulating pair, making repeated coughs and touching them. Sometimes the copulation is broken off as the copulating male drives them away. Very small males (less than 1 year) sit and watch copulations and then go to the same female and make mounting intentions themselves, of which a swollen female is usually tolerant, if not encouraging. It is difficult to avoid the impression that imitation learning is taking place.

In this wild population, consort pairs were usually short-lived, a day or less. Swollen females left the male and wandered alone and might be mounted by juvenile males, to which nearby adult males rarely objected. Some adult males mated frequently, others not at all; but these differences could not be correlated with other differences in behavior, except that two easily recognized males that stayed noticeably more with their respective troops' nursery groups than others were not frequent maters.

In the wild, 33 per cent of seventy-six mating attempts by adult males were avoided by swollen females. Only one out of twenty-nine attempts by large juveniles was avoided. Attempts to mount other classes of females were very few (Table 3), and no avoiding was seen.

In the cages, attempts by the adult male were avoided least by swollen females (25 per cent), and most by deflating females (50 per cent). Other states were between 30 and 40 per cent. As in the wild, the juvenile male was avoided considerably less—between 20 and 30 per cent of attempts for all states. Apparently a relatively constant proportion of mounting attempts are avoided.

Presenting.—This is a familiar "courtship" pattern of female primates. Here "true presenting"—that is, with the rump toward the presentee—is distinguished from "lateral presenting," where the posture is the same but the flank is to the presentee. The latter is one of a great variety of invitation-to-groom postures employed by baboons. True presenting also very frequently elicits grooming, especially from females. Presenting was never used in either habitat to stop aggression. The nearest approach to this was "guilty conscience" presenting by a juvenile which has just attacked another and then returned and presented to an adult which had observed the attack. Presenting to an animal which had shown any signs of attacking the presenter did not occur. The most frequent context of presenting was the brief pause when walking past a seated male, waiting till he acknowledged the gesture by looking at or touching the rump, and then walking on.

Table 4 shows the incidence of presenting between all classes of the wild population. One-fourth of the presenting was by inflating or swollen females to potent males, far above the expected number on a random distribution. This made a very striking behavior pattern for such females. In both habitats adult males were presented to far more than juvenile males. Figure 6 shows the frequency of presenting in the cage to the adult and the juvenile male.

Tables 5 and 6 show the responses of the male presentees. A very small proportion of presentings was responded to by mounting or intention mounting, and the proportion was if anything higher for nonswollen females. The most

TABLE 4

PRESENTING IN WILD TROOPS

PRESENTER	PRESENTEE									TOTAL
	A Male	BJ Male	SJ Male	SJ Fem.	BJ Fem.	FA Fem.	PSA Fem.	SA Fem.	P Fem.	
Adult male (17.5)	17	2	3			14		6	1	43
Big juv. male (22)	28					16			2	46
Small juv. male (11)	10					8			1	19
Small juv. female (11)	14	2				4	1	1	2	24
Big juv. female (11)	12	3				4		1	1	21
Adult females: (17.5)										
Flat (10)	66	7	1			11		1	2	88
Partly swollen (1)	18					2			1	21
Swollen (1.3)	92	10	3			2			1	108
Pregnant (5.7)	24	6	7			0			0	37
Total	281	30	14			61	1	9	11	407

The rough percentage of each class of animal in the population (excluding babies) is given in parentheses.

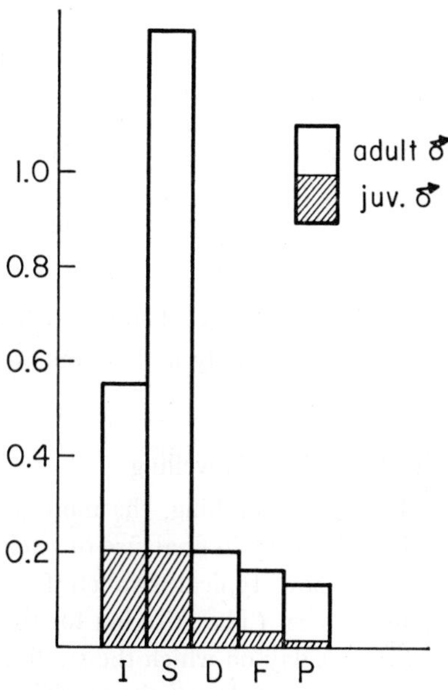

FIG. 6.—Caged females. Frequency of presenting to the two males (interactions per female day) in each cycle state. Symbols as in Figure 1.

frequent response was a friendly gesture—lip-smacking, touching, or grooming; almost as often there was no apparent response by the male. Juvenile males responded more frequently by mounting, although there were fewer presentings made to them. The proportion of different types of response remains about the same for all cycle states, but since swollen females present more, they are also mounted more.

TABLE 5

INCIDENCE OF PRESENTING TO POTENT MALES IN CAGED FEMALES AND THE RESPONSES ELICITED

Cycle State	Total No.	No/Female Day	Responses (%)			
			None	Friendly	Intention Mount	Mount
Inflating	68	0.55	37	56	3	4
Swollen	248	1.3	33	50	6	11
Deflating	19	0.2	37	37	15	11
Flat	42	0.16	38	36	16	10
Pregnant	36	0.13	33	44	8	14

TABLE 6

PRESENTING TO POTENT MALES BY WILD FEMALES AND THE RESPONSES ELICITED

Cycle State	Total No.	Per Females in Population (adjusted)	Responses (%)			
			None	Friendly	Intention Mount	Mount
Part-swollen	18	6	44	33	6	17
Swollen	102	25.5	42	38	7	13
Flat	74	2.5	42	23	20	5
Pregnant	30	2	23	40	30	7

The most usual pattern of copulation is when the male initiates the interaction, going to the female who stands only as he mounts—too late to provide a visual stimulus by her posture.

Behavior Changes within the Period of Swelling

The axiom that "the bigger the swelling, the more sexual behavior" only holds within each cycle. Females vary in final size of swelling a great deal, but each female only begins to show typical swollen female behavior as she approaches her own maximum size. (The criterion for the change between Inflating and Swollen was the first day on which the swelling was not noticeably bigger than the day before.) The last day of the Swollen stage was marked by a change in behavior. The females' appearance did not change, but they became inactive and interacted little with males. (According to Gilman and Gilbert, ovulation occurs 1 or 2 days before deflating, so this would be a postovulatory day.) Figure 7 shows the total interaction on 17 last days of the swollen state compared with their 17 preceding days, and the middle recorded day of the same swollen periods. Comparison with all male-female interactions in other cycle states (Fig. 7, second histogram) shows that on the last day the female behaved, and was responded to, almost like a deflating female.

The Swellings of Rhesus Monkeys

Normal adult rhesus monkeys do not show cyclical swelling on the whole, although there are exceptional individuals. Adolescent females swell between about 2 and 4 years of age. (Plate 2.1*D*). Changes in this ill-defined swelling are rather difficult to measure exactly, and study is further complicated because in young animals the cycles are also sometimes irregular. Figure 8 shows the relation of cycle to swelling in two females about 3 years old (one with a very

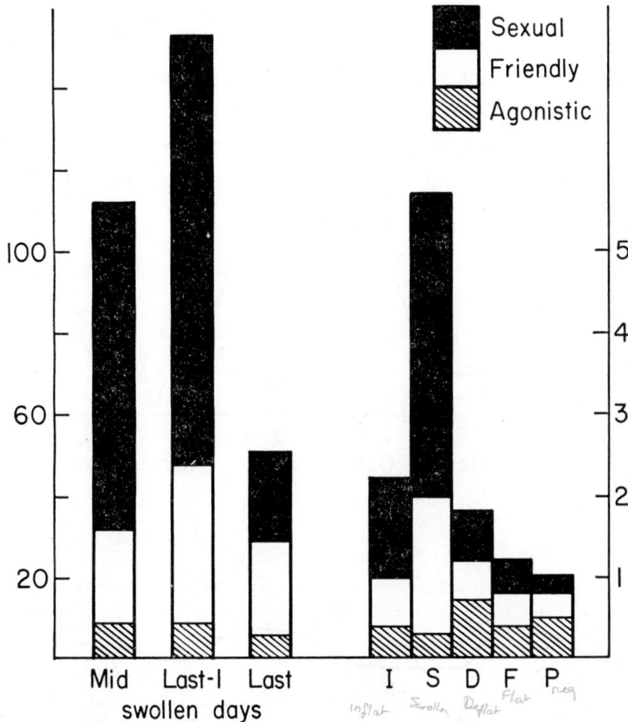

FIG. 7.—Caged baboons. A comparison of female-male behavior on the last day, the next-to-last day, and a middle day of the full swelling, based on seventeen complete cycle records. The figures are the total of interactions observed. For comparison, the right-hand histogram shows interaction per female day in each cycle state, the two males summed in both cases. Symbols as in Figure 1.

pronounced edema), having regular cycles, and caged with a male. The picture they give corresponds with remarks made by Zuckerman, van Wagenen, and Gardener (1938). Swelling begins a few days after menstruation and increases until just before the next menstruation. The male never responded to these females' swellings by looking or touching. One female (upper diagram) appeared irritated by hers and clutched at it often. The function of these adolescent swellings is obscure, since the females were not of reproductive age. In most cases the female has stopped swelling before she becomes pregnant. It would seem, therefore, that they could not be the result of any sexual selection.

Zuckerman *et al.* regarded the swellings of baboons and macaques as homologous, since both are produced by estrogen injection, differing only in that the swellings of the rhesus "mature"—that is, stop occurring after a certain time—whereas those of baboons do not. It seems to me that there are other differences which should be taken into account.

1. The swellings are not in the same place (see Plate 2.1*A–D*). The rhesus swelling involves mainly the interfemoral membrane, the tail root, and the hips. These areas are never swollen in normal baboons, where the swelling is confined to the perineum, which in turn is only slightly swollen in rhesus.

2. The baboon does not swell until slightly before she is fertile. The rhesus stops swelling about this time.

3. The baboon swelling is almost entirely restricted to the follicular phase

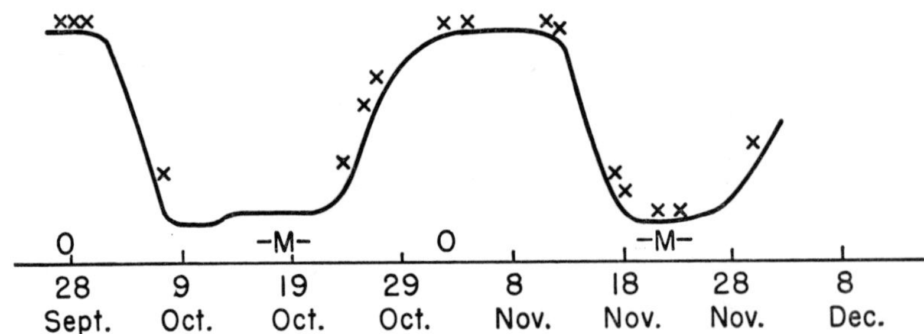

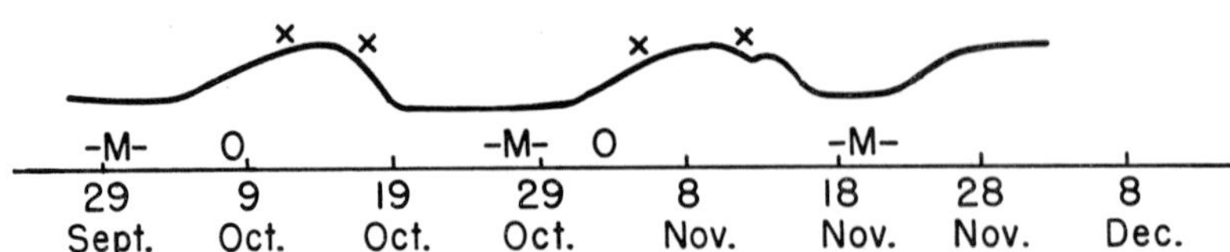

FIG. 8.—Swelling-size changes in two adolescent rhesus monkeys related to their menstrual cycles.—*M*—, menstruation; *O*, presumed ovulation; *X*, mating on that day. The vertical scale is subjective.

of the cycle. That of the young rhesus begins in the follicular phase, continues to grow through ovulation, and reaches maximum size toward the end of the luteal phase. This suggests that either the cyclical hormonal changes in the two species, or the hormonal changes which govern the swelling size, are different.

On the other hand, the duration and size changes of the swelling do appear to correspond to cyclical mating paterns of caged adult rhesus monkeys (Rowell 1963). Mating in the colony studied occurred throughout the cycle except the days preceding and during menstruation; there was, however, a steady rise in mating frequency through the cycle until a sudden drop just before the next menstruation (Fig. 9).

DISCUSSION

It is clear that sexual behavior is highly correlated with the presence of a (large) perineal swelling in the adult female baboon; perhaps a similar correlation exists in the rhesus money. There is a general assumption that a causative relation exists between the presence of the swelling and the occurrence of sexual behavior, but such a relationship has not in fact been shown to exist. There are two ways in which a swelling could cause sexual behavior—either by acting as a visual stimulus to the male or by irritating the female in such a way as to change her behavior.

If the swelling provides a stimulus to the male, it is acting in a very general way. It might be expected that such a sign stimulus would be responded to as it was perceived, but the male baboons reacted to the presentation of a swollen rump in the same way as they did to a nonswollen one; they usually initiated sexual behavior when the swelling was *not* visible. On the other hand, the female gives ample clues about her estrous condition by behavior changes, as has been shown, and a male would have no difficulty in detecting these changes

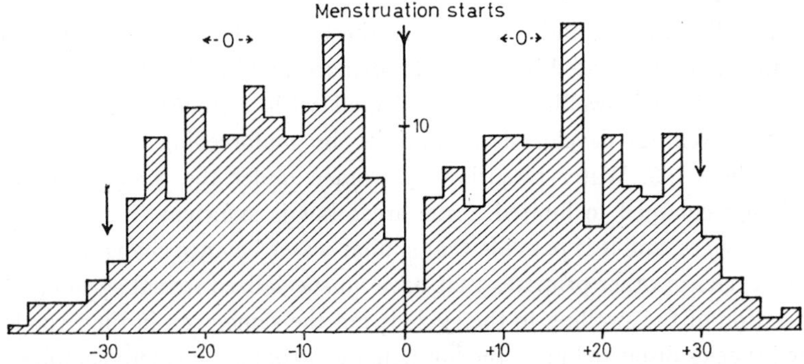

FIG. 9.—Copulations on each day of the menstrual cycle in a colony of rhesus monkeys (from Rowell 1963). Arrows mark median cycle length. Period during which ovulations probably occurred is marked *O*. Each column is the mean of 2 days. (From the *Journal of Reproduction and Fertility,* by permission.)

even if the swelling were not visible—in the same way as males of related, nonswelling species do. When a fully swollen female does not *behave* like an estrous female, the males ignore her, a fact that suggests that they are not reacting primarily to her changed appearance.

This is not, of course, to suggest that an animal as intelligent as a baboon does not notice the correlation between willingness to mate and perineal swellings. Perhaps the increased selectivity of males as they get older reflects such a learning process. But it seems likely that this learned cue is secondary in importance.

The second possibility, that the swelling irritates the female and thus changes her behavior, has been suggested also for the rhesus monkey and the chimpanzee (Birch and Clark 1950). Hinde and Steel (1964) have shown that hormones affect skin sensitivity in breeding canaries. Baboons with swellings do not show irritation directly by their behavior (unlike rhesus monkeys which frequently feel their swellings with their hands, and unlike menstruating females of both species, which often investigate their vulvas with their hands, showing that they will respond to irritation in this area in an observable way). The effects of estrogen in the normal cycle, apart from causing the swelling, seem to be similar to the effect of estrogen on most mammals; so it does not seem necessary to invoke the intervention of swelling irritability to explain the increase in activity, willingness to copulate, and so on.

There is of course the possibility that the swelling has no social function, that it is a by-product of some other adaptation. This is rather difficult to accept, but on the other hand some experimental study seems called for before one can confidently assert otherwise. There is a general tendency among primates to respond with edema to changes in hormone balance. Thus rhesus monkeys swell during adolescence but before full maturity; the vulva of a patas money (which have no "sexual skin") was also observed to swell at puberty. Some mature rhesus monkeys swell a little after a seasonal or lactational acyclic interval, and some elderly rhesus females swell excessively over the whole body. Castrated baboons, chimpanzees, and rhesus monkeys of both sexes show sexual swellings when injected with estrogens. The relatively enormous swellings developed in captivity by nonbreeding animals may be relevant here too. Water retention by women under the influence of estrogens is not confined to limited tissue areas but is sufficient to cause marked weight changes, and it is possible that similar water balance changes occur during the cycles of other primates that have no specialized sexual swelling. Since the relation of the swelling to the cycle is not the same even in the two closely related species discussed here, the hormonal control may be more complex than has been supposed; it may be that it is from the angle of water balance physiology rather than social behavior that the sexual swelling can best be understood.

The interactions between swollen females and males stand out immediately as being different, even after the most perfunctory baboon watching. Yet the details of behavior patterns, as so far analyzed, did not change very much with the females' cycle state. The effect seemed to be produced by rather general changes in the females' over-all activity level and their spatial relations and frequency of interaction with males. Very little direct response to the females' swelling was observed, but a general long-term response by males was not excluded.

SUMMARY

The behavior of caged and wild baboons was analyzed in relation to the cycle states of the females.

Wild swollen females were more active than usual and stayed with the males. Caged, swollen females interacted with other females as frequently as at other times, but the proportion of agonistic to friendly interactions changed with cycle state.

Interactions with males increased in inflating and again in swollen females; the increase was almost entirely composed of behavior which could be called "sexual." (This term included all male mounting and all female presenting to the opposite sex.) Fully adult and large juvenile males showed rather different response patterns. The same trends were seen in both wild and caged groups but were more exaggerated in the wild.

Only males mounted. A comparison of observed with expected mountings by males of different ages shows that an increase in the selection of inflating and swollen females could be seen as the male gets older. A fairly constant proportion of mounting attempts were avoided by females in all cycle states.

Presenting was most frequent by swollen females to adult males. The response of the male did not vary with cycle state, and a very low proportion of presenting was followed by mounting.

On the last day of the full swelling, the females' behavior and the males' responses to them resembled those of, and to, a deflating animal.

The swellings of adolescent female rhesus are discussed, as are differences between them and those of adult baboons.

Although there is a correlation between swelling and sexual behavior, there is as yet no evidence for a causative relation, and the function of the swelling may be primarily physiological rather than social.

This work was carried out on a D.S.I.R. research assistantship. I am also extremely grateful to Miss Unity Stack, who did most of the paper work in this analysis.

REFERENCES

Birch, H. G., and Clark, G. 1950. Hormonal modification of social behavior. IV. The mechanism of estrogen-induced dominance in chimpanzees. *J. Comp. Phys. Psychol.* 43:181–93.

Bopp, P. 1953. Zur abhängigkeit der Inferioritätsreaktion vom Sexualzyklus bei weiblichen Cynocephalen. *Rev. Suisse Zool.* 60:441–46.

Carpenter, C. R. 1942. Sexual behaviour of free-ranging rhesus monkeys (*Macaca mulatta*). *J. Comp. Psychol.* 33:113–62.

Gillman, J., and Gilbert, C. 1946. The reproductive cycle of the chacma baboon with special reference to the problems of menstrual irregularities as assessed by the behaviour of the sex skin. *S. Afr. J. Med. Sci.* 11 (Biol. Suppl.):1–54.

Hinde, R. A., and Steel, E. 1964. Effect of exogenous hormones on the tactile sensitivity of the canary brood patch. *J. Endocrinol.* 30:355–59.

Kaufmann, J. 1965. A three-year study of mating behaviour in a free-ranging band of rhesus monkeys. *Ecology* 46:500–512.

Rowell, T. E. 1963. Behaviour and reproductive cycles of female macaques. *J. Reprod. Fertility* 6:193–203.

———. 1965. The habit of baboons in Uganda. East African Academy Symposium 2, Nairobi.

———. 1966. Forest living baboons in Uganda. *Proc. Zool. Soc. Lond.* (in press).

van Wagenen, G. 1945. Optimal mating time for pregnancy in the monkey. *Endocrinology* 37:307–12.

Washburn, S. L., and DeVore, I. 1961. The social life of baboons. *Sci. Am.* 204:62–71.

Zuckerman, S. 1930. The menstrual cycle of the primates. I. *Proc. Zool. Soc. London* 100:691–754.

———. 1931. *Ibid.* 101:325–43; 593–601.

———. 1932. *The social life of monkeys and apes.* London: J. Paul.

———. 1937. The duration and phases of the menstrual cycle in primates. *Proc. Zool. Soc. London* 107:315–29.

———. 1937. Cyclical fluctuation in oestrin threshold. *Nature* 139:628.

Zuckerman, S., and Fulton, J. F. 1934. The menstrual cycle of the primates. VII. *J. Anat. London* 69:38–46.

Zuckerman, S., and Parkes, A. S. 1952. The menstrual cycle of the primates. V. *Proc. Zool. Soc. London* 102:139–91.

Zuckerman, S., van Wagenen, G., and Gardener, R. H. 1938. The sexual skin of the rhesus monkey. *Proc. Zool. Soc. London* 108:385–401.

3

LABORATORY OBSERVATIONS OF EARLY MOTHER-INFANT RELATIONS IN PIGTAIL AND BONNET MACAQUES

LEONARD A. ROSENBLUM AND I. CHARLES KAUFMAN

Considering the widespread interest in social behavior and its development throughout the primate order as evidenced by the diversity of work included in this book, we can see the need for close comparisons between primate species studied under comparable and relatively controlled conditions. As part of a laboratory program designed to assess the role of varying patterns of mother-infant relations upon the behavioral development of the young, during the past 4 years we have been studying homospecific groups of two species of macaque, the pigtail (*Macaca nemestrina*) and the bonnet (*M. radiata*) under identical social and environmental conditions.

The pens in which the animals are housed are each approximately 8 feet wide, 7 feet high, and 13 feet deep; the walls are opaque except for two one-way vision observation screens set into the front wall, through which all observations are carried out (Fig. 1). The observations are recorded by dictating into a continuously running tape recorder all of the behaviors in which a given subject engages, together with the social partner or partners involved. The time of occurrence of each behavior is retrievable from the tape on transcription and is entered on the typed records; these times are then transferred to punched cards and, through appropriate programing, the calculation and tabulation of behavior durations (as well as tabulations of frequencies and other related dimensions) are carried out by high-speed computers. The general findings presented in this chapter are based on a small segment of the

Leonard A. Rosenblum and I. Charles Kaufman, State University of New York, Downstate Medical Center, Brooklyn, New York.

observations of three bonnet and two pigtail groups; that portion of the currently available computer analysis of data from one pigtail and one bonnet group, (each including four mother-infant dyads), which deals with mother-infant relations, comprises the bulk of the quantitative data which are included.

To provide a heterogeneous social grouping for our observations, each of these groups, when originally formed, contained one adult male, four adult females, and, to provide intermediate aged partners with whom the first group of infants could interact, a male and a female adolescent. Members of our pigtail groups were generally dispersed around the pens during the day, coming into contact usually only to interact in some overt fashion, such as that involved in grooming, sexual behavior, or aggressive behavior. Bonnets, on the other hand, have regularly been observed to spend long periods of time during both the day and the evening in close "passive contact" with one another,

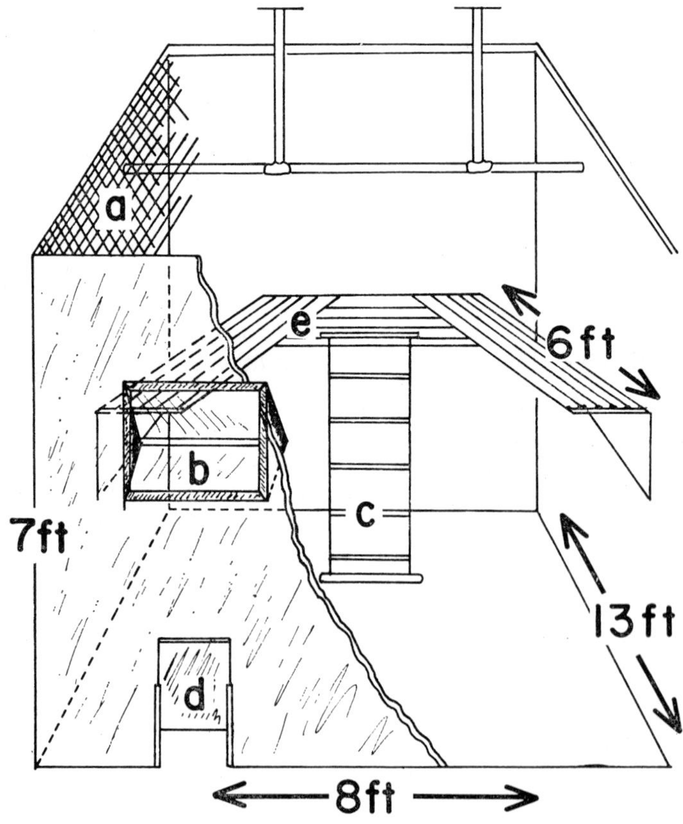

FIG. 1.—Scale diagram of the group pens. (*a*) Overhead mesh; (*b*) one-way vision screen; (*c*) mesh-covered ramp; (*d*) guillotine animal door; (*e*) aluminum shelves.

often in large groups reminiscent of group sleeping huddles sometimes observed in other species. It is significant that there is virtually no overlap in the distribution of scores for passive contact between the two species, and in addition, that the two species also show a somewhat similar difference in the tendency simply to remain in close proximity to one another (1 foot or less) particularly during the late night hours (Rosenblum, Kaufman, and Stynes 1964).

It increasingly has become our opinion that this species difference in characteristic individual distance patterning is but one dramatic reflection of a basic difference in the whole tone of social behavior in these two species as we observe it in the laboratory; and furthermore, that this difference has important bearing on a number of aspects of the dyadic interaction and the ultimate behavioral development of the young.

As is true of many other primates, births in our animals normally occur during the evening hours when the laboratory is quiet and illumination is low; all but one of our forty-seven births have occurred between 6:00 P.M. and 6:00 A.M. The placenta, eaten after all pigtail births and after almost all bonnet births, when observed, has been devoured eagerly by the mother even before licking and grooming of the infant has progressed. Although our sample of primipara is somewhat small ($N = 6$), they seem to eat the placenta with about the same regularity as multipara.

The interest of the group in the arrival of a newborn seems marked in both species. However, the reactions of the bonnet and pigtail mothers is in no way comparable. Pigtail mothers show marked reluctance to engage socially, generally withdrawing from voluntary participation in group activities and, as much as possible, thwart the interest in the young by "protective withdrawal" or "protective threat," "pursuit" or actual "attacks." The bonnet mothers, on the other hand, almost immediately return to close contact with other females. Although failing to allow the extensive freedom of access and handling of the infant by members of the group reported for langurs (Jay 1963), bonnet mothers frequently permit others to explore, handle, and groom their newborn without removing it from them (Plate 3.1). These early indications of species differences in maternal relaxation regarding the familiar social group (that is, return to passive contact and tolerance of others' inspection of the infant), as we shall see below, continues to be manifest throughout the period of dyadic involvement.

Turning now to specific patterns observed between mother and infant, we have seen among the earliest behaviors a striking difference between species in the frequency with which the mother (and in bonnets, others) inspects the perineal region of male and female infants. During the first week or so of life, the male infant repeatedly is lifted and turned over or has his hind legs pulled aside or apart to allow close visual and olfactory inspection of the genitalia, whereas such behaviors are directed at female infants only rarely. It is our opinion that this and perhaps related differential treatment of male and female neonates may play a significant role in developing behavioral dimorphisms and would merit greater general scrutiny among primatologists working in both the field and the laboratory.

In both of our species the infant in the early weeks of life, when not being examined or groomed by the mother or others, generally remains in close ventral-ventral contact with its mother. Although occasionally loosening its grip and falling back somewhat away from the mother's ventrum into the "ventral-ventral hold" position, in the early weeks tight "ventral-ventral cling" is far the more common behavior (Fig. 2). This behavior is initially the predominant one on the part of the infant, gradually declining during the first year of life; however, even at the end of the fifteenth lunar month, the infant is still spending approximately 20 per cent of its time clinging to the mother's ventral surface. (See Fig. 3.) Similarly, time spent in oral contact with the mother's

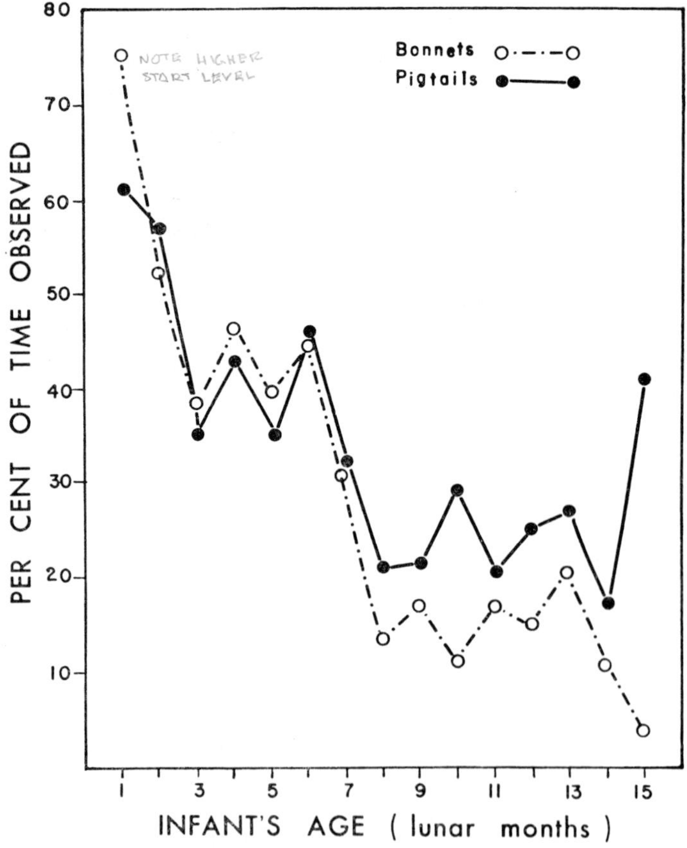

Fig. 2.—The development of "ventral-ventral cling" in bonnet and pigtail infants

PLATE 3.1.—"Passive contact" of bonnet mother with newborn and other females

nipple (without actual nursing) declines sharply during this period, but even at the end of the first year, the infant still spends most of the time that it is on the mother's ventrum holding the nipple in its mouth.

During the periods of close contact with their infants, both bonnet and pigtail mothers assume one of three basic postures toward the infant. During the early postpartum days, the mother will often "cradle" the infant, actively grasping it and pressing it to her. However, in almost all animals, this maternal behavior virtually disappears in the second half-year of life, and in some animals, "cradle" is rare even with regard to a newborn except during periods of stress. By far the more frequent behavior is that of "enclose," in which the mother, while seated, surrounds the clasping infant with the flexed forearms and/or hind limbs without actually making palmar contact with the infant's body. This maternal behavior also shows a decline as the infant ages but at a much more gradual rate than for "cradle." The third type of support posture assumed by the mother, almost never seen in the early days of life and still rather infrequently thereafter, is "passive support." In this behavior the mother does nothing to aid in support of the infant clinging to or holding her ventrum. This most frequently occurs at times when the mother is eating, exploring inanimate objects, or engaging in a social grooming interaction.

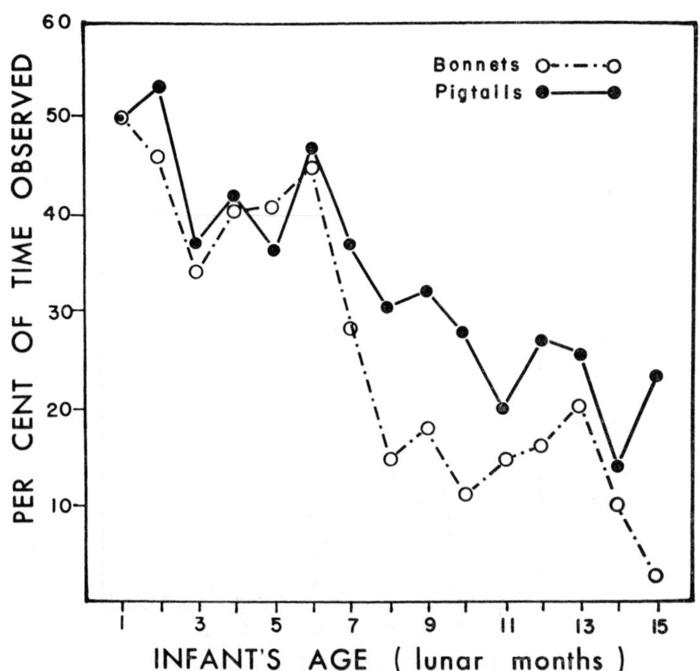

FIG. 3.—The development of "nipple hold" in bonnet and pigtail infants

During the early weeks of life, the mother not only maintains close contact with her infant while remaining stationary but she maintains this contact during locomotion as well. Although occasionally providing momentary support for her infant during carriage, prolonged "cradle carriage" is only rarely observed, as all healthy infants even on the first day of life are well able to support their own weight even when their mothers are in very active locomotion. In both species the time spent transporting the infant on the mother's ventral surface rapidly drops to almost zero by the end of the fifth lunar month. In both species under our laboratory conditions we have only rarely observed "dorsal-ventral carriage" in which the infant, supporting itself, sits or lies on the mother's back. It is our hypothesis that this behavior, frequently seen in many primate forms in the wild, may facilitate the long journeys which feral groups make, but is never required for the limited trips which occur in the confines of our laboratory pens.

By the third week of life, the previous total dependence of the infant on the mother begins to show signs of a gradual transition toward independence. The infant at this time begins to attempt to break contact with the mother. In the pigtails particularly, the mother initially shows considerable "departure

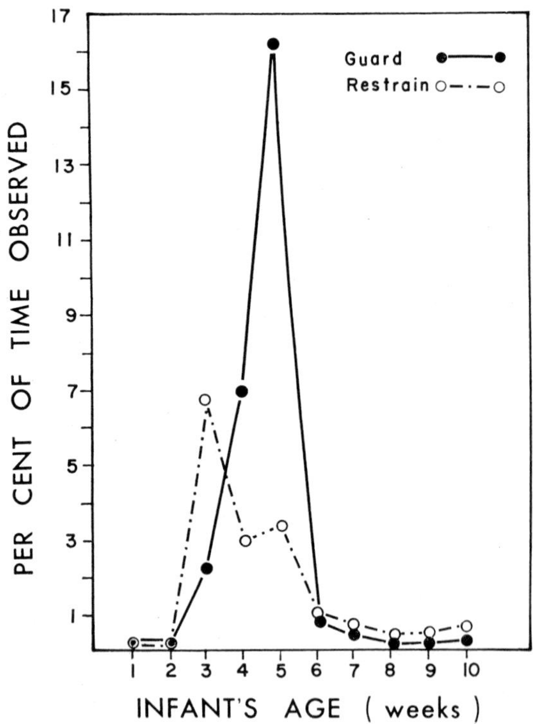

Fig. 4.—The development of "departure restraint" and "guard" in pigtail mothers

restraint" at this stage and generally prevents the infant from leaving. The infant's continued attempt to leave the mother is increasingly successful, as the mother's tendency to restrain departure decreases. However, the mother by no means simply disregards the separated infant at this stage. During the next several weeks she continues to show considerable "guarding" behavior as the infant walks about near her. As seen in Figure 4, pigtail mothers initially restrain their infants at the end of the first month, guard them during the second month, and show little obvious protectiveness thereafter. It is at this point that visual checking and watching of the infant by the mother, and vice versa, becomes the primary protective tie between the dyad during separations. The infant, during this period, frequently returns to the mother, and she shows repeated retrieval of the separated infant (Fig. 5). Except in obvious instances of social or environmental danger, it is not clear what stimulates the return by the infant or retrieval by the mother. It often seems that as the duration of separation increases, both partners grow uneasy and respond to some subtle, perhaps internal cues.

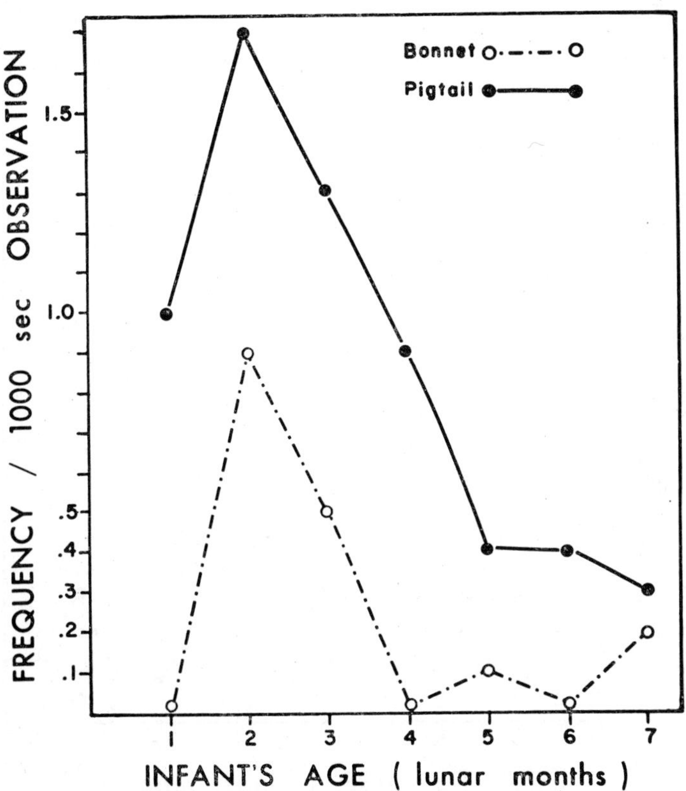

Fig. 5.—The development of "retrieval" in bonnet and pigtail mothers

Before leaving the discussion of this transitional stage of maternal hesitancy regarding the tendency of the infant to seek brief and repeated separations from her, we note that bonnet mothers show considerably less departure restraint and guard behavior of their infants than do the pigtail mothers. They do, however, as seen above, show the same peaking of retrieval behavior, although at a lower level, during this transitional stage. We speculate at this time that the continuity of close physical contact between adult members of the bonnet groups and the apparently related placidity of ambient dominance relations, which seems characteristic of their social interactions, may lead to less hesitancy on the part of the mothers, not only to allow other females to touch, explore, and handle their newborn, but also to allow their infants to separate from them more readily as they get older.

The mother's initial hesitancy to allow separation by the infant is followed, among the pigtails at least, in the fifth to eighth lunar months by behaviors which actively encourage the infant to break contact. By this time, mothers of both species actively engage in one or another forms of nipple withdrawal (Plate 3.2). This behavior may include a sudden raising of the arms, pulling the nipple from the infant's mouth; the use of the elbow to press the infant's head aside and off the nipple; or, as with one female, sliding the middle finger along the nipple into the infant's mouth and allowing it to suck the finger instead.

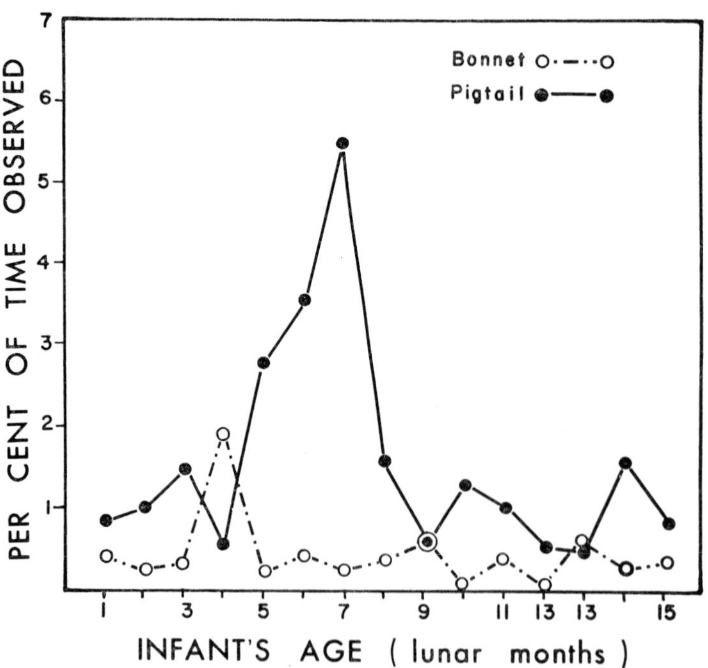

FIG. 6.—The development of "punitive deterrence" in bonnet and pigtail mothers

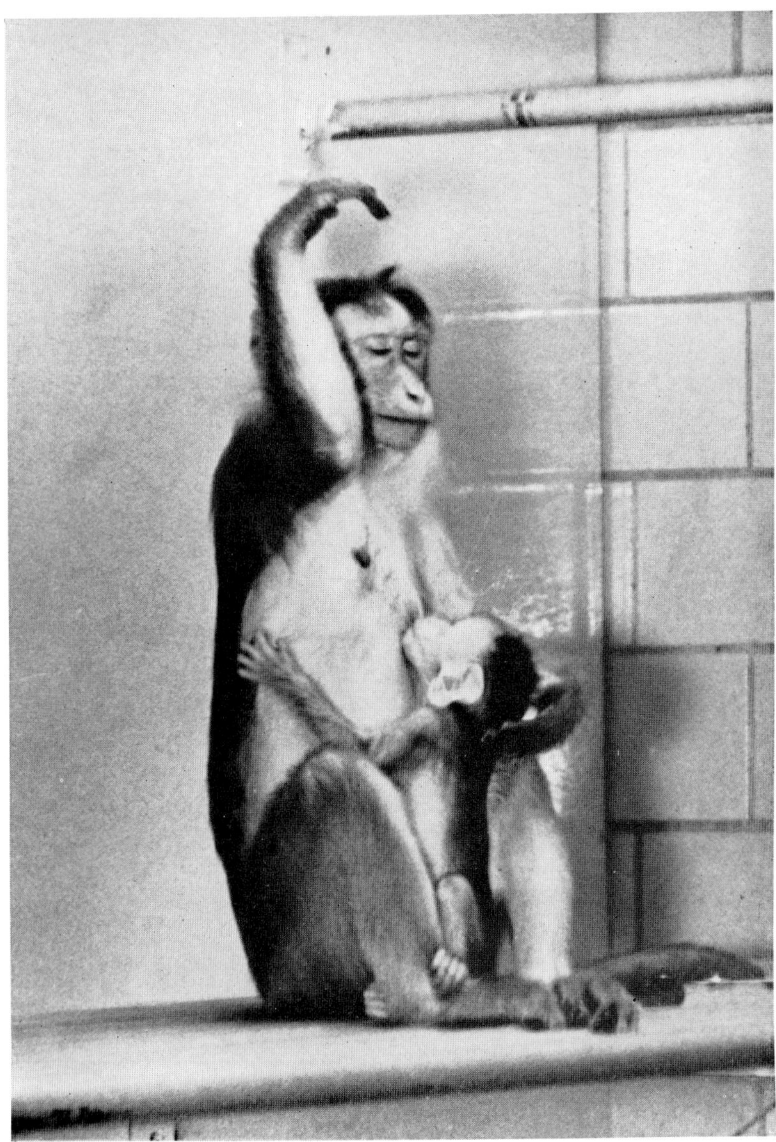

PLATE 3.2.—A frequent form of "nipple withdrawal" seen here in a bonnet mother

PLATE 3.3.—"Punitive deterrence" directed by a pigtail mother toward her infant

During this period of relative rejection, mothers not only engage in removing the infant from the nipple, but from time to time also show complete "infant removal."

"Punitive deterrence" (Plate 3.3) involves restrained biting of the limbs, shoulders, and head of the infant, or violent shaking of the infant's body (Fig. 6). Although seen to some extent in both species, it is engaged in to a considerable degree, particularly by pigtail mothers, from the fifth to the seventh month. This punitive behavior, occurring at times when the infant is attempting to reestablish or to maintain contact when the mother does not wish to, after some repetition and despite the cries and struggles of the infant, usually succeeds in preventing or breaking contact between the pair. As the infant grows older, this overt behavior wanes and a simple touch of the infant's side or the sudden shifting of the mother's posture suffices to cause the infant to release his grasp. It is noteworthy that the bonnet females show far lower levels of punitive deterrence than the pigtails, thus enhancing further the picture of differential maternal treatment in the two species.

Thus the macaque mother-infant dyad during the first 9 months of the infant's life passes through a marked evolution. The initial mutuality of maternal and filial dependence gradually gives way to an increasing independence of activity as the infant's and the mother's peer group involvements take precedence, culminating in a dramatic step toward dissolution of involvement upon the birth of the mother's next offspring.

REFERENCES

Jay, Phyllis. 1963. Mother-infant relations in langurs. In *Maternal behavior in mammals,* ed. H. L. Rheingold. New York: Wiley.

Rosenblum, L. A., Kaufman, I. C., and Stynes, A. J. 1964. Individual distance in two species of macaque. *Animal Behav.* 12:388–42.

4

THE DEVELOPMENT OF MUTUAL INDEPENDENCE IN MOTHER-INFANT PIGTAILED MONKEYS, *Macaca nemestrina*

GORDON D. JENSEN, RUTH A. BOBBITT, AND BETTY N. GORDON

The attachment relationships or affectional bonds among primates are the basic processes of the structure of their social behavior. The mother-infant relationship is the first such attachment and perhaps the prototype of all later such bonds. The following is a description of some preliminary observations of the loosening of this attachment in *Macaca nemestrina*, the pigtailed macaque. You might call this the process of untying the apron strings. We are investigating it in detail under highly controlled and simplified conditions in the experimental laboratory.

Our general question is: What roles are played by the mother and the infant in the detachment process? We are also interested in the ways in which differences in environment affect this process. Such environmental differences occur naturally among social groups of the same species. There is also reason to believe that environmental differences within a group result in many of the individual differences of its members. In order to test the effects of environment on the interactions of mother and infant, we are raising mother-infant pairs in two types of environment. One is a privation environment.[1] It is essentially devoid of changing extraneous stimuli or opportunities to manipulate or climb (Plate 4.1). These infants and mothers live in bare cages in soundproof

Gordon D. Jensen, Ruth A. Bobbitt, and Betty N. Gordon, Department of Psychiatry and Regional Primate Research Center, University of Washington, Seattle, Washington.

[1] We distinguish between privation and deprivation in terms of environmental stimulation. A privation environment is devoid of stimuli; deprivation occurs when available stimuli are removed.

rooms. The second environment is a rich one. Although confinement in a cage is a basic feature of this environment also, it is more representative of the natural setting, where there are other monkeys to see and hear, and a variety of occasions for climbing and manipulation (Plate 4.2).

For the study of these developmental processes, a quantitative, reliable observational method has been developed (Bobbitt, Jensen, and Kuehn 1964). It permits the analysis of behavior patterns as they emerge and change in relation to other patterns. It also permits us to define empirically the contingencies of behavior. In brief, most behavioral events are recorded as they occur, in sequence (Bobbitt, Jensen, and Gordon 1964). One 10-minute observation is made daily at random times. During this time, an average of three hundred behavioral units are noted for a mother-infant pair. Continuous reliability testing results in 80–85 per cent agreement between observers (Bobbitt, Gordon, and Jensen, in press).

When the study was about half completed, we perused the data on the animals already observed to derive empirical hypotheses about the nature and development of the interactive patterns. This process was essential to eventual analysis of all data because it enabled us to define the questions we wished to program for testing with the computer. The present discussion of behavioral trends is based on this preliminary perusal of patterns and thus must be viewed as indicating tendencies. In the main, these trends seem significant, but the statistical tests to confirm this will not be performed until data for the animals in the remainder of the study are available.[2]

Of course, there are many facets to the development of mutual independence. Only two of these will be discussed here: the mother's cradling the infant and the mother and the infant approaching and leaving each other. These behaviors will be described in terms of the position of the mother and infant in relation to each other. Relative position is basic because we define independence operationally according to the degree of closeness or separation of the two subjects.

We have defined four different positions. Position 1 is the closest, essentially the mother and infant situated in a ventral-ventral position with the infant grasping the mother with his arms and legs. Position 2 is defined as the infant still within the mother's "lap" area, but holding on with only one arm or not holding on with arms at all. In Position 3 the animals are in contact, but the infant is outside the mother's lap area. In Position 4 the two animals are completely separated and space can be observed between them.

Figures 1, 2, and 3 show the developmental trends of relative position changes. As with all curves presented here, the linear and curvilinear functions are determined by inspection of empirically derived points and the line

[2] At the time of going to press the trend analysis of individual behavior and relative positions on the full study of three groups of four pairs each is nearly complete. With one exception, to be noted later, the trends of the preliminary data have persisted.

PLATE 4.1.—The privation environment. Cage with no climbing facilities or manipulative objects situated in a soundproof room.

PLATE 4.2.—The rich environment. Toys for climbing and manipulation; visual and auditory access to other animals and availability of extraneous laboratory stimuli.

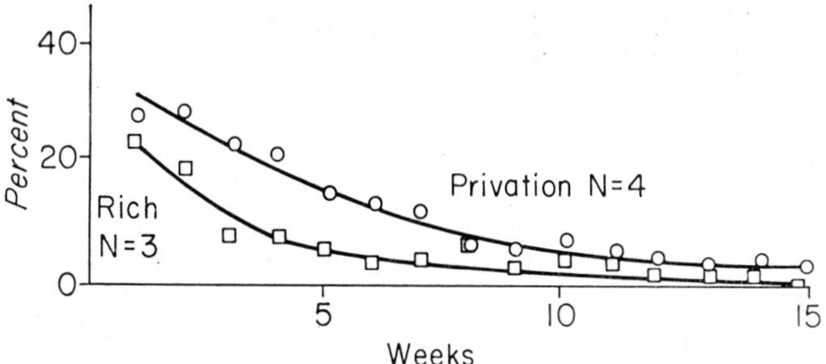

FIG. 1.—Relative frequency of changes to Position 1. (Transitions to Position 1/total number of transitions.)

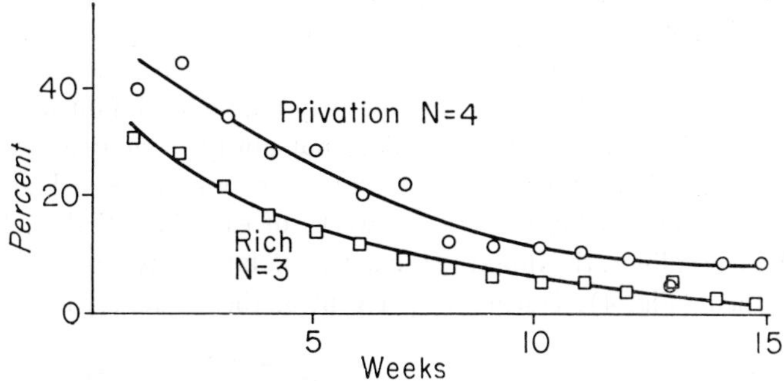

FIG. 2.—Relative frequency of changes to Position 2

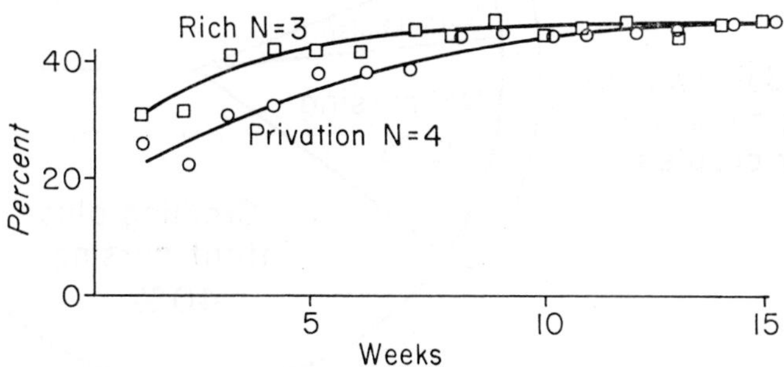

FIG. 3.—Relative frequency of changes to Position 3

smoothed graphically. Figure 1 shows the trends of the frequency with which the mother and her infant change into Position 1. This curve drops fairly rapidly, indicating that while the close position is relatively common during the first few weeks of life, it becomes much less frequent by the time the infant is 3 weeks old. Note the slower decrease in Position 1 for animals in the privation environment. In these curves, as in most of those to follow, this difference between pairs of monkeys in the two environments, whatever its significance, is remarkably consistent. Figure 2 shows the trend for Position 2. It is very similar to the trend of Position 1, with consistent differences between the two environments. In Figure 3, showing the trend for Position 3, the curve rises. Clearly, the infant increasingly ventures outside the mother's immediate ventral area but remains in contact with her. In this position he may be playing at his mother's side or may be actively climbing on her; the mother may or may not be actively in contact with him.

One of the most common groups of patterns that occur in close positions of 1, 2, and 3 is cradling. It is probably the most typical maternal behavior observed, and is not momentary but lasts for a measurable period of time. For purposes of analysis, cradling is called a duration behavior. It occurs primarily in Positions 1 and 2, although the mother may cradle her infant in Position 3. Cradling in a given relative position does not always comprise the total behavior pattern. About four thousand cradlings were observed among five mother-infant pairs during the first 15 weeks of each infant's life. In 44 per cent of these cradlings, the infant was inactive so far as any sustained behavior was concerned (Fig. 4). During 40 per cent of the cradlings, the infant was

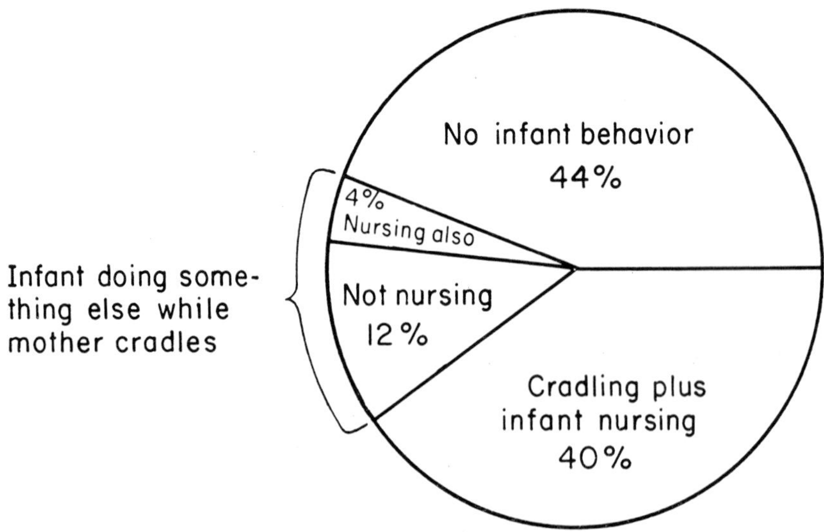

Fig. 4.—Patterns in which mother cradling occurs; with reference to concurrent infant behaviors.

only nursing. In 16 per cent of the cradlings, however, the infant was doing something else, either with or without nursing. Thus it is apparent that cradling is a very inactive situation for the infant, except for nursing, and that cradling and nursing tend to go together. We may hypothesize that cradling has an inactivating effect on the infant.

In connection with these behavior patterns and this hypothesis it is interesting to look at the variations in the types of cradling. We have defined three types: (1) the tight cradles in which the mother holds the infant tightly with both arms; (2) the usual cradle (Plate 4.3) in which the infant is held in the ventral area with his mother's arms closely surrounding him or with the infant outside the ventral area but still with at least one of the mother's arms around him; and (3) the loose cradle in which the mother appears to be passively inclosing the infant, her arms loosely surrounding him. The first two graphs in Figure 5 indicate the correlation of these three kinds of cradling with the infant's concomitant behaviors. The decreasing frequency with which the infant nursed is associated with loosening of the cradle. Non-nursing activity by the infant also increased as the cradle became looser. The third graph in Figure 5 shows the average duration of each type of cradle. Note that the tight cradle was held for longer periods of time. These results lead us to hypothesize that the variations of the cradle serve to modulate infant closeness. In other words, as the mother loosens her cradle, the infant decreases his nursing and increases his activity; also, he remains in the position of being cradled for a shorter period of time. We suspect that the changes in the mother's cradling have a definite communicative significance for the infant, that the infant's changing from nursing to active behavior has a definite communicative significance to the mother, or that both kinds of communication occur. In our final analysis we will look at the sequences of the individual behaviors to identify the contingencies of the individual behaviors and thus to define more precisely the communicative significance of each.

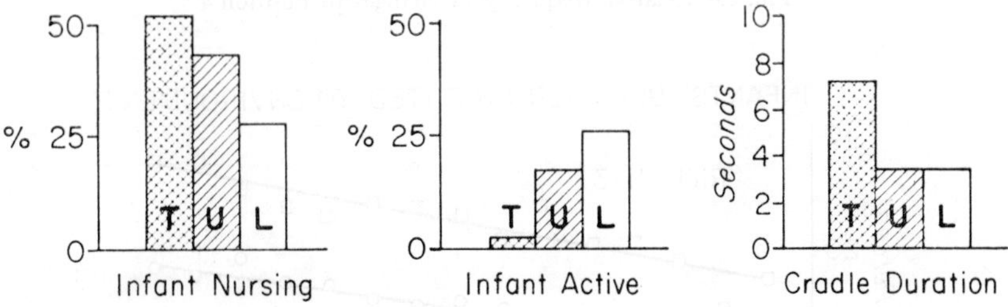

FIG. 5.—Infant nursing, infant activity, and average duration of cradling relative to intensity of mother cradling.

T—Tight Cradle, N = 107
U—Usual Cradle, N = 3996
L—Loose Cradle, N = 716

Turning again to the developmental trends of position changes, we can see in Figure 6 a gradual rise in the proportion of changes to Position 4. Again, there is a difference between the trends for the infants in the two environments. The infants in the privation environment have not attained the level of Position 4 behavior seen in the infants in the rich eironment.

Another aspect of the development of mutual independence, the proportions of the infant's behavior oriented to his physical environment, is illustrated in Figure 7. Environmentally oriented behavior increased at the same rate for infants in both environments but is greater throughout for infants in rich environments. These curves complement the evidence of the relative position curves in showing a gradual loosening of the mother-infant attachment and a strengthening of the infant's interaction with his environment. This supports our concept of the infant's becoming detached from his mother as he becomes reattached to other objects. The basic question, of course, is what roles the mother and infant play in the detachment-reattachment process.

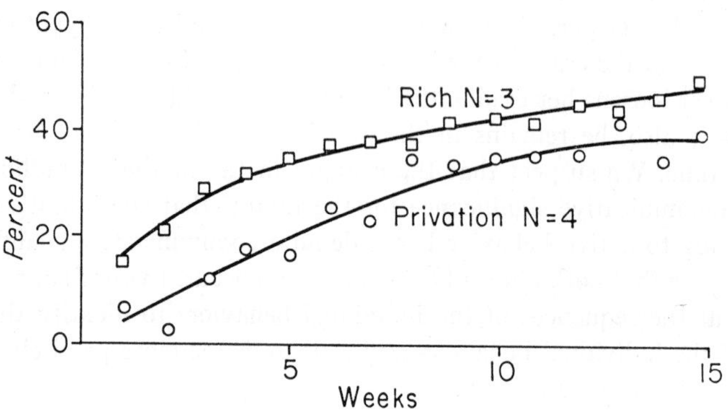

Fig. 6.—Relative frequency of changes to Position 4

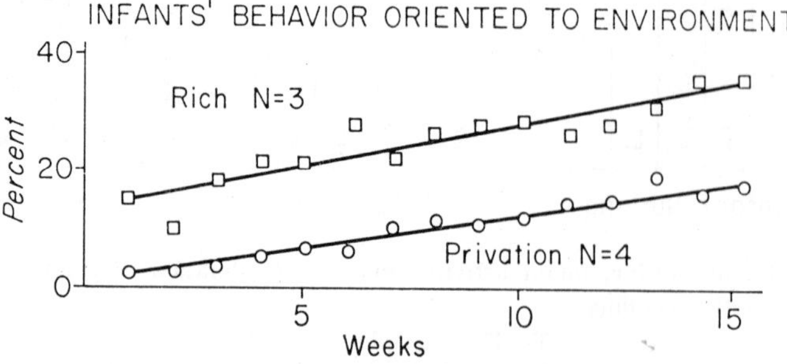

Fig. 7.—Relative frequency of infants' behavior oriented to physical environment

PLATE 4.3.—Mother cradling infant with a usual cradle

PLATE 4.4.—Infant approaching mother; the predominant behavior in his interactive locomotion.

PLATE 4.5.—Mother leaving infant; the predominant behavior of her interactive locomotion.

One possible method for elucidating these roles is to study patterns produced by the animals' approaching (Plate 4.4) and leaving (Plate 4.5) each other. The trends for the mothers' interactive locomotion, their approaching and leaving their infants, are displayed in Figure 8. Again we see a higher level for animals in the rich environment, at least initially. Gradually, however, the privation mothers' interactive locomotion exceeded that of the rich. The peak or bulge in trend for mothers in the rich environment, indicated by the dotted line, is concurrent with an increase in the infant's activity on his mother (Jensen and Bobbitt 1963; Bobbitt, Jensen, and Kuehn 1964).[3] The interactive trends for the infants are shown in Figure 9. It is obvious that the lag in infant interactive locomotion is a function of the maturation process. In interpreting the curves for interactive locomotion by both mother and infant, we find them consistent with our general hypothesis about the effects of environment. Initially, the rich pair engages in more interactive locomotion than does the deprived pair. As the infant in the rich environment becomes capable of doing more in his wider environment, however, he becomes more interactive with it. In contrast, the privation infant and mother, who can interact only with each other, gradually attain high levels of interactive behavior.

The proportion of this interactive locomotion that consists of one animal's leaving the other is plotted in Figure 10. The mothers' interactive locomotions have been almost exclusively movements away from their infants. In contrast, the majority of the infants' movements have been toward their mothers. There were essentially no differences in the curves for the animals in the two different

[3] In the final analysis for the entire study, the trends of interactive locomotion were the same for mothers in both environments. Although the trend for the group in the rich environment starts from a higher level, the difference between curves is not significant at a high level of confidence.

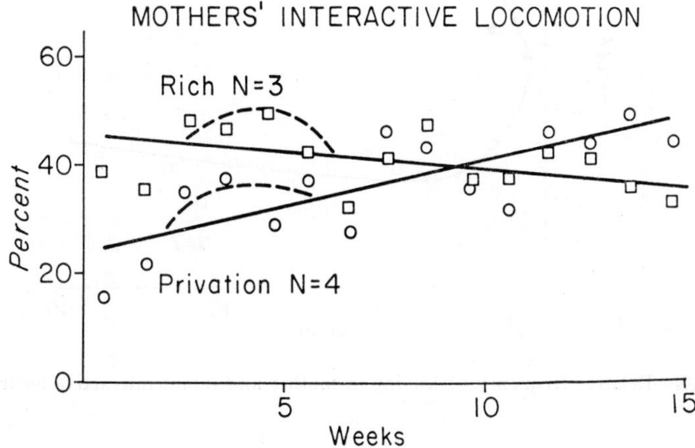

FIG. 8.—Relative frequency of mothers' locomotion that is interactive

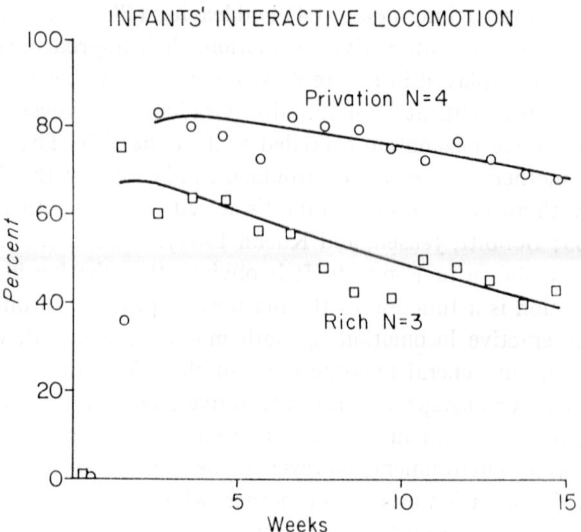

FIG. 9.—Relative frequency of infants' locomotion that is interactive

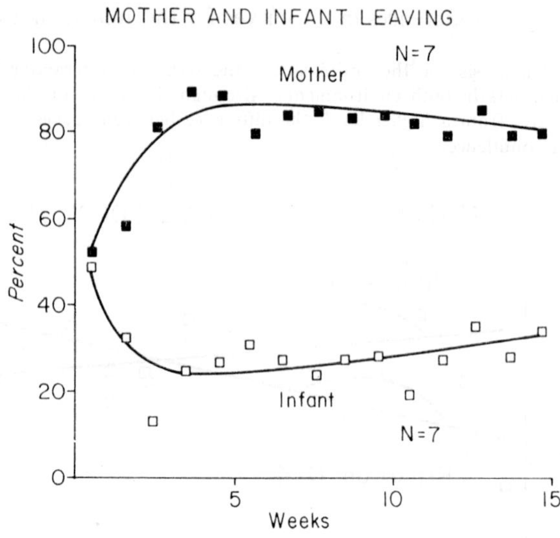

FIG. 10.—Relative frequency of leaving in mother and infant interactive locomotion

environments. Thus, mother-leaving-infant and infant-approaching-mother types of behaviors are probably constants irrespective of differences in environments. This is strong evidence for the active role that mothers play in instigating infant independence and their own independence.

To round out this picture of the development of interactive locomotion, the proportion of the total leaving done by the mothers and by the infants is shown in Figure 11. Although the data in Figure 10 indicates that 90 per cent of the mothers' interactive locomotion is leaving, this leaving is only about half the total leaving after the infant is a few weeks old. Furthermore, even though environment has had no apparent effect on leaving as a percentage of the mothers' or infants' interactive locomotion, such a difference does appear in the distribution of leaving between mother and infant. It seems that although mothers in general are predominantly "leavers," the mothers in the rich environment do a greater proportion of the total amount of leaving than do the mothers in the poor environments during the early period of developing independence.

During our observations of rhesus monkeys in a seminatural state on the island of Cayo Santiago, in the Caribbean, infants who were playing at a distance from their mothers regularly seemed to come immediately when their mothers began to move away. It seemed that the mother's leaving somehow functioned as a communicative signal for the infant to come to her. In the laboratory, 50 per cent of the mother's leavings are associated with immediate infant approaches. This percentage is quite significant, since the infants'

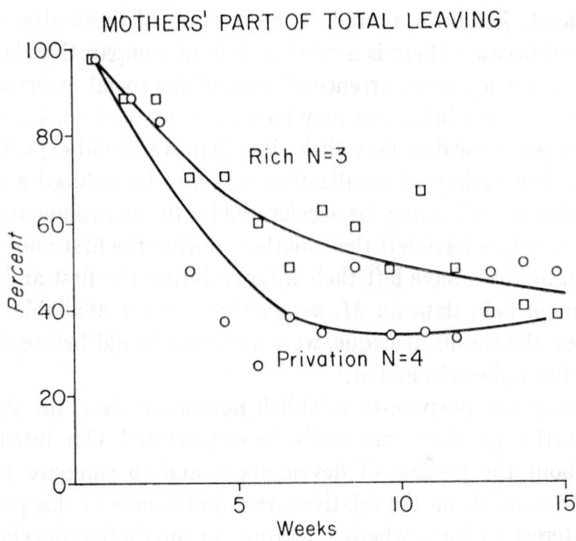

Fig. 11.—Proportion of mother-leaving behavior in relation to total-leaving behavior

approaches to their mothers constitute only 9.4 per cent of their behavior in general.

Some indications of sex differences in leaving have been derived from the sample data mentioned earlier in the discussion of cradling. Among the male infants in the privation environment, 22 per cent of the interactive locomotion was leaving, whereas among the female infants it was 18 per cent. The mothers of the male infants do 92 per cent leaving and the mothers of female infants 85 per cent. In other words, for the female infants and their mothers a smaller proportion of their interactive locomotions was leaving. Itani's (1959) field observations of Japanese macaques seem consistent with these data. He noted a difference between the sexes in the process of social maturing. Young males left their mothers to form groups of male peers at an age when young females were remaining with their mothers.

DISCUSSION

Our initial analyses of data for the early period of life begin to suggest some of the ways that environment affects development and behavior performance. These differences seem to support our general hypotheses that environmental privation, without any maternal deprivation, retards the mutual independence or detachment process and the process of reattachment to other stimulus objects, and that this degree of early environmental privation critically affects the animals' later social performance.

Any laboratory environment is a restricted environment when compared with the natural one, so that the course of development of animals in each of our environments has been altered. The direction of these alterations cannot yet be assessed because there is a relative lack of comparative data from field studies. The frequency of occurrence of some of the social repertoire may have been reduced and some behaviors may have been accelerated. For example, our mother-infant pairs vocalize very little, less than we would expect in a natural environment. The variety of vocalization may also be reduced and altered. In contrast, leaving behavior may be accelerated by the environmental restriction. In our studies, babies have left their mothers during the first and second weeks of life; often, mothers have left their infants during the first and second days of life. Although field data on *M. nemestrina* are not available, infants of a related species, the rhesus macaque, were 3 to 4 weeks old before they left their mothers in a natural environment.[4]

Since it is not our purpose to establish normative data, the significance of these comparative problems can easily be exaggerated. Our intent is to learn something about the process of development and to compare the effects of experience differentials on the relative speed and course of this process. It will also be of interest to learn whether features of interactive development other

[4] C. B. Koford, personal communication.

than mother-leaving and infant-approaching locomotion are differentially affected by environmental differences. If they are not, this may mean that there exist constants in social interaction. Such "robust" behavior patterns, as we might call them, may constitute the basic structure on which the interactive process is built. Further, these patterns could be common to other nonhuman primate species and even to the human primate. From our knowledge of evolution we should expect some behavior patterns in human primates similar to those in subhuman primates, as with many physiological structures.

In summary, we have presented empirically derived measures of the development of mutual independence of mothers and infants. These data and those derived from aspects of the study not presented here enabled us to form hypotheses about the communicative dynamics of these situations. Computer analysis for trends in the behavioral measures of the entire study have since substantiated the hypotheses about the development of mutual independence set forth here.

This research was supported by a PHS Research Grant No. HD 00883-03 from the National Institute of Child Health and Human Development, Public Health Service.

REFERENCES

Bobbitt, R. A., Gordon, B. N., and Jensen, G. D. Development and application of an observational method: Continuing reliability testing. In press.

Bobbitt, R. A., Jensen, G. D., and Gordon, B. N. 1964. Behavioral elements (taxonomy) for observing mother-infant-peer interactions in *Macaca nemestrina*. *Primates* 5(3–4):71–80.

Bobbitt, R. A., Jensen, G. D., and Kuehn, R. E. 1964. Development and application of an observational method: a pilot study of the mother-infant relationship in pigtail monkeys. *J. Genet. Psychol.* 105:257–74.

Itani, Junichiro. 1959. Paternal care in the wild Japanese monkey, *Macaca fuscata fuscata*. *Primates* 2:61–93.

Jensen, G. D., and Bobbitt, R. A. 1963. On observational methodology and preliminary studies of mother-infant interaction in monkeys. In *Determinants of infant behavior,* Vol. III. ed. B. M. Foss. New York: Wiley.

DISCUSSION OF REPRODUCTIVE BEHAVIOR

STUART A. ALTMANN

The fundamental process of evolution is change of gene frequency in a population. So far as is known, such changes can be brought about by only four processes: (1) selection, that is, differential reproduction and differential survival (genetically, the latter is a special case of the former), (2) gene migration, (3) mutation pressure, and (4) genetic drift. Consequently, reproductive behavior is of central importance in the relation between behavior and evolution. The chapters that have been presented in this section are related to two quite distinct aspects of reproduction: sexual behavior and parental care. In either case, the question of basic evolutionary impact turns primarily on the way in which such behavior affects the relative contribution of each individual in the society to the gene pool of the next generation.

In what follows, the problems that are involved in several such relations will be singled out, with emphases on possible means of empirical solution. I make no pretext that these problems are exhaustive or even the most important in understanding the relations between reproductive behavior and evolution. They are included because they impressed me and because they are relevant.

Seasonal breeding has been indicated for *Lemur catta* by Jolly and for several other species of primates as well (Lancaster and Lee 1965). In those instances in which seasonal breeding apparently does occur, one wonders about the adaptive significance of this phenomenon. In general, it seems unlikely that there is any striking disadvantage to sexual behavior per se being carried out at any time of the year. But seasonal breeding leads to seasonal births, and judging by studies on other animals, it is much more likely that survival of infants is at stake.

Selection leads not only to the development of characters, but also to their maintenance. Consequently, we can expect that if seasonal breeding is selected on the basis of infant survival, it will work against any processes that tend to

Stuart A. Altmann, Yerkes Regional Primate Research Center, Emory University, Atlanta, Georgia.

increase the temporal variability of parturition. A difficult but crucial study would compare the survival of those infants born very early and very late in the season with those born at less deviant times.

The available data on seasonal breeding have led Jolly, as well as several other authors, to criticize the theory of Zuckerman (1932) that the sexual bond is responsible for the year-round social groupings of primates; this theory has also been criticized on the basis of the disruptive effect of sexual behavior on the group, because it does not account for the sociability of prepubescent individuals, and because even an adult female is in estrus during only a small portion of her life, with "time out" for periods of anestrum, diestrum, pregnancy, and lactation (Altmann 1962, DeVore 1962, Jay 1962).

For the sake of argument, let me play the devil's, or rather, Sir Solly's advocate. First, the data that have been published on seasonal breeding are anything but conclusive (Lancaster and Lee 1965), with the exception of those for the rhesus on Cayo Santiago (Koford 1965) and for the Japanese macaques (Lancaster and Lee 1965), nor is there any reason to believe that we can extrapolate from the breeding of these macaques to other species of primates. And even if, for some populations of some species, breeding is in fact seasonal, this does not preclude that the animals are conditioned to this powerful incentive during those times when it does occur. The same can be said in answer to the argument that the female is periodically in anestrum or diestrum.

As for the data that indicate a marked increase in aggressive interactions, intergroup movements, and so forth during the breeding season, these are perhaps the data that best show what a powerful incentive such sexual behavior is! For just the same reason, fights break out when a group of hungry monkeys are given food.

Regarding prepubescent primates, two other remarks should be made. For one thing, sexual behavior is, in fact, quite common among young primates of many species. Second, no one has denied the existence of other social bonds. Certainly, that between mother and infant is a powerful one. But this bond is present in all mammals and hence cannot account for the fact that a primate social group, unlike that of most (but not all) other mammalian social groups, consists of members of both sexes and all ages.

These various arguments, pro and con, have been presented not because any of them is very convincing, but because they illustrate a very real dilemma for students of primate behavior, which, in broadest aspect, stems from the problem of relating the behavior of individuals to the processes of the group. The question is, Just how does one evaluate the relative potencies or strengths of the various cohesive factors that keep the members of a group of primates together? To my knowledge, this problem has never been squarely faced anywhere in the literature on primate behavior. The best available discussion of primate sociability is that of Mason (1964).

Any attempt to take a social group apart, either by setting up partial groups, as has been done, for example, by Jensen, Bobbitt, and Gordon, and by Rosenblum and Kaufman, or with an entire group, by any partial sampling of behavior, as is the inevitable procedure in field studies, runs the risk that, as a result of emergent properties, studies of the parts will not, when put together, reconstitute the whole. Perhaps the first demonstration of this was Maslow's 1936 study, which indicated that the agonistic relations within a group could not be predicted from the interactions when the monkeys were put together in pairs.

A problem similar to the last is that of disentangling from the total configuration of stimuli that impinge on the organism those that constitute the social signal. The studies of Jensen, Bobbitt, and Gordon and of Rowell illustrate the great difficulty of this task, a difficulty that is compounded by the fact that primates, like mammals in general, often rely upon multisensory information input. Certainly, the obvious assumption about sexual skin swellings has been that they evolved as a means of communication by estrous females. Indeed, they have all the hallmarks of "good" social signals for this purpose: they are particularly prominent during just those phases of the menstrual cycle when conception is most likely; they are absent during diestrum, pregnancy, and early lactation; they are made more conspicuous by the sexual presenting of the female; and they can hardly be "missed" by the males as they put their muzzles to the perinea of the females. Yet, as Rowell has realized, none of this is direct evidence that the sexual skin swellings have any communicative significance. What would be? Perhaps the closest thing to decisive evidence could be obtained through small, localized administrations of progesterone—enough to induce a detumescence of the sexual skin, but not enough to affect behavior or reproductive physiology. Studies on the responses given to such females, compared with those to normal females, should help clarify the communicative significance of the sexual skin.

What is the adaptive significance of interspecific and intraspecific differences in maturational rates such as are indicated by the data for bonnet and pigtail macaques presented by Rosenblum and Kaufman and by Jensen, Bobbitt, and Gordon? The sensitive techniques that have been developed have enabled Jensen and his colleagues to pick out consistent developmental differences in both maternal and infantile behavior between those pigtail macaques in a *rich* environment and those in a *privation* environment. Yet, by comparison with the natural environment, even the so-called *rich* environment is highly impoverished. How much greater, then, must be the differences between these animals and those in the wild? But more important, if maternal care and infant development are so dependent upon environment phenomena, how do these differences affect the behavior and consequently the social organization of animals that are raised in various environments, and to what extent are these differences adaptive to these particular environments? Comparable questions about basic adaptive relations can be posed about the converse influences of social structure on mother-infant relations such as those that Rosenblum and Kaufman are studying.

The task of relating a process of mother-infant behavior to the social structure is neither simple nor obvious. For example, by comparison with pigtails, bonnet macaque mothers restrain their infants less; Rosenblum and Kaufman have suggested that this may result from the greater placidity of dominance relations in which the mother and infant bonnets find themselves. Had the results for infants of these two species come out in the reverse direction, could one not have thought of an equally plausible explanation? These studies of interspecific and intraspecific differences in developmental processes raise many questions about the corresponding natural situations. It is difficult, however, to pose the questions in such a way that the results of any research that attempts to answer them will not be equivocal.

If we are going to increase the relevance of our research to an understanding of how primate behavior has evolved, and how it has affected other evolutionary processes, we must continue to bring field and laboratory studies into greater congruence. There are many ways in which this can be, and has been, accomplished. One way is brought to mind by the two studies of pigtail and bonnet macaques: just as field workers can benefit enormously by observations on captive animals, particularly in conjunction with an ongoing laboratory study of behavior, so it would seem advisable for any laboratory study that involves many years of intensive research on the behavior of a species to be preceded by, or at least to include, several months of field observation. When such field observations could be carried out in collaboration with a fieldworker, particularly near the end of a field study—at the time when the fieldworker's perceptive abilities and background information on a local population are at their greatest, much could be accomplished in a relatively short time.

Among the cercopithecine primates, the group for which we have the most abundant research results, a common core of behavior patterns is becoming apparent, albeit with some species-specific modifications. Doubtless, many of these behavior patterns have a common evolutionary background. In contrast, the social structures in this group seem much more variable and divergent. Thus, it appears that in this group of primates social evolution has been not so much an evolution of social behavior patterns as it has been an evolution of the uses to which they were put.

REFERENCES

Altmann, Stuart A. 1962. A field study of the sociobiology of rhesus monkeys, *Macaca mulatta*. *Ann. N.Y. Acad Sci.* 102 (Art. 2):338–435.

DeVore, Irven. 1962. The social behavior and organization of baboon troops. University of Chicago doctoral dissertation.

Jay, Phyllis C. 1962. The social behavior of the langur monkey. University of Chicago doctoral dissertation.

Koford, Carl B. 1965. Population dynamics of rhesus monkeys on Cayo Santiago. In *Primate behavior: field studies of monkeys and apes,* ed. I. DeVore, pp. 160–74. New York: Holt, Rinehart and Winston.

Lancaster, Jane B., and Lee, Richard B. 1965. The annual reproductive cycle in monkeys and apes. In *Primate behavior: field studies of monkeys and apes,* ed. I. DeVore, pp. 486–513. New York: Holt, Rinehart and Winston.

Maslow, A. H. 1936. The role of dominance in the social behavior of infrahuman primates: IV. The determination of hierarchy in pairs and in a group. *J. Genet. Psychol.* 49:161–98.

Mason, William A. 1964. Sociability and social organization in monkeys and apes. *Adv. Exptl. Sociol. Psychol.* 1:277–305.

Zuckerman, S. 1932. *The social life of monkeys and apes.* London: Kegan Paul, Trench, Trubner.

Part II

AGONISTIC BEHAVIOR

5

TRIPARTITE RELATIONS IN HAMADRYAS BABOONS

HANS KUMMER

Analyzing the interrelation of three rapidly unfolding individual sequences of behavior is a task which, if properly carried out, requires recording methods of higher sensitivity for timing minute events than those available so far in field studies. While tripartite behavior in other baboons and macaques has been repeatedly described in qualitative terms (especially by DeVore 1962 and Altmann 1962), its interpretation on the basis of mere description by others seems too daring to attempt. This chapter, therefore, will be limited to the species which I could observe in the Zurich Zoo from 1955 to 1958 and near Erer-Gota and at other locations in eastern Ethiopia in 1960 and 1961. Fortunately, hamadryas baboons display all the major tripartite patterns observed in other species. This was especially true in a captive hamadryas colony in which more complex tripartite patterns had developed than were found in free-living troops (Kummer 1965).

In this chapter, no emphasis will be placed on minute descriptions of the behavioral sequences, since most of them have been described earlier, together with definitions of the component behavioral units (Kummer 1957). The aim here is to link the main facts together in ontogenetical order and to give a working hypothesis of their genesis—namely, that the first and basic triangular relation in a primate's life is the protection of the infant by its mother against another group member's curiosity or aggression, and that many of the tripartite relations among adults stem from it. For the sake of brevity, terms like "threat" and "agression" will be used as labels for sets of behavioral units commonly considered to have those functions.

A tripartite relation as it is understood in the following is not merely a behavioral sequence in which three monkeys participate, since hardly any

Hans Kummer, Delta Regional Primate Research Center, Covington, Louisiana.

interaction between two animals remains unaffected by the presence of others. Rather, it is composed of sequences in which three individuals *simultaneously* interact in three *essentially different roles* and *each of them aims its behavior at both* of its partners. Thus, many of the more common scenes in primate social life are excluded. If a subadult male is attacked by an adult male and afterward redirects his counteraggression onto a low-ranking female, this is not a tripartite relation, since the events occur at different times.[1] The cooperation of two animals in threatening a third one or in searching for a lost infant also must be excluded, since the cooperators do not appear in essentially different roles. All tripartite behavior described in this chapter is displayed in agonistic contexts. It is certain that other tripartite and more complex relations exist, for example, among the males directing the troop's movement, but they will not concern us here.

The relation of an infant, an aggressor, and a protective mother, though transitory, fulfils the above conditions. It is known from many primate species that a mother will slap at a female trying to touch her infant, or that she will threaten an aggressive playmate when her infant screams. Mason (1964) recently has noted again that the filial attachment of the infant may later be transferred from the mother to another adult (for instance, a dominant male). The evidence of this shift is especially convincing in hamadryas groups. The question arises whether the triangular relationship of an infant, its protector, and an antagonist is equally preserved into the juvenile and adult life. According to the evidence gathered on the hamadryas groups, the answer is yes, although the originally simple behavior takes on a number of forms as the infant grows and sexes differentiate.

A hamadryas infant frightened by another will run to its mother and grasp her fur, whereupon the mother may threaten the aggressor. But a baboon, at 1 year of age, very often plays with other juveniles in a group out of sight of its mother. Such a play group of hamadryas juveniles usually forms around a subadult or young adult male. He does not take part in the play, but now and then screaming, frightened players run into his arms, whereupon he threatens the aggressor, acting as a substitute for the several mothers in succession. Instead of just clinging to the male's fur, the infant sometimes turns around and looks at the aggressor, still screaming, and now it seems to learn that sitting close to a male and screaming at another animal is likely to induce submission in the latter and provoke attack by the male protector. Thus, the infant or young juvenile is able to *establish* the triangle with himself in the role of the protégé and with the male in the original role of the protective mother. However, the male is not that juvenile's protector exclusively, as its mother was, for he will also attack it in certain situations in favor of another protégé.

[1] In rhesus macaques (Altmann 1962) and hamadryas baboons such redirected aggression is sometimes aimed into empty space instead of at a third animal. Hamadryas also threaten while sleeping, or present to nobody.

Therefore, the large male becomes an ambivalent figure, mainly protective toward the infant sitting close to him, but on the other hand aggressive if another infant manages to sit closer. In contrast to the certainty of the roles in the infant-mother-aggressor situation, it is no longer certain which will be the protected, and which will be the attacked. The consequences of this situation are easily observed in behavior. First, the juvenile wavers between fleeing toward the adult male and fleeing away from him. Second, the place in the arms of or close to the protective male is now an object of competition among the participants in an agonistic encounter, and this situation is new for the infant who, until now, was the only protégé of its mother. In spite of these complications, juvenile and subadult hamadryas baboons continue to run toward an adult male in most serious encounters, and the females even do so throughout their lives.

The conflict about the two flight pathways results in several displaced and redirected forms of grooming the male's coat (Kummer 1957), which need not concern us here, except for the fact that even adults, not only infants (Harlow 1961), may under stress be strongly attracted by fur. The other difficulty, namely, the competition of two potential protégés and the need for protection even against the protector, is the cause of the social techniques which we are to consider now. In the strongly cohesive one-male groups of hamadryas baboons, several females are conditioned to follow one male at all times. Females wandering away are brought back by the group leader's neck bite. For each female, thus, there exists but one protector, and seeking out another male would lead to severe fights in which the female would be physically torn back and forth between the two males. This protector, however, is also her main potential aggressor, and thus the situation is basically the same as it was in the play group around the subadult male.

In some one-male groups the competition of the different females for the male's protection is an almost constant matrix of interactions. The females in this respect are especially resourceful. The most simple form of becoming the protégé is to arrive first at the male's side, as shown in Plate 5.1. This, however, is far from a safe means; therefore in addition to screaming, even a 2-year-old female will present to the male. It is almost certain that presenting protects against attacks by a dominant male.

At this stage, the competitive atmosphere of the situation is quite obvious: the female closer to the male now tries to threaten her opponent away from the male, staying as much as possible between the two, while presenting to the male. Both females may sometimes run in circles around the male, the closer one always holding her place on the radius of the more distant one while presenting to the male. This whole pattern of so-called protective threat in front of the group leader is the fundamental form of aggressive encounter between adult hamadryas females.

In the next most complex step, the experienced adult female encounters an opponent of roughly equal or lower dominance status than herself. She now proceeds to perform her own attacks away from the male, aiming her bites typically at the tail of the other female. The latter will not flee, but will stay as close to the male as the attacking female will let her. Between the attacking, even a very dominant female will each time withdraw and present before the male.

This whole technique probably must be learned, as the following shows. A juvenile female sometimes runs such attacks away from the male against an adult female much above her in the dominance order. In these instances, when the juvenile aggressor is safely away from the protector, the dominant female may counterattack and displace the juvenile from her central position, thus reversing the original pattern.

What is the reaction of the male to all this? Usually none. But if he attacks, it is usually the more distant of the opponents which is bitten. The following sequence observed in Ethiopia may exemplify the effectiveness of staying close to the male—even if he does not interfere at once. A very aggressive old female ran attacks against a subadult female who was clinging to the back of the group leader. The aggressor passed the two very closely and tried to slap at the young female, but did not reach her. After 8 seconds the male rose to his feet and stared at the aggressor, the young female still clinging to his flank. Now the aggressor ceased to attack and withdrew.

During phases of changing group structure in the zoo colony, females sometimes would initiate protective threat without apparent reason against an opponent which was not even aware of the aggressor. Now, it was never observed that a male attacked an unaware opponent. The technique developed by several females of the colony to overcome this problem was pulling the opponent's tail before threatening.

Adult female IV quietly sat cradling her infant. Subadult female 3a lingered around without being noticed. Suddenly 3a started screaming, rushed past IV pulling her tail violently, and ran on to their group leader. Female IV jumped to her feet, screamed, and ran toward the male, where 3a was already in the protected threat position. The male aimed a slight brow-lifting threat at IV.

The last step of development, in which the female protégé practically took over the male's role, was only once observed in captivity. The old, experienced female Vecchia had recently entered a young and inexperienced male's group. In situations which usually provoke a group leader's attack and neck bite, this male consistently failed to react. The old female usually was sitting behind the male. In most instances where the male failed to attack, she ran forward and carried out "his" attack with a typical neck bite, which normally is the group leader's privilege. Immediately she would withdraw behind the male again. Vecchia also mounted females when they presented to the male and bit them when they mated with the male. The following sequence is an illustrative example:

Plate 5.1.—A group leader holds the first of two quarreling females to approach him and stares at her opponent while two other females cooperate in his threat (Ethiopia).

Without apparent reason, a female, Sora, of the inexperienced male's group attacked the subadult female of the senile group leader nearby. Two seconds later, both females were attacked, but while the subadult female was chased by her senile group leader, Sora drew a neck bite from Vecchia. The subadult female then followed the senile male and groomed him; Sora ran to her inexperienced group leader, but there Vecchia constantly got between the two and finally took Sora's grooming.

Let us examine briefly the probable mechanisms underlying this entire development.

First, the growing female infant seems to transfer the mother's role onto a male. Throughout adult life, a female under extreme stress will cling to the male's back or be embraced by him in the way of the infant. Another piece of evidence supporting the transfer hypothesis is that the hamadryas female enters the consort relation with her group leader not as a fully mature consort, but as an infant of about 1 year of age. These initial groups, which make up almost one-fifth of all one-male groups in our Ethiopian population, consist of one immature female and a young adult male. They exhibit typical maternal behavior, whereby the group leader takes on the mother's role toward his young consort. The initial groups seem to develop into normal one-male groups as the female grows up.

Future studies may support the speculation that the hamadryas one-male group evolved on the basis of a transferred mother-infant relation. This transfer, however, is not sufficient to keep the adult female closer and more constantly within the male's neighborhood than even a 1-year-old infant stays with its mother. Additional means of attraction are needed. First, the "following" response of the female is reinforced again and again by the male's attacks and neck bites. In the field these attacks occur several times per hour in the first days of the male's relation with the female infant, while later they occur once every few days. Besides this, the hamadryas male's coat may have evolved as a supernormal stimulus (see Tinbergen 1951) aimed at a persisting infantile attraction to fur. In captivity, the tendency of all subadults and adults to inspect or groom the male's coat during stress was so pronounced that this behavior even was redirected to the ground when inhibited by strong flight motivation, and again when the group leader became senile and his hair became short and brown.

A second mechanism underlying the phenomena of protected threat is social facilitation. The probability that a primate threatens or attacks an object is higher if another group member has already done so. Hamadryas baboons sometimes just stare and slap in the direction of their neighbor's threat, even if they cannot see the object of this threat because of an obstacle. In protected threat, the central female provokes the male's threat partly by means of social facilitation. Altmann suggests that threats away from the dominant animal appease the latter.

Third, the aggression away from the dominant animal may partly be redirected aggression originally aroused *against* the dominant male.

A fourth mechanism is also probably responsible for the behavior of the experienced female acting in place of her group leader. Imagine an animal A clearly aiming communicative gestures at B who, however, does not react. Now, a third animal, C, with a very low threshold for exactly this reaction, stays near B. It is likely that under certain circumstances C will react instead of B. Two examples may illustrate this. A mother did not react to the screams of her infant which had slipped into a crevice in front of her. Another female having no infant of her own, and sitting close by, after a while grasped the infant and gently pushed it to the mother's breast. Another time, an adult female ran to groom the male when a juvenile female belonging to the same group looked away from the group leader who threatened her for staying too far away. The adult female thus did what the younger one, which was probably her daughter, should have done. Similarly, the female applying neck bites in place of the inexperienced group leader probably may have had a lower threshold for this male reaction than even the male himself. That she performed as a male is not unusual, since there are no gestures strictly limited to one sex in hamadryas baboons.

There is a further aspect of the protected threat situation. By placing itself exactly in front of the dominant male, the protégé makes it almost impossible for the opponent to threaten or attack back, since any such gesture would automatically be aimed at the male also. It looks as if the protégé would graft its own gestures upon the protector, thus exploiting the protector's physical features of dominance. Even if the dominant figure is most unlikely to act at all, as in the example of the neck-biting female, its features still can be used as an impressive backdrop.

Finally, there is a possibility that one monkey, for instance the inexperienced group leader, can by slight intention movements induce another monkey, for example the old experienced female, to carry out an attack against a third. There is no evidence at all for this in our material, but experimental investigation should not fail to consider the possibility, especially since it is of importance in human tripartite relations (see Russell and Russell 1961).

So far, we have considered the development of tripartite relations in hamadryas females. Although the males follow the same course up to an age of about 2 years, they never develop the full pattern of protected threat. As subadults, the males often show another kind of tripartite behavior when under threat from an adult male. It is again the triangle of mother, infant, and aggressor which is simulated, but typically, the subadult male adopts not the infant role, as the females do, but the role of the mother, while the dominant male appears as the aggressor.

In a typical sequence the frightened subadult male grasps an infant and embraces it, turning sharply away from the adult male. He also may invite an infant or juvenile to jump on his back by lowering his hindquarters, and then carry it in front of the adult male or away from him. A 5-year-old subadult may thus carry a large 4-year-old. Often, the infant which is embraced or carried is not at all frightened or even aware of the threat, but is definitely pulled onto the stage by the subadult male, as in the following sequence from the zoo colony:

> Subadult male 4a saw that adult male α looked at him. Male 4a started lip-smacking and approached α, finally grimacing and screeching intermittently with knees bent. Male α merely stared at him. Two feet from them a 1-year-old female explored the wire net, turning her back to the males. Suddenly, 4a looked at her, pulled her down, and embraced her. Gradually, his screeching diminished; the female jumped off and he ran away.

Three-year-old females often mated with males of the same age—always hidden by some rock from their group leader's view. At intervals, both would run to a place where they could see the group leader and there the male sometimes embraced the female with repeated grasps on her flank, grimacing at the group leader; or he carried her on his back to the group leader's place.

The evidence is not sufficient to decide whether such maternal behavior reduces the probability of the dominant male's attack against the subadult male, in spite of the fact that adult males attack two animals simulating a mother-infant pair only half as often as they attack single animals. Itani (1963) gives some evidence which supports the idea of such a protective function, describing a subleader in a troop of Japanese macaques who "by hugging an infant succeeded in being tolerated by the females and leaders" in the central part of the troop. While the original function of maternal behavior is to protect the infant which is embraced or carried, a protective effect here is sought in the maternal role. This is especially evident when a subadult hamadryas male flees *toward* a dominant male. He will not dare to groom the male or to cling to him as a female would, nor will he present. He is trapped in front of the male, looking at his face and screaming more intensively every second until finally he may draw a bite from the adult male. If, however, he manages to invite a nearby infant to jump onto his back in this situation, he will at once run off with it. Carrying an infant triggers the flight, and thus the carried infant breaks the magic circle of attraction toward the dominant male.

Subadult males never used the protected threat, and females under threat never picked up an infant. One reason for the strictness of this rule is that subadult males are not members of the one-male groups. They are never attacked for leaving a group leader, and their staying close and presenting to adult males has little, if any, appeasing effect. Their tendency to flee *away* from an adult male is much stronger than the tendency to flee *to* him—so much so that they do not even touch him. If they choose the flight pathway toward the male, they groom the ground near him or merely look at his fur and chew. According to Mason (1964), clinging and perhaps grooming have a "stress-reducing" effect. Both are impossible for the subadult males, at least in the infant role. Their resort, that is, mothering an infant, may have a similar effect in reducing stress, especially since embracing allows both animals to cling to something. It is interesting to note that group leaders when under stress themselves during fights within the troop, would embrace one of their females.

Since hamadryas males begin their career as group leaders in a maternal role toward their infant females, it would be expected that the males of this species are more strongly motivated toward maternal behavior than the males of other species. This is in fact observed: from sexual maturity onward, subadult hamadryas males frequently catch very young infants, cuddling and carrying them up to half an hour. Also, motherless infants are invariably picked up and "adopted" by young adult males having no females yet. These patterns are never observed in females, and in males they abruptly vanish once they have their own females. It is not surprising, therefore, that subadult males under threat reduce their stress by maternal behavior, for which they have a strong motivation anyway.

The most obvious class of tripartite relations in hamadryas baboons, then, can be described by the roles of a protected, a protector, and an antagonist. First in ontogeny, and probably in evolution, these roles are taken on by the infant, its mother, and an aggressor. As the infant grows, these roles are preserved, but they shift to other actors. The situation is at once complicated by the fact that several individuals transfer the role of a protective mother to the same dominant male and that thereby they are forced to compete for his protection. This situation is first encountered by the juvenile when he plays with others around a single subadult male. It is again experienced by the females sharing the protection of one group leader. The genesis of the one-male groups suggests that a transferred mother-infant relation was an important root of their evolution. Consistent with this view is the strong maternal motivation of hamadryas males, which is revealed throughout the subadult and early adult age until they meet the needs of their first consorts.

This hypothesis barely covers more than one aspect of the complex phenomena. Two points, however, emerge from the phenomena themselves: it is, first, astonishing how often the behavior patterns of the mother-infant relation appear in tripartite relations of subadult and adult hamadryas baboons. And second, the tripartite behavior in some instances comes close to exploitation: a primate may learn to use for his own protection, or for increasing the effect of his aggression, another one who primarily is not involved in the events.

REFERENCES

Altmann, Stuart A. 1962. A field study of the sociobiology of rhesus monkeys, *Macaca mulatta. Ann. N.Y. Acad. Sci.* 102:338–435.

DeVore, Irven. 1962. The social behavior and organization of baboon troops. University of Chicago doctoral dissertation.

Harlow, Harry F., and Harlow, Margaret K. 1963. A study of animal affection. In *Primate social behavior,* ed. Charles H. Southwick, pp. 174–84. Princeton, New Jersey: Van Nostrand.

Itani, Junichiro. 1963. Paternal care in the wild Japanese monkey, *Macaca fuscata.* In *Primate social behavior,* ed. Charles H. Southwick, pp. 91–97. Princeton, New Jersey: Van Nostrand.

Kummer, Hans. 1957. Soziales Verhalten einer Mantelpaviangruppe. Beiheft Scheiz. *Z. Psychol.* 33: 1–91.

———. 1965. A comparison of social behavior in captive and free-living hamadryas baboons. In *The baboon in medical research,* ed. Harold Vagtborg, pp. 65–80. Austin: University of Texas Press.

Kummer, Hans, and Kurt, Fred. 1963. Social units of a free-living population of hamadryas baboons. *Folia Primatol.* 1: 4–19.

Mason, William A. 1964. Sociability and social organization in monkeys and apes. *Adv. Exptl. Soc. Psychol.* 1:227–305.

Russell, Claire, and Russell, William M. S. 1961. *Human behaviour.* London: André Deutsch.

Tinbergen, Niko. 1951. *The study of instinct.* Oxford: Clarendon Press.

6

SOCIAL RELATIONS OF ADULT MALES IN A FREE-RANGING BAND OF RHESUS MONKEYS

JOHN H. KAUFMANN

INTRODUCTION

This chapter describes the social interactions of adult male rhesus monkeys (*Macaca mulatta*) of different social ranks and at different seasons of the year. These monkeys were all members of the largest social group (about one hundred fifty members) in the free-ranging rhesus colony on Cayo Santiago, a 37-acre islet off the east coast of Puerto Rico (see Altmann 1962 for description and history of the colony). This colony, where every adult monkey is marked and recognizable on sight, provides an unusual opportunity for quantifying the social relations of the members of a large band. The information thus obtained can serve as a standard of comparison for studies on other species, or for studies of rhesus monkey behavior under different conditions (for example, the behavior of monkeys suffering from experimentally produced brain damage).

METHODS

The data in this report were obtained from field observations totaling 342 hours, made in four periods from December, 1962, to September, 1963.

Period I—75 hours in December, 1962, at the end of the 1962 mating period.

II—100 hours in February, March, and April, during the 1963 birth period.

III—75 hours in July, during the interim between the birth period and the mating period.

IV—92 hours in September, during the 1963 mating period.

John H. Kaufmann, Zoology Department, University of Florida, Gainesville, Florida.

Period III was apparently a true interim period, with a minimum of influence from either the birth or mating periods. Strong influence from the 1962 mating period was evident in period I, however, with the behavior intermediate between that typical of the mating period and that observed in period III. Therefore, the data from period I will be included only where necessary to illustrate specific points.

Although the monkeys in this colony are free ranging and eat some natural foods, they are dependent on commercial monkey food from a few artificial feeders. Of the total observation time, 18 per cent was spent at these feeders. Because intragroup conflict is exaggerated around the feeders, the data obtained there were kept separate from those obtained elsewhere. The feeder data provide much quantitative information on agonistic behavior, whereas the other data give a truer over-all picture of the monkeys' natural behavior.

During each day's observations, I concentrated on one or two males at a time, recording as many as possible of the social interactions of each male observed. This procedure provided a record of the fraction of each male's total social interactions that was accounted for by each category of behavior (mounting, grooming, and so forth), even though all males were not observed for the same length of time. Because some acts are more conspicuous than others, these percentages are valid only for comparing individuals or seasons, not as absolute measures of the relative frequency of each act. The latter can only be obtained by observing each male for long uninterrupted periods, a laborious task successfully accomplished for three males by Dr. George Fisler (in preparation).

Special observations were made to determine the males' spatial relations to the rest of the band in "stable" situations. Stable situations were those in which (1) the band as a whole was not traveling and (2) neither the male in question nor any monkey within 20 feet was traveling or engaged in agonistic behavior. No observations were made near the artificial feeders or water sources, because stable conditions rarely if ever existed there and crowding was the rule. Each time I saw a male in a stable situation, I noted down all adults within 10 feet and within 20 feet, and also recorded whether the male was in the central group, on the fringe of the central group, or outside the central group. In order to have a single standard by which all males could be placed, I defined the central group with respect to the females. A male was within the central group if he was inside an imaginary line drawn to include all adult females in the band, with the exception of isolated females accompanying the male in question. A male outside this imaginary line, but still within immediate potential visual and vocal contact with the group, was on the fringe of the central group. A male cut off from the central group by topography or distance was outside the central group, even though other males were present. For example, the peripheral males were often together, but away from the rest of the band. Although such judgments in the field are often somewhat subjective, comparisons should be valid if all males are classified in a like manner. The consistency of the results supports their validity. The conclusions drawn here about spatial relations are based on 2,284 observations; each male was observed five to fifty-six times in each period, most of them twenty to forty times in each period.

RESULTS

Organization

Divisions of the hierarchy.—The adult males in the band constituted a clearly defined, linear dominance hierarchy, expressed through spatial displacements and the exchange of aggressive and submissive signals. This hierarchy included three discrete divisions: the central hierarchy, the 4-year-old males, and the peripheral males. The central hierarchy formed a continuum of rank but can be arbitrarily subdivided for convenience into high-, medium-, and low-ranking groups. The high-ranking males were most active in breaking up intraband disputes, but lower-ranking males also did this occasionally. Most of the males born in 1959 were becoming integrated into the central hierarchy in 1963, but still acted chiefly as a distinct subgroup. The peripheral males, mostly orphans or sons of females in other bands, ranked below all other males in the group and constituted another distinct subgroup, usually staying on the fringe of the band. The rank groups at the end of each observation period are shown in Table 1.

Changes in rank.—Although the ranks in the male hierarchy were fairly stable, there were a few important changes during the course of this study. Male 14, the alpha male since 1961, began to show signs of weakening authority in July, 1963, when he received a neck wound. By September, 1963, he had slipped to third place, below DW and DV. During his tenure as alpha male, 14 was the main focal point for band activity. He was usually followed by a group of females and their young, and although he seldom initiated group movements, such movements were rarely successful unless he participated. Male 14 was deferred to by all other members of the band, and consistently had more associates within the band than any other male. In period IV, after his drop to third position, 14 was passively displaced by other males (DW, DV) for the first time (Table 2). Male 14 as not actively displaced more often in period IV, nor did he receive more active threats, and he continued to take the lead in breaking up intraband fights. Except for DW and DV, who now dominated him, 14 retained his favored social position for at least 1 month after his drop in rank. Furthermore, DW and DV showed no immediate changes in their social relations, except with respect to 14.

The males (EY, IA, DS, KU, EP, KX) born in 1959 to mothers in this band were still with their mothers in the central group in period I. DK and KN were also 3 years old at this time, but their mothers were not present and they

TABLE 1

RANK OF ADULT MALES AT THE END OF PERIODS I, II, III, IV

Four-year-old males EY, IA, DS, KU, EP, KX began integrating into the central hierarchy in period II.

Rank Groups	Period I	Period II	Period III	Period IV
CENTRAL HIERARCHY				
High	14 DW DV 56 26	14 DW DV 56 EY 26	14 DW DV 56 EY 26	DW DV 14 56 EY 26
Medium	08 79 121 S05	08 79 TA 121 DS S05 KU EP	08 79 IA 121 DS, KU, EP, KX S05	08 79 121 S05 IA[2] EP[2]
Low	AL BA DX?	AL KX BA BZ DX	AL BA BZ DX	AL BA DX
PERIPHERAL MALES	113 AV EG AN KN 53? KB CZ DK	113 AV EG AN KN 53? KB CZ DK	AV AN 53 KB EG KN CZ DK 113[1]	AV AN 53 KB EG KN CZ DK 113[3]

[1] Attacked and driven from band July 17. Semisolitary thereafter.

[2] Rank unstable, falling.

[3] Occasionally seen with band, but avoided interactions.

TABLE 2

RANK-CORRELATED CHANGES IN PASSIVE DISPLACEMENTS

Males 14 and DW are compared with the average for all high-ranking central hierarchy males in the per cent of instances in which they passively displaced other monkeys vs. the per cent of instances when they were passively displaced. For example, in period III, 14 displaced other monkeys in all of the displacements in which he was involved. In period IV, after he was deposed, 14 displaced the other monkeys 74 per cent of the time and was displaced 26 per cent of the time. The numbers in parentheses indicate sample sizes.

PERIOD	MALE 14		MALE DW		ALL HIGH-RANKING MALES	
	Displaces (%)	Displaced (%)	Displaces (%)	Displaced (%)	Displaces (%)	Displaced (%)
I	(38) 100	0	(26) 96	04	(157) 87	13
II	(46) 100	0	(60) 95	05	(318) 91	9
III	(23) 100	0	(19) 100	0	(145) 94	6
IV	(19) 74	26	(24) 100	0	(194) 93	7

were already peripherals. The others had ranks roughly proportional to their mothers' ranks in the central group, but these young males were not yet included in the central male hierarchy. In period II these males were just 4 years old, and beginning to infiltrate the central hierarchy. They tended to act together as a group, joining in a common defense against other males, and because of this it was difficult to assign them individual ranks. EY moved up rapidly in the central hierarchy, an example of a son of a high-ranking mother achieving high rank early and easily (Koford 1963). EY was the brother of DW and the son of 119, the band's highest-ranking female. The other 4-year-olds were of medium to low rank in the central hierarchy. In period III these

TABLE 3

CHANGE IN SOCIAL SPACING OF 113 AFTER BEING DEPOSED BY OTHER PERIPHERAL MALES

Male 113 is compared with the average for all peripherals. The numbers in parentheses indicate sample sizes.

PERIOD	MALE 113					AVERAGE FOR PERIPHERALS				
	Per Cent of Observations			Avg. No. of Assoc. within 20 ft.	Observed Alone (%)	Per Cent of Observations			Avg. No. of Assoc. within 20 ft.	Observed Alone (%)
	CG	FG	OG			CG	FG	OG		
II....	(29) 0	14	86	0.86	48	(204) 1	43	56	1.17	36
III...	(5) 0	20	80	0	100	(124) 4	81	15	0.69	50
IV....	(8) 0	25	75	0	100	(165) 37	61	2	0.70	54

CG = in central group.
FG = on fringe of central group.
OG = outside central group.
Observed alone means no adults were within 20 feet.

4-year-olds continued to act as a group. EY, IA, and DS had reached plateaus in their climb in rank, but the others increased their standings slightly. By period IV, DS, KU, and KX had left the group, and IA and EP were of unstable but lower rank. Only EY maintained his high position and was apparently completely integrated into the central hierarchy. The details of the behavior of the 4-year-olds may be followed in the tables and graphs throughout this chapter. As a group, their behavior was usually intermediate between that of the high- and medium-ranking males in the central hierarchy.

Male 113 was the highest-ranking peripheral male until he was attacked and driven from the band by BZ, KB, and CZ on July 16 (period III). Thereafter he was semisolitary, occasionally entering the band but avoiding close contact with the members. His sudden change in status is shown clearly in Table 3, which compares the data for male 113 to the average for all peripherals in (1) the per cent of time spent within, on the fringe of, and outside the central group; (2) the average number of adults within 20 feet; and (3) the percentage of observations during which no other adults were within 20 feet. Rank changes were more common within the peripheral group than within the central hierarchy or within the 4-year-old group.

Several changes occurred during the course of this study in rank group and band affiliation. During period I, DX and AL occasionally joined the peripheral males. Male 98, originally a regular peripheral male, was seen only four times—once with the peripherals and three times alone. Male 53 was also usually alone during this period. During period II, AL continued to associate at times with the peripherals, as did (less often) 121, 08, and 79. Male HC, a 3-year-old from another band, was seen several times with the peripherals. In period III, HC was seen oftener with the peripherals, while 98 was now living entirely alone. Male 53 was with the peripherals part of the time, and in the central group part of the time. Period IV was marked by widespread instability among the colony's social groups, with many changes in band affiliation among the males. HC was now a constant companion of the band's peripherals, and newcomers 127, S15, R15, ET, EX, and CY all acted as peripherals. Four-year-olds DS, KU, and KX had joined other bands.

TABLE 4

Agonistic		Nonagonistic (Friendly)
Aggressive	Submissive	
Break up fights	...	...
Attack	Attacked	Groom, groomed
Displace, active	Displaced, active	Play
Displace, passive	Displaced, passive	Hold young in arms
Threaten (no displacement)	Threatened (no displacement)	Sit within 1 foot
Mount	Mounted	Follow, followed
Appeased	Appease	Copulate

Social Acts

Description of acts.—In this chapter a social act is defined as any incident in which the behavior or presence of one animal affects the behavior of another. The social acts observed for each male were recorded according to the classification in Table 4.

Breaking up intraband fights was a special case of threatening, directed at two or more individuals, in which the interfering male ran toward the combatants and usually gave some auditory or visual threat. Of the ninety-eight fights which I saw broken up, the high-ranking males in the central hierarchy broke up ninety-one (Table 5).

An attack was any hostile encounter involving physical contact. At least 10 to 20 per cent of the attacks were redirected; that is, a monkey attacked a lower-ranking one when a higher-ranking monkey approached. Many attacks began when a female (especially one in estrus) or an immature screeched at a peacefully approaching male, who then attacked the screeching monkey. In defense of their young, adult females sometimes attacked higher-ranking males.

Displacements were the most common social interaction observed, and 75 per cent of them occurred at the feeders. Active displacements, in which one monkey caused another to move aside with an overt threat, were only one-half as common as passive displacements, in which no overt threat was involved. Adult females displaced males in 11 per cent of the active displacements involving females, but in only 3 per cent of the passive displacements. This was

TABLE 5

NUMBER OF FIGHTS BROKEN UP BY VARIOUS MALES

Rank Group	Males	No. of Fights Broken Up
High	14	53
	DW	7
	DV	12
	56	10
	26	9
Medium	08	1
	79	1
Low	BA	1
4-year-olds	EY	3
	IA	1
Total		98

because males tend to ignore higher-ranking females unless overtly threatened by them, while females retreat more readily from higher-ranking males even when there is no overt threat. Displacements were highly reliable indicators of rank; alternation of the dominant role occurred only when social rank was changing or not yet established.

Threat usually meant simply staring or running toward another monkey, often with the mouth open and with a threat call; but two special cases of threat were also seen. Tree shaking was apparently a long-distance threat, often directed at foreign objects or members of other bands. Tree shaking was seen eighteen times—fourteen times high-ranking males in the central hierarchy were responsible. In the other special form of threat, a high-ranking male would walk very close to another monkey, or even around it in a tight circle, brushing against it. At the same time he would thrust his face close to the other's face and protrude his lips. The threatened monkey responded by sitting very still and giving a fear grin, and these rare rituals were never followed by an attack. Threat, in general, was a far less reliable indicator of dominance than was displacement.

A single, brief mounting, with or without pelvic thrusts, was interpreted as an agonistic act, while "copulation" was reserved for repeated mountings of a female culminating in ejaculation. Simple mounting occurred throughout the year, and often both participants were males. Mounting was not a very reliable dominance indicator, since the mounted animal was the dominant one in ten of the thirty-nine observations.

Play by adult males included the following acts: inhibited biting (twenty-one times), wrestling (thirteen times), grabbing (eleven times), and chasing and jumping at (six times). Of the twenty-nine play sessions involving adult males, twenty-one were with immatures, seven were with other adult males, and one was with an adult female in the mating period. Four-year-old males played about twice as frequently as older males, and play was most common in period II.

Following and copulation are diagnostic of the consort relations with estrous females and were restricted to late period III (a few cases) and period IV.

A total of 5,763 separate social acts involving adult males was recorded during this study. The over-all rate of observed social interaction (acts observed per hour) was highest in the mating period and lowest in the interim period. Away from the feeders with the interim period as the standard, 24 per cent more social activity of all kinds was observed in the birth period and 56 per cent more in the mating period. At the feeders, where the observed rate of social interaction was six times as great, 2 per cent more was observed in the birth period than in the interim period, and 23 per cent more in the mating period.

Effect of rank.—There was only a rough correlation between rank and the proportions of agonistic and friendly behavior away from the feeders (98 per cent of all acts observed at the feeders were agonistic). Within the central hierarchy the proportion of agonistic behavior was generally higher with monkeys of lower rank, though DV, 26 and BZ had unusually high proportions of agonistic behavior (Table 6). The 4-year-olds generally had a low proportion of agonistic behavior, but EY was a conspicuous exception. The peripherals as a group were about as friendly as the high-ranking males in the central hierarchy but 113, KB, and CZ were conspicuous exceptions.

It should be emphasized here that frequent agonistic behavior is not an indication of high rank. Thus, many times no acts were recorded when alpha male 14 approached or fed near crowds of females and immatures, because they simply ignored him, even when he brushed against them. There was less reaction to 14's approach than to that of any other male in the central hierarchy. When DV approached, however, the females and immatures habitually dashed away. Females rarely sat around DV at the feeders as they did round 14;

TABLE 6

COMPARISON OF ALL MALES IN PER CENT OF AGONISTIC VS. NONAGONISTIC SOCIAL ACTS AWAY FROM THE FEEDERS

For example, 29 per cent of male 14's 289 observed social acts were agonistic, while 71 per cent were nonagonistic.

Male	Sample Size	Agonistic Acts (%)	Nonagonistic Acts (%)
High			
14	289	29	71
DW	240	43	57
DV	222	53	47
56	159	40	60
26	154	50	50
Average	....	43	57
Medium			
08	137	47	53
79	101	48	52
121	97	46	54
S05	98	55	45
Average	....	49	51
Low			
AL	77	56	44
BA	119	59	41
BZ	16	80	20
DX	54	49	51
Average	....	61	39
4-year-olds			
EY	93	51	49
IA	59	32	68
DS	53	37	63
KU	48	21	79
EP	40	27	73
KX	24	07	93
Average	....	29	71
Peripherals			
113	35	64	36
AV	65	47	53
EG	38	38	62
AN	38	35	65
KN	32	24	76
53	44	41	59
KB	38	63	37
CZ	45	53	47
DK	24	46	54
Average	....	46	54

often they watched from a distance as DV fed, then rushed to feed when 14 displaced him. Similar but less extensive fear responses were evoked by DW, and by EY in periods III and IV, when he was asserting his position in the central hierarchy. Some males (for example, DV and EY) that had recently worked their way up in the rank order were unusually assertive and "touchy," with relatively high agonistic scores. Other males, regardless of their rank, were easygoing and caused a minimum of disturbance. Males 14 and 56 were good examples of the latter.

Looking now just at agonistic behavior, we see that the proportion of aggressive acts to submissive ones is more obviously correlated with rank. Within the central hierarchy and the 4-year-old group, aggressiveness at the feeders was more frequent among males of higher rank (Table 7). Most of 14's submissive behavior occurred in period IV, after he was deposed. Males 121 and DX were unusually aggressive for their ranks; but 121 had a relatively high proportion of interactions with 4-year-olds and peripherals, and DX had a high proportion of interactions with peripherals. As a group, the peripherals were submissive, but the scores of individual peripherals are not comparable because of their frequent shifts in rank and infrequent, erratic appearances at the feeders. Away from the feeders, there was a similar close positive correlation between rank and aggressiveness. However, the ratio of aggressive to submissive behavior was lower than at the feeders, because the males had fewer contacts with females and immatures away from the feeders.

Some monkeys made a good living from bold submissiveness. Some of the lower-ranking, central hierarchy males like S05 and BA avoided interactions at the feeders, lurking at the edge of the crowd and rushing in to grab food only when higher-ranking males were absent. DK, however, the lowest ranking of the peripheral males, boldly stayed around the feeder. Submissive and appeasing to all he met, he was tolerated, and fed easily with no need to be furtive.

Table 8 compares the different rank groups in the relative frequency of displacements, mounting, grooming, and sitting within 1 foot. Displacements and grooming were by far the commonest acts for all males, and there were consistent correlations between rank and the relative frequencies of these acts. All rank groups, except the peripherals, were involved in mounting about equally, with the central hierarchy males doing most of the mounting. The highest-ranking males sat in contact with other monkeys much more frequently than did any other males.

Seasonal differences.—Seasonal differences were evident in the proportions of agonistic vs. nonagonistic and aggressive vs. submissive behavior for all males (Table 9). The males were relatively less agonistic in the birth and mating periods, when they had increased associations with females and immatures. Looking only at the agonistic behavior, we see that the males were relatively more aggressive (and thus correspondingly less submissive) in the mating period, when they had a very high proportion of interactions with females. In a consort relation, the male is typically aggressive and the female submissive, regardless of their relative ranks. The over-all trend of more frequent agonistic behavior with lower rank was not followed in the mating period; at this time all males had a uniformly low proportion of agonistic acts. However, the over-all trend of less frequent aggressiveness with lower rank was followed in all periods.

Table 10 shows the seasonal changes for all males in the relative frequency of displacements, mounting, grooming, sitting within 1 foot, and following and

TABLE 7

COMPARISON OF ALL MALES IN PER CENT OF AGONISTIC BEHAVIOR AT THE FEEDERS THAT WAS AGGRESSIVE AND SUBMISSIVE

For example, 95 per cent of male 14's agonistic behavior was aggressive and 5 per cent was submissive.

Male	Sample Size	Aggressive Acts (%)	Submissive Acts (%)
High			
14	257	95	5
DW	214	99	1
DV	491	92	8
56	166	85	15
26	268	84	16
Average	....	91	09
Medium			
08	168	86	14
79	182	66	34
121	200	87	13
S05	122	58	42
Average	....	74	26
Low			
AL	96	57	43
BA	173	48	52
BZ	75	25	75
DX	93	52	48
Average	....	45	55
4-year-olds			
EY	120	91	9
IA	76	70	30
DS	63	60	40
KU	72	48	52
EP	85	51	49
KX	56	23	77
Average	....	57	43
Peripherals Average	225	21	79

TABLE 8

Comparison of Rank Groups in Relative Frequency of Displacements, Mounting, Grooming, and Sitting within 1 Foot

The figures given are percentages of the total of each group's social acts away from the feeders. For example, the high-ranking males displaced other monkeys in 23 per cent of their social acts and were displaced in 3 per cent of their acts, for a total of 26 per cent for displacements among all of their social acts. The numbers in parentheses indicate sample sizes.

Rank Group	Displace (%)	Displaced (%)	Total Displacements (%)	Mount (%)	Mounted (%)	Total Mounting (%)	Groom (%)	Groomed (%)	Total Grooming (%)	Sit within 1 Foot (%)
High (1064)	23	3	26	4	1	5	10	26	36	13
Medium (433)	16	14	30	3	1	4	14	21	35	5
Low (266)	21	24	45	3	1	4	11	13	24	6
4-year-olds (317)	12	7	19	2	3	5	26	25	51	9
Peripherals (359)	8	27	35	1	1	2	20	24	44	7

TABLE 9

Seasonal Differences in Per Cent of Agonistic vs. Nonagonistic, and Aggressive vs. Submissive Behavior for All Males

For example, in period II, 40 per cent of the males' social acts away from the feeders were agonistic. At the feeders, 56 per cent of their acts were aggressive and 42 per cent were submissive, for a total of 98 per cent agonistic acts. The numbers in parentheses indicate sample sizes.

Period	Away from Feeders		At Feeders		
	Agonistic (%)	Nonagonistic (%)	Aggressive (%)	Submissive (%)	Total Agonistic (%)
II	(612) 40	60	(1,588) 56	42	98
III	(436) 53	47	(624) 57	42	99
IV	(869) 42	58	(659) 65	33	98

TABLE 10

Seasonal Differences for All Males in Relative Frequency of Displacements, Mounting, Grooming, Sitting within 1 Foot, and Following and Copulation

The figures given are seasonal averages, for all groups, of social acts observed away from the feeders. For example, in period II, the males displaced other monkeys in 12 per cent of their social acts, and were displaced in 13 per cent of their acts, for a total of 25 per cent for displacements among all of their social acts.

Period (Sample Size)	Displace (%)	Displaced (%)	Total Displacements (%)	Mount (%)	Mounted (%)	Total Mounting (%)	Groom (%)	Groomed (%)	Total Grooming (%)	Sit within 1 Foot (%)	Following, Copulation (%)
II (612)	12	13	25	4	3	7	21	22	43	14	0
III (436)	18	20	38	1	<.5	1+	18	18	36	4	1
IV (869)	17	13	30	3	1	4	10	25	35	5	17

copulation. Period II is marked by a high proportion of grooming and sitting within 1 foot, period III by a decrease in these friendly acts and a corresponding increase in spatial displacements, and period IV by the appearance of consort behavior.

Grooming

Grooming is probably the most important social act for revealing rhesus band organization, since it is an important part of the relationships between all sex and age categories and is involved in both mating and dominance behavior. The total amount of grooming observed (number of pairs per hour) did not vary much from season to season, though the central hierarchy males did much less grooming in period III than in periods II and IV. The chief seasonal changes were in the identities of the grooming partners (Fig. 1). All ranks

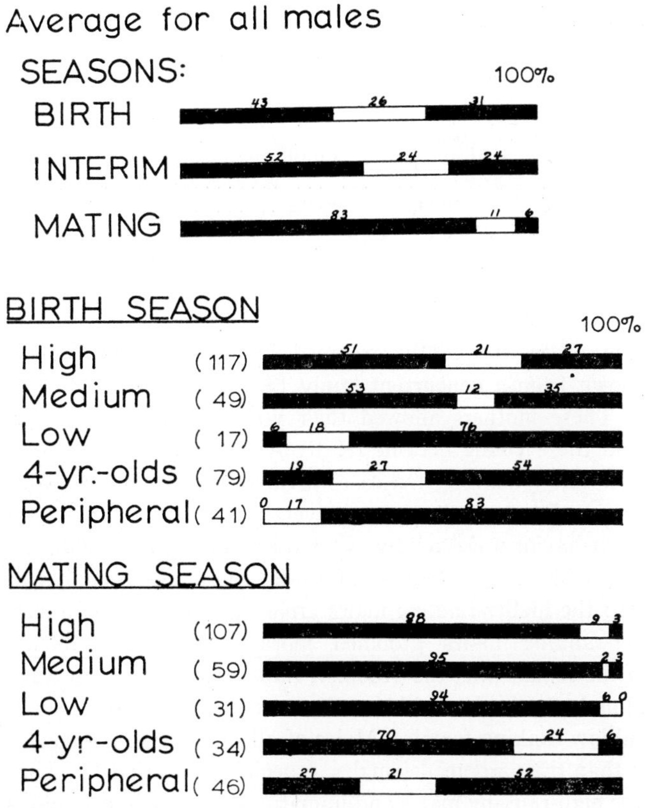

FIG. 1.—Seasonal differences in grooming partners. Each bar represents 100 per cent of the grooming partners for a given rank group in a given season. Three categories of partners are distinguished: Adult females on the left, immatures in the center, and adult males on the right. The numbers in parentheses indicate the sample sizes.

showed a consistent relative increase in grooming with females and a relative decrease in grooming with males as the mating season approached. The central hierarchy males showed a consistent decrease for periods II to IV in the per cent of grooming with immatures, but the 4-year-olds groomed immatures about equally in all seasons. The peripherals devoted about 17 per cent of their grooming in period II to only one immature (3-year-old male HC), decreased this proportion to 4 per cent in period III (though HC was with them more then), and increased their grooming of immatures to 21 per cent in period IV as they moved into the central group and groomed various immatures there.

The immatures groomed by adult males fell into several significant categories (Table 11). Close relatives (siblings, first cousins) and orphans were most frequently groomed. The males also groomed the young of their close female associates. Male 56 was especially active in grooming JM, the daughter of female 106, with whom 56 spent much of his time in periods I, II, and III.

TABLE 11

CATEGORIES OF IMMATURES GROOMED BY ADULT MALES
N = 115

Category	%
Relatives	20
Orphans, or with mothers in another band	19
Orphaned relatives	18
Offspring of close female associates	17
Offspring of known permissive mothers	17
Other	9
	100%

Most of the other, frequently groomed immatures were the offspring of mothers known from a concurrent study (Kaufmann 1966) to be unusually permissive. These mothers allowed their young greater than average social freedom, and these young became relatively independent of their mothers at an early age. The remaining ten cases involved the young of two females on which I have no data regarding permissiveness.

Grooming behavior was strongly affected by rank relations (Plate 6.1). Central hierarchy males tended to groom females whose rank roughly corresponded to their own: the higher-ranking males groomed more higher-ranking females, while lower-ranking males groomed more lower-ranking females (Table 12). When the female was the groomer, however, this pattern was not maintained, and males in the central hierarchy were groomed about equally by high-ranking and low-ranking females. Males of all ranks were groomed by females more often than they groomed females, especially in period IV. The same pattern held for high-ranking males and immatures, but medium-ranking and low-ranking males groomed and were groomed about equally by immatures. In 77 per cent of the cases where males groomed males, the groomer was the lower ranking of the two. This relation was consistent in all rank groups and at all seasons.

PLATE 6.1.—Four-year-old males KU and DS groom DS and EY in order of ascending rank.

PLATE 6.2.—Alpha male 14 consorts with a low-ranking female in the mating period

Use of Trees

To determine whether or not arboreal activity was correlated with rank, I included observations of the elevation above ground of all males seen in stable situations in periods III and IV. As shown in Figure 2, there were no consistent rank-oriented differences among the males in the central hierarchy. Fisler (in preparation) made similar observations of three males in the central hierarchy, with similar results. The 4-year-olds were erratic, staying on the ground more than the central hierarchy males in period III, and less in period IV. The

TABLE 12

RANK OF FEMALES GROOMED BY, AND GROOMING, MALES IN DIFFERENT RANK GROUPS (PERIODS II THROUGH IV)

For example, high-ranking males groomed high-ranking females nineteen times and high-ranking males were groomed by high-ranking females forty-three times. There were equal numbers of each rank of females.

MALES	FEMALES		
	High	Medium	Low
High			
Groomer	19	25	12
Groomee	43	51	44
Medium and Low			
Groomer	7	4	28
Groomee	30	17	42
4-year-olds			
Groomer	12	7	5
Groomee	13	8	9
Peripherals[1]			
Groomer	0	0	3
Groomee	0	6	3

[1] Groomed with females in period IV only.

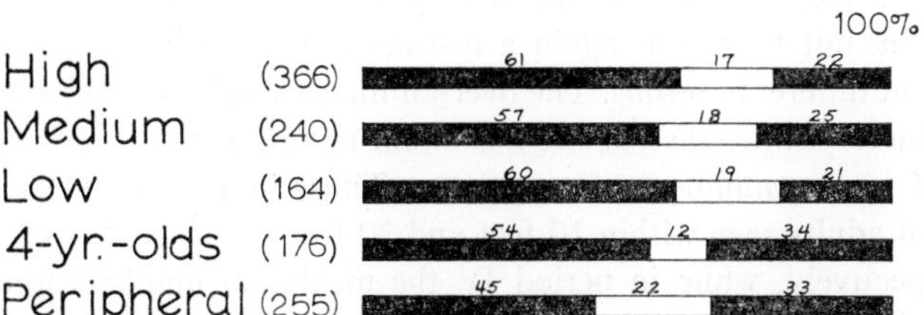

FIG. 2.—Frequency of occurrence above ground level for different rank groups in periods III and IV. Each bar represents 100 per cent of the observations in "stable situations" for each group. Three heights are distinguished: less than 1 foot on the left, 1 to 5 feet in the center, and more than 5 feet on the right. The numbers in parentheses indicate the sample sizes.

peripherals were consistently above ground more frequently than the central hierarchy males. All males showed an average increase of 10 per cent in use of trees in period IV. The pattern of arboreal activity could not be definitely correlated with either weather conditions or mating activity, but followed the increase in mating activity more closely.

Spacing

With few exceptions, males of all rank groups consistently had more adults within 10 feet of them than they had between 10 and 20 feet, for an over-all average of 25 per cent more associates within 10 feet. Since the doughnut-shaped space between 10 and 20 feet from the males has three times the area of the circle with a radius of 10 feet, the observed distribution indicates a definite clumping tendency as opposed to a random or even distribution of adults within the group.

Effect of rank.—There were definite correlations between social spacing and rank (Table 13). The tendency to remain in the central group was lower among males of lower rank, and the number of adults within 20 feet was also lower for males of lower rank in the central hierarchy. The peripherals had about as many associates within 20 feet as did the low-ranking males in the central hierarchy, but there was a distinct difference in the nature of their associates (see following section on associates). Males of higher rank tended to sit alone (no adults within 20 feet) less often than lower-ranking males, and male 14 was outstanding in that he was sitting alone in only 5 per cent of the observations. All other males were sitting alone in more than 10 per cent of the observations.

Seasonal differences.—There were also seasonal differences in social spacing. The tendency to remain in the central group rose sharply in the mating period (Fig. 3), especially for peripherals. This was not true of the highest-ranking males, however, which were consistently in the central group throughout the year. For males in the central hierarchy the numbers of associates within 20 feet were highest in the mating period, though also high in the birth period (Table 14). The peripherals had more associates within 20 feet during the birth season, but there was again a distinct difference in the nature of their associates at different seasons. The over-all increase in associates within 20 feet was not due to an increase in the maximum number, but to an increased frequency of higher numbers of associates. Thus, in period II the maximum numbers of adults seen within 10 feet and 20 feet of any male were seven and eight, respectively, while in period IV the maximum numbers were five and eight. There is a suggestion here that the increased tendency for social clumping in the mating period meets resistance at distances of less than 10 feet. This is further supported by a comparison at different seasons of the per cent of all associates within 20 feet that were within 10 feet. Thus, although the average number per male of associates within 20 feet rose from 1.26 in the birth period to 1.80 in the mating period, the per cent of these that were within 10 feet dropped from 62 per cent in the birth period to 52 per cent in the mating period.

Position in marching columns.—Apart from their spacing when the group was at rest, males of different ranks tended to occupy characteristic positions

TABLE 13

SPACING OF ADULT MALES, PERIODS II THROUGH IV

MALES	SAMPLE SIZE	PER CENT OF OBSERVATIONS			AVG. NO. OF ASSOC. WITHIN 20 FT.	OBSERVED ALONE (%)
		CG	FG	OG		
High						
14	163	97	3	0	3.38	5
DW	127	95	5	0	1.99	21
DV	113	90	10	0	1.68	26
56	128	86	13	1	2.29	13
26	128	81	18	1	1.52	22
Average	....	90	10	+	2.17	17
Medium						
08	108	73	26	1	1.15	31
79	113	51	49	0	1.05	46
121	103	48	46	6	1.35	31
S05	88	47	53	0	1.07	37
Average	....	55	43	2	1.16	37
Low						
AL	88	30	70	0	.60	40
BA	91	50	49	1	.84	52
DX	80	24	70	6	.96	33
Average	....	35	63	2	.80	42
4-year-olds						
EY	91	68	27	5	1.29	15
IA	85	74	22	4	1.34	29
EP	59	40	45	15	1.58	22
Average	....	61	31	8	1.40	22
Peripherals						
113	42	0	20	80	....	....
AV	64	10	60	30	0.83	46
EG	68	2	64	34	.92	43
AN	58	12	65	23	.85	45
KN	64	3	59	38	.98	37
53	53	32	52	16	.81	47
KB	47	12	72	14	.63	59
CZ	66	20	64	16	.95	35
DK	52	21	65	14	0.62	59
Average	....	14	62	24	.82	47

CG = in central group.
FG = on fringe of central group.
OG = outside central group.
Observed alone means no adults within 20 feet.

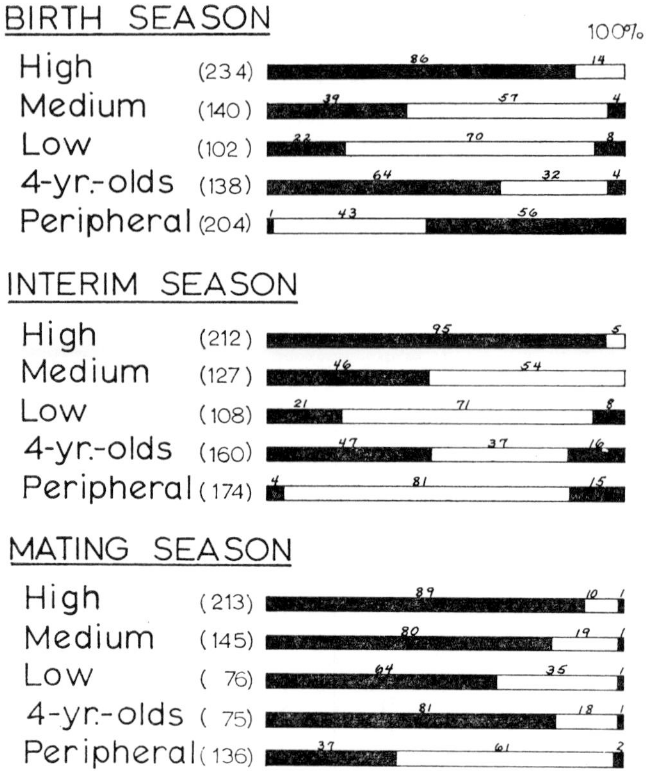

BIRTH SEASON

100%

High (234) 86 14

Medium (140) 39 57 4

Low (102) 22 70 8

4-yr.-olds (138) 64 32 4

Peripheral (204) 1 43 56

INTERIM SEASON

High (212) 95 5

Medium (127) 46 54

Low (108) 21 71 8

4-yr.-olds (160) 47 37 16

Peripheral (174) 4 81 15

MATING SEASON

High (213) 89 10 1

Medium (145) 80 19 1

Low (76) 64 35 1

4-yr.-olds (75) 81 18 1

Peripheral (136) 37 61 2

FIG. 3.—Seasonal differences in position with reference to the central group. Each bar represents 100 per cent of the observations in "stable situations" for a given rank group. Three positions are indicated: inside the central group on the left, on the fringe of the central group in the center, and outside of the central group on the right. The numbers in parentheses indicate the sample sizes.

TABLE 14

SEASONAL DIFFERENCES IN NUMBER OF ADULT ASSOCIATES WITHIN 20 FEET

The numbers in parentheses indicate the sample sizes.

RANK GROUP	PERIOD II		PERIOD III		PERIOD IV	
	Avg. No. of Assoc. within 20 ft.	Observed Alone (%)	Avg. No. of Assoc. within 20 ft.	Observed Alone (%)	Avg. No. of Assoc. within 20 ft.	Observed Alone (%)
High	(234) 1.95	22	(212) 1.52	29	(213) 3.03	1
Medium	(140) 0.85	47	(127) 0.78	47	(145) 1.84	16
Low	(102) 1.02	35	(108) 0.63	63	(76) 1.36	14
4-year-olds	(138) 1.40	22	(160) 1.40	22	(75) 2.10	11
Peripherals	(204) 1.07	36	(174) 0.69	50	(206) 0.70	54
Average, all groups	1.26	32	1.00	42	1.80	19

"Observed alone" means no adults within 20 feet.

during extended marches. On these marches, a group of adult males typically marches at the head of the column (Unit 1); the females and immatures, with more males, come next (Unit 2); and another group of males brings up the rear (Unit 3). An analysis of twelve such marches shows that high-ranking and medium-ranking males in the central hierarchy tended to stay in Unit 2, among the females and immatures (Table 15). Low-ranking males in the central hierarchy and especially 4-year-olds tended to be in Unit 1, while peripherals were usually at the rear, in Unit 3.

TABLE 15

COMPARISON OF RANK GROUPS IN POSITION IN COLUMN DURING EXTENDED MARCHES

The figures show the per cent of times members of each rank group were seen in each unit. Numbers in parentheses indicate sample sizes.

Rank Group	Unit 1 (%)	Unit 2 (%)	Unit 3 (%)
High (58)	7	91	2
Medium (38)	18	66	16
Low (34)	48	26	26
4-year-olds (27)	59	41	0
Peripherals (66)	14	12	74

The lead male in the column was low-ranking male BA on six of the twelve marches, high-ranking males 14 and 56 once each, medium-ranking male 121 once, 4-year-old EP once, and a peripheral male twice. Male BA was in Unit 1 on ten of the twelve marches, male 56 was in Unit 1 three times, and alpha male 14 was in Unit 1 only once.

Associates

The social status of their associates is just as important as the males' spatial relations within the band. In the following discussion, "associates" are those monkeys located within 20 feet of the males during stable situations; there were no important differences in the status of associates within 10 feet and between 10 and 20 feet. The per cent of adult associates that were female and male is given in Table 16 for different rank groups and different seasons. The proportion of female associates was lower among males of lower rank, and the 4-year-olds scored about the same as low-ranking males in the central hierarchy. All rank groups except the highest-ranking males increased the proportion of their female associates as the mating period approached. Even more clear-cut was the increase in period IV in the proportion of females among monkeys of all ages which sat within 1 foot of the males (Table 17). Because of the small samples here, the raw data are presented instead of percentages.

Female associates.—There was no consistent correlation between the males'

rank and rank of female associates (Fig. 4). Male 14 associated predominantly with high-ranking females, even after slipping to third position in the hierarchy, but there were exceptions (Plate 6.2). The other high-ranking males associated with high-ranking and low-ranking females about equally except in the mating period, when they associated more with high-ranking females. Medium-ranking and low-ranking males tended to associate with low-ranking females except in the mating period, when they associated about equally with high-ranking and low-ranking females. These generalizations do *not* indicate that high-ranking females were involved in more mating activity; they simply tended to sit near the males. A 3-year study of mating behavior in this band (Kaufmann 1965) showed no correlation between mating activity and rank

TABLE 16

SEASONAL DIFFERENCES IN SEX OF ADULT ASSOCIATES

Numbers in parentheses indicate sample sizes.

Rank Group	Period							
	II		III		IV		Aver. (II–IV)	
	♀ (%)	♂ (%)	♀ (%)	♂ (%)	♀ (%)	♂ (%)	♀ (%)	♂ (%)
High	(343) 78	22	(274) 84	16	(498) 79	21	80	20
Medium	(88) 53	47	(80) 62	38	(199) 73	27	63	37
Low	(62) 16	84	(41) 33	67	(66) 65	35	38	62
4-year-olds	(152) 25	75	(179) 19	81	(116) 67	33	37	63
Peripherals	(169) 0	100	(102) 9	91	(67) 37	63	15	85
Average, all groups	34	66	41	59	64	36	......	......

TABLE 17

NUMBERS OF ADULT FEMALES, IMMATURES, AND ADULT MALES SITTING WITHIN 1 FOOT OF THE MALES IN DIFFERENT RANK GROUPS AT DIFFERENT SEASONS

Rank Group	Period II			Period III			Period IV		
	♀♀	Imms.	♂♂	♀♀	Imms.	♂♂	♀♀	Imms.	♂♂
Central Hierarchy									
High	15	10	12	7	7	0	21	3	0
Medium	5	1	4	1	0	0	9	0	0
Low	0	0	6	0	2	0	5	0	0
Total	20	11	22	8	9	0	35	3	0
4-year-olds	4	5	6	2	2	4	4	0	0
Peripherals	0	0	12	0	0	2	4	0	1
Totals, All Males	24	16	40	10	11	6	43	3	1

of females. The 4-year-olds associated with high-ranking and low-ranking females about equally except in the interim period. Peripherals associated with females only in the mating period, when they associated about equally with high-ranking and low-ranking females.

Male associates.—Central hierarchy males, 4-year-olds, and peripherals all showed a strong tendency to associate with males in their own division of the hierarchy (Fig. 6). The high-ranking males associated often with 4-year-olds, but rarely with peripherals. Medium-ranking and low-ranking males associated slightly more often with peripherals than with 4-year-olds. Peripherals and

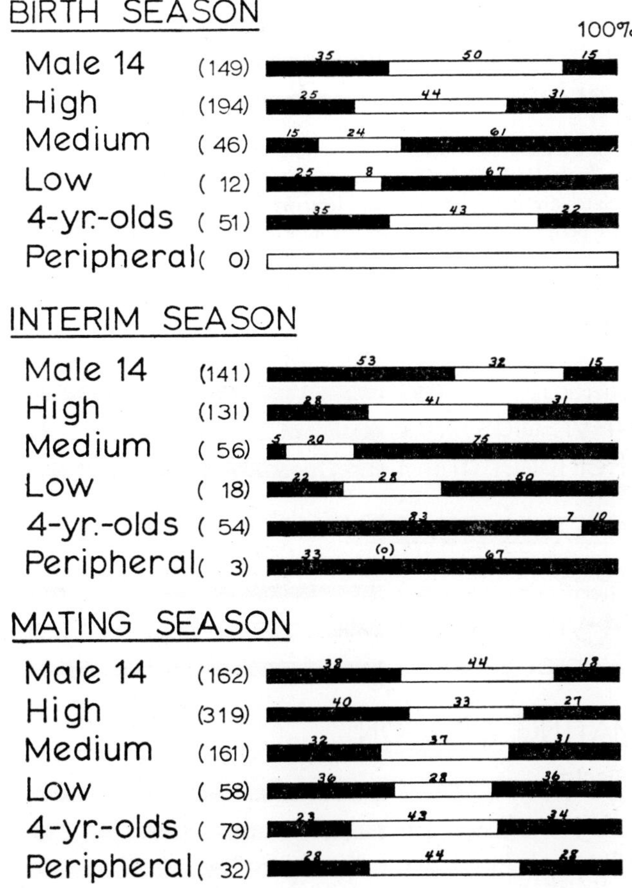

Fig. 4.—Rank of female associates. Each bar represents 100 per cent of the adult female associates for that season of the male in a given rank group. High-ranking females are indicated on the left, medium-ranking females in the center, and low-ranking females on the right. The data for male 14 are given separately from those of the other high-ranking males. The numbers in parentheses indicate the sample sizes.

4-year-olds rarely associated with each other. The central hierarchy males collectively showed no important seasonal changes in their male associates, but the highest-ranking males associated with peripherals only in the mating period (Fig. 6). The 4-year-olds associated almost exclusively with the central hierarchy males in the mating period, since by this time the three remaining 4-year-olds were members of the central hierarchy and no longer acted as a

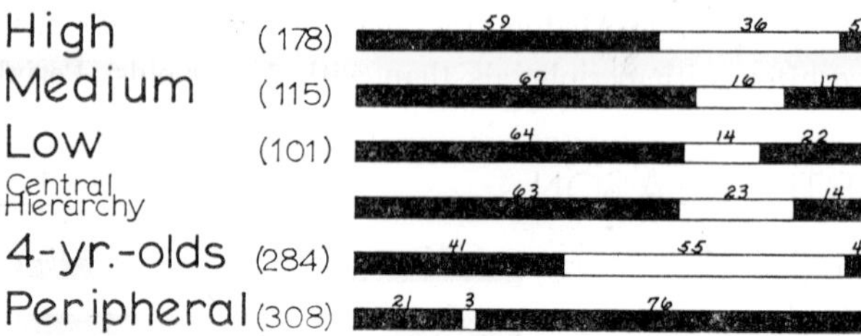

Fig. 5.—Rank of male associates. Each bar represents 100 per cent of the adult male associates in periods II through IV of the members of a given rank group. Three categories of associates are indicated: central hierarchy males on the left, 4-year-old males in the center, and peripheral males on the right. The numbers in parentheses indicate the sample sizes.

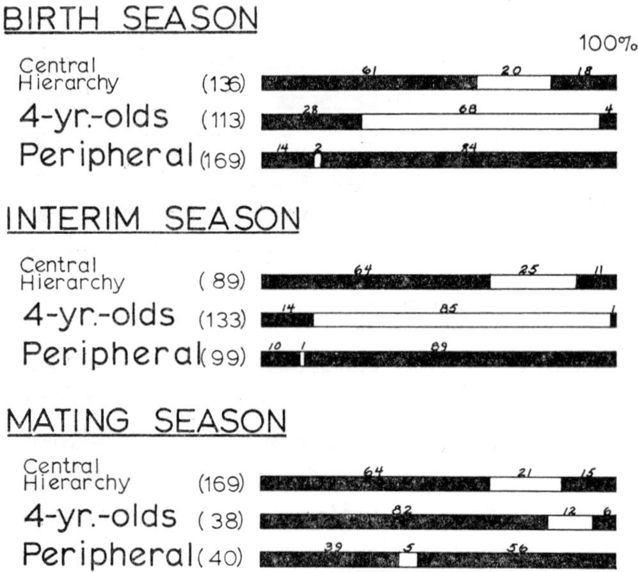

Fig. 6.—Seasonal differences in rank of male associates. Each bar represents 100 per cent of the adult male associates for that season of given categories of males. Three categories of associates are indicated: central hierarchy males on the left, 4-year-old males in the center, and peripheral males on the right. The numbers in parentheses indicate the sample sizes.

group. The peripherals greatly increased their associations with central hierarchy males in the mating period. This increase is correlated with the increased tendency of the peripherals to be in the central group at this season.

Although rough generalizations can be made concerning the associates of males of different rank, it should be kept in mind that each male associated predominantly with particular *individuals,* not indiscriminantly with members of a certain sex or rank category. Frequent male associates were almost all of nearly equal rank, or else were close relatives. Outside of consort relations with estrous females in the mating period, persistent associations with females were less common than those with males. Those female associations that existed again indicated a tendency for males to associate with relatives, or with females whose rank roughly corresponded with their own. These tendencies were not apparent in the consort relationships.

DISCUSSION

To understand and document fully the activity of free-ranging monkeys, one would need a continuous record of the activity of all monkeys at all seasons. This is obviously impossible, forcing us to rely on various sampling procedures. This chapter and the work of Fisler (in preparation) are complementary attempts to describe the activity of adult males in the Cayo Santiago colony. This chapter gives an over-all view of the social relations of all of the adult males in one band at different seasons. Because this study necessarily relies on many brief observations of each male, there are inevitably unmeasured sampling errors, and the percentages for each social act are valid only for comparing seasons or males of different ranks. The spacing data, however, are valid in absolute terms.

Fisler obtained a more accurate account of the total activity of three males (14, 26, BA) during the early part of the interim period between the birth and mating periods. By following each male continuously for about 40 daylight hours, he put social behavior in proper perspective with other types of activity and showed that social acts took up less than 15 per cent of male 14's daylight hours and less than 5 per cent of 26's and BA's daylight hours. Fisler also got accurate counts of the social acts performed by each male, and these are more valid in absolute terms than my comparative percentages. But because some acts were performed so infrequently during his observation periods, and some not at all, this procedure is also open to sampling errors. And, of course, it gives information on only three males during a period of less than 3 weeks.

Despite the different approaches and different sources of error, the two studies support each other nicely. There are no real contradictions in the results, since the differences can easily be accounted for by the expected sampling errors and the different ways in which we handled our data (for example, Fisler lumped all of his data, while I separated data obtained at the feeders from those obtained elsewhere). In all cases where one can make direct comparisons between Fisler's data and mine for males 14, 26, and BA in the interim period, the same conclusions are indicated.

These two studies, combined with more intensive studies of mating behavior (Conaway and Koford 1964, Kaufmann 1965) and the behavior of mothers and infants (Kaufmann 1966), give us a detailed picture of the social relations of adult males in this band. These studies also shed considerable light on the social relations of adult females and immatures. Both males and females have dominance hierarchies. That of the males is clearly defined, while the female hierarchy is less obvious and perhaps not completely linear. The complex interrelations between male and female hierarchies in such a large band are still not clear, though Donald S. Sade (chap. 7 of this volume) has been able to work out a single linear hierarchy which includes both males and females in a small band of about thirty monkeys. With this information we can speculate on the types of natural selection that might involve social rank in rhesus bands.

Presumably there are at least two types of selection in which rank may be involved: (1) selection for those monkeys best adapted physically and psychologically to survive in their general physical and biological environment and (2) selection to maintain the intraband rank system. The first type of selection is, of course, very general, operating through differential survival among all age, sex, and rank categories. But it is probably accelerated by the rank system, since it seems likely that "superior" (for example, healthier, more vigorous) monkeys would tend to achieve higher rank. It is well-known that a monkey's rank at any given time is largely determined by experience, but presumably inherited characteristics also play a part. If this is true, then an increased number of offspring from higher-ranking animals would tend to accelerate selection for desirable characters. There is a positive correlation between the males' rank and the amount of their mating behavior, and thus presumably their success in fathering offspring (Carpenter 1942, Kaufmann 1965). No correlation has been found, however, between the rank of females and their mating success or birth rate. Selection may operate here through the increased survival of the offspring of high-ranking females. Such young benefit from their mothers' protection in being able to feed more freely than the offspring of lower-ranking females, and in being less subject to attack from other monkeys. These favored young may also tend to achieve high rank themselves, thus perpetuating the advantage, as shown by Koford (1963) for males. Finally, higher-ranking females may survive and breed longer on the average than lower-ranking females. More studies of free-ranging females are needed to prove or disprove these suppositions.

Beyond selection on this general level, of which social rank is but one of many facets, there may also be selection on a more restricted level which acts to maintain the rank system itself. In loosely organized mammalian social groups, there may be no apparent dominance hierarchy, as with coatis (*Nasua narica*), which have a relatively limited repertoire of social signals, and which commonly settle intraband disputes with actual fighting (Kaufmann 1962). However, in large, tightly organized groups such as those of rhesus monkeys, there is a definite advantage in a well-defined rank system which is expressed chiefly through harmless signals. Thus, the activities of the members of a closely knit band can be integrated with a minimum of biologically wasteful conflict. Such a system might depend on selection (increased survival) for monkeys which, regardless of their rank, achieve the highest degree of social integration. Such selection might involve the efficiency of sense organs: sight and hearing are especially important in the social integration of rhesus monkeys. Also involved must be the physical and psychological abilities to give and respond properly to social signals—to the extent that such traits are inherited. Unfortunately, we have no accurate knowledge yet of the extent to which heredity and experience interact to determine a monkey's social responses.

SUMMARY

The social relations of the adult males were studied at different seasons in a band of free-ranging rhesus monkeys on Cayo Santiago, Puerto Rico. These males were organized into a clearly defined, linear dominance hierarchy which included a central hierarchy, a group of 4-year-olds just becoming integrated into the central hierarchy, and a group of peripheral males. Although the ranks were fairly stable, a few important changes occurred during this study. Most important was the drop in rank of alpha male 14 to third position near the end of the study. Except for his subordination to the two males which surpassed him, 14's social relations were essentially unchanged for at least 1 month after he was deposed.

An analysis of the social acts performed by the males revealed that within the central hierarchy there was a roughly negative correlation between rank and the proportion of agonistic behavior. Peripherals, however, considering their low rank, engaged in comparatively little agonistic behavior, because they associated chiefly with each other. There was a positive correlation between aggressiveness and rank. The males showed less agonistic behavior in the birth and mating periods than in the interim period, but this agonistic behavior consisted of a higher proportion of aggressive acts during the mating period.

Grooming relationships varied with both rank and season. All ranks groomed proportionately more females as the mating season approached, and central hierarchy males tended to groom females whose rank roughly corresponded to their own. Males of all ranks were groomed by females more often than they groomed females, and in 77 per cent of the cases where males groomed males, the groomer was the lower in rank.

Peripheral males tended to stay in trees more than central hierarchy males, but within the central hierarchy there were no consistent rank-oriented differences in the amount of arboreal activity.

Social spacing also varied with rank and season. The tendency to remain within the central group of females and immatures was directly related to rank, as was the average number of adults sitting within 20 feet. Except for the high-ranking males in the central hierarchy, the tendency to remain in the central group rose sharply in the mating period. For central hierarchy males, the number of adults within 20 feet also rose in the mating period. The proportion of female associates was directly related to rank and rose for most ranks as the mating period approached. There was no consistent correlation between the males' rank and that of their female associates, but males in each rank group tended strongly to associate with males in their own division of the hierarchy. Each male associated chiefly with certain individuals, not indiscriminately with members of certain sex or rank categories.

REFERENCES

Altmann, S. A. 1962. A field study of the sociobiology of rhesus monkeys, *Macaca mulatta*. *Ann. N.Y. Acad. Sci.* 102:338–435.

Carpenter, C. R. 1942. Sexual behavior of free ranging rhesus monkeys (*Macaca mulatta*). *J. Comp. Psychol.* 33:113–62.

Conaway, C. H., and Koford, C. B. 1964. Estrous cycles and mating behavior in a free ranging band of rhesus monkeys. *J. Mammal.* 45:577–88.

Fisler, G. F. Nonbreeding activities of three adult males in a band of free-ranging rhesus monkeys. In preparation.

Kaufmann, J. H. 1962. Ecology and social behavior of the coati, *Nasua narica*, on Barro Colorado Island, Panama. *Univ. Calif. Publ. Zool.* 60(3):95–222.

———. 1965. A three year study of mating behavior in a free ranging band of rhesus monkeys. *Ecology* 46:500–12.

———. 1966. Social relations of infant rhesus monkeys and their mothers in a free ranging band. *Zoologica* 51:17–28 + 3 plates.

Koford, C. B. 1963. Rank of mothers and sons in bands of rhesus monkeys. *Science* 141:356–57.

7

DETERMINANTS OF DOMINANCE IN A GROUP OF FREE-RANGING RHESUS MONKEYS

DONALD STONE SADE

What determines the rank of monkeys in hierarchies of aggressive dominance within free-ranging social groups? The few reports available from long-range studies of groups of recognizable individuals point to the rank of the mother as being the most important factor in determining the rank of young monkeys, and show that rank established early in life tends to persist into adulthood. Studies of the Koshima (Kawai 1958) and Minoo B (Kawamura 1958) groups of *Macaca fuscata* in Japan show that in paired competitions over food items in limited areas away from other monkeys the successful monkey is usually the offspring of the higher-ranking mother. Koford's (1963) comments on six groups of *M. mulatta* in the Cayo Santiago, Puerto Rico, colony indicate that the adolescent sons of the highest-ranking female in each group held, at least for a time, high rank in the hierarchy of adult males within the mother's group.

The following report is based on about 550 hours of observations made between 1961 and 1963 on Cayo Santiago Group F, a rhesus monkey group similar in size to the Japanese macaque groups mentioned above. This report confirms the general conclusion of the above studies: that an offspring's rank is in part determined by its mother's rank, even though I use different criteria of dominance than do Kawai and Kawamura.

CRITERIA OF RANK: FIGHTS

My criteria for ranking individuals in a hierarchy of aggressive dominance derive from the aggressive interaction itself, rather than from the causes, con-

Donald Stone Sade, Department of Anthropology, University of California, Berkeley. Present address: Department of Anthropology, Northwestern University, Evanston, Illinois.

text, or consequences of that interaction. These are the most unambiguous criteria, since the behaviors that mark the winning monkey usually are not also used by the losing monkey in the same interaction, and are independent of the context of the interaction, whether it be competition over an infant, food, or a female, or simply an act of malevolence.

In an observed intense fight the attacking monkey charged, roaring and batting at another, then grabbed and held the victim while biting him on the back. As the attack began, the victim cowered away, grimacing and shrieking, presenting his hindquarters to the attacking monkey at the same time. Almost immediately the victim leapt away and fled, still shrieking and grimacing, and finally escaped after being bitten.

It is convenient to analyze both attack and flight into components of locomotion and posture, head and face movements, and vocalizations. In less intense fights the same components occur in milder or incomplete forms; some may be lacking entirely or at least may be expressed too subtly for the human observer to recognize.

Gestures of Attack

The attacking monkey often halts its charge before reaching the victim: The charge is expressed as a lunge. In a still milder expression of the charge, the attacking monkey faces the victim and bobs his head and shoulders up-and-down, as if bounding toward the victim. In a yet milder expression only the head bobs toward the victim. In its mildest form the charge is expressed simply by staring at the victim, perhaps drawing the head slightly toward the shoulders. The attacking monkey, instead of bobbing or lunging, may walk, head slightly lowered, toward the victim. Such an approach grades into a stalk: The body is held more rigidly, and other components of attack (such as open mouth) may be obvious. The stalk grades into the lunge and into the charge.

Instead of holding the victim, the attacking monkey may grab him hard and let go. Instead of grabbing, he may simply bat or cuff the victim with his open hand without actually grasping skin or hair. Instead of batting, he may push the other monkey down or away, either gently or not.

While biting, although the mouth opens and closes, he does not draw back his lips in a grimace. Any of the incomplete attacks, from the mild stare to the fierce charge, may be accompanied by an incomplete bite: the open-mouth gesture. The mouth may barely open, or may open as widely as in an actual bite. During the bite the eyelids almost close; probably many of the muscles of the face, scalp, and neck tense, and the ears flatten against the head. The open-mouth gesture is usually accompanied by a lowering of the eyelids and often by exaggerated ear-flapping and raising and lowering the brows. In some individuals, brow and ear movements may be independent, and a few can appress the top of the ear against the head independently of the bottom part of the ear.

When expressed less intensely, the roar is given as a long or short grunt, which may be loud or barely audible. The short grunt may be repeated rapidly in short series.

The more intense expression of the elements of locomotion and posture, head and face movements, and vocalizations tend to combine in a single attack, as do the less intense expressions of the same elements. The tendency is not an invariable rule, however, and the elements may combine in different degrees of intensity in a single attack. For instance, a monkey may bite without charging, roar while bobbing, or charge without vocalizing.

Gestures of Flight

Instead of fleeing, the monkey under attack may hop aside, move aside slowly, or not move away at all. If he stays, he may cower to the ground on one shoulder, his face toward the attacker, his hindquarters up toward the attacker, his tail down and twitched to one side. He may cower mildly, leaning away and hunching slightly. He may merely glance toward the attacking monkey and then glance quickly away.

A grimace may accompany any of the above movements. The lips may be drawn far back exposing most of the clenched teeth, or the lips may be drawn back only slightly.

A shrill vocalization often accompanies the grimace, in which case the teeth are not clenched. In its mildest form the vocalization grades from a barely audible hiss, through a chirp or squeak, to a long and loud shriek.

The more intense forms of the postures, gestures, and vocalizations of flight tend to combine as do the less intense forms. However, as with the elements of attack, a mild form of one component may combine with an intense form of another component.

The behaviors of the attacking monkey and of the fleeing monkey are summarized in Figure 1. The arrows point from the less intense to the more intense forms of the behaviors. The names of the behaviors refer to regions of continua rather than to discrete units of behavior. As the intensity of attack or flight increases, the magnitude of the movements and the volume of the vocalizations increase and the number of components increases. Some elements of attack and flight are illustrated in Figures 2 and 3. There is agreement between the above descriptions of rhesus fighting behavior and the descriptions in Altmann (1962), Chance (1956), Hinde and Rowell (1962), Rowell (1962), and Rowell and Hinde (1962).

Definition of a Fight

Agreeing with Harrison Matthews (1964, p. 30) that overt and ritual fighting cannot be sharply distinguished because they form a continuum of behavior, I define a fight as an interaction in which an attack of any intensity is followed by a flight of any intensity. Within the group, fights were usually simple: The attack was followed immediately by the flight. Sometimes the attacked monkey fought back. In such a fight the loser was the last to display gestures of flight. If neither combatant eventually responded with flight gestures, I did not attempt to judge a winner, but such sequences were very rare. I did not try to judge individual winners in fights of more than two monkeys, except when more than one monkey responded with flight gestures to the same attack of a single monkey. Then I judged that the single attacking monkey had won a fight with each of his victims.

Since mounting, lip smacking, grooming, and yawning by one monkey may be followed by the same gestures from the other monkey in an interaction, and since these behaviors seem not to derive from the behaviors used in extreme fights, I exclude them from the criteria used to judge winners and losers in

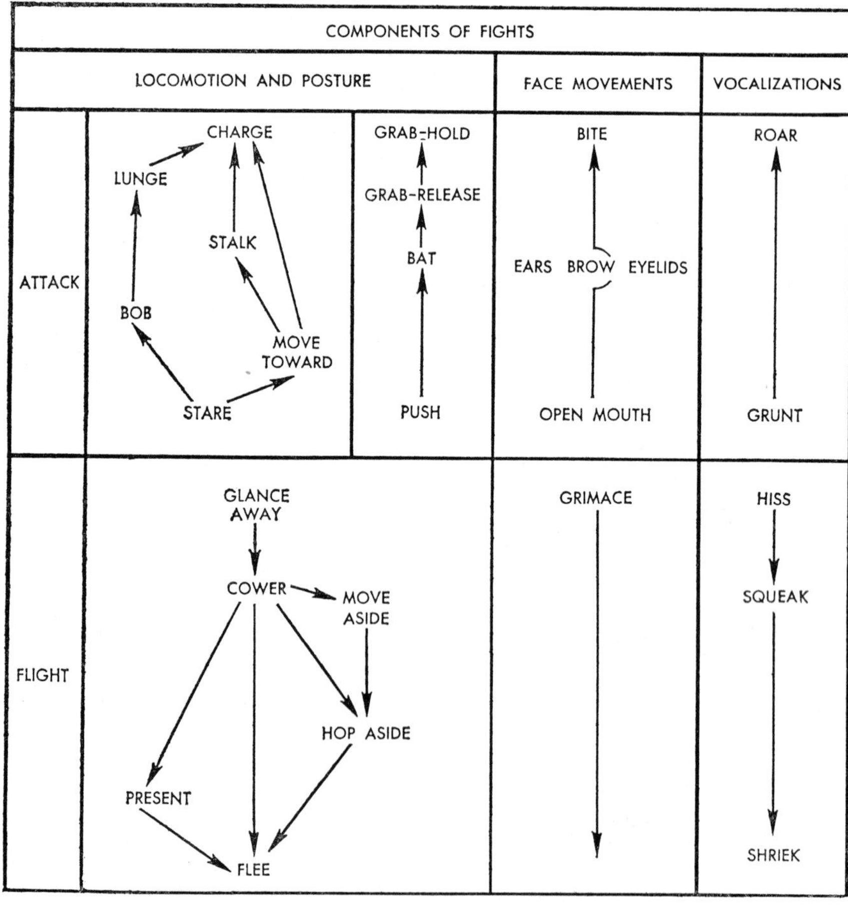

FIG. 1.—Summary of attack and flight behavior. Arrows lead from less intense to more intense forms of the behavior. The names indicate regions of continua rather than discrete units of behavior.

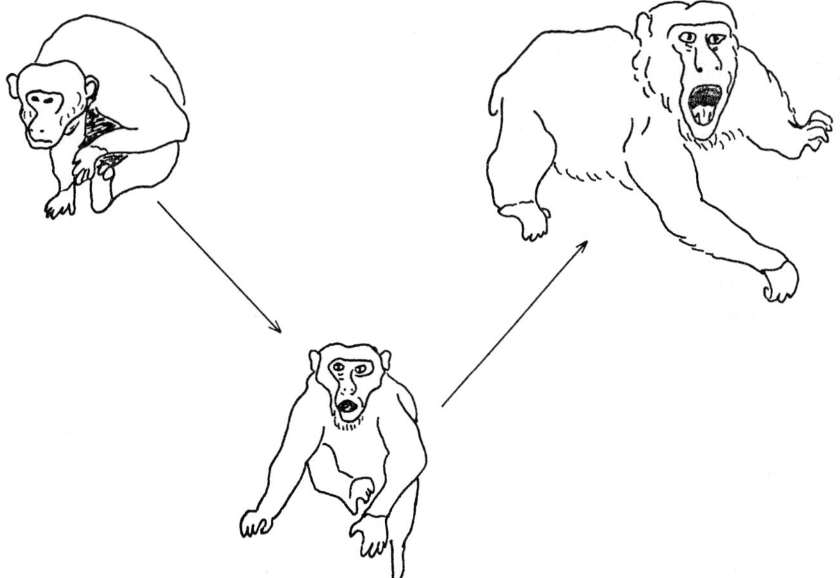

Fig. 2.—Gestures of attack. Arrows lead from less intense to more intense forms of the behavior. Sketched from 35 mm transparencies taken by the author. All are adult males.

Fig. 3.—Three grimaces showing increase of intensity in direction of arrow. Monkey in center is an adult male; the others are adult females. Sketched from 35 mm transparencies taken by the author.

agonistic interactions. Nor do I base judgments on tail positions. Monkeys raise their tails in sequences of behaviors that are clearly not fights, and although monkeys may judge dominance by tail position, I cannot. I also do not use success at obtaining food as a criteria for winning a fight, although I sometimes use food to start fights between monkeys. Too often the loser of the fight fled with the prize or some other monkey got it while the competitors were

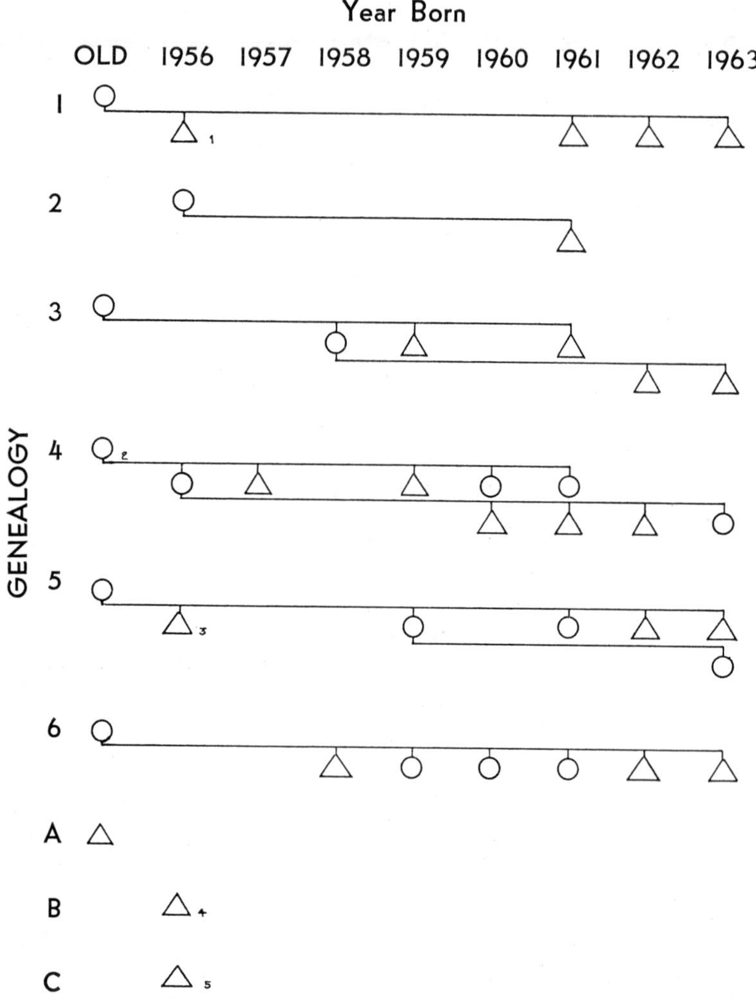

FIG. 4.—Diagram of group structure. Circles represent females, triangles males. Lines of descent are shown by heavy lines. Notes: (1) *1956-Male-1* left group late in 1962. (2) *Old-Female-4* died at end of 1961. (3) *1956-Male-5* castrated near end of 1960. (4) *1956-Male-B* left group near end of 1961. (5) *1956-Male-C* joined group after end of 1962 observations.

fighting. The tabulations on which the rest of this report is based include only fights as narrowly defined above. Winning fights and losing them are my only criteria of dominance.

DOMINANCE RELATIONS

This section includes the description of some dominance relations within Group F in the summer of 1963, with some comments on the history of the relations. In the text and tables individuals are identified by the year of their birth, sex, and genealogy (males of unknown descent are given a letter instead of a genealogy number). Monkeys born before 1956 are called "Old." The structure of the group and the descent relations of the individuals are shown in Figure 4.

TABLE 1

FIGHTS BETWEEN ADULT FEMALES IN 1963

Only females with yearling or older offspring included. Numbers in boxes are fights won by female in column from female in row. Figure in parentheses indicates a sequence that possibly was not a fight.

WINNER	LOSER						
	Old-Fem.-1	1956-Fem.-2	Old-Fem.-3	1958-Fem.-3	1956-Fem.-4	Old-Fem.-5	Old-Fem.-6
Old-Fem.-1	—	7	3	9	6	4	1
1956-Fem.-2		—	6	6	13	1	(1)
Old-Fem.-3		1	—	1	5	5	2
1958-Fem.-3				—	12	6	1
1956-Fem.-4					—	1	2
Old-Fem.-5						—	6
Old-Fem.-6							—

Adult Females

Females with yearling or older offspring can be ranked in a linear hierarchy (Table 1). Only once during this period did a female of lower rank defeat a female of higher rank.

Yearlings

Fights between the yearling offspring of the females ranked in Table 1 are tabulated in Table 2. All of the fights in Table 2 occurred while the mothers of the fighting yearlings were either far away or were clearly not paying attention. Sometimes the loser of a fight ran to huddle against his mother, who threatened the pursuing yearling away. Sometimes the mother of the losing yearling paid no attention, even though the fight took place beside her. It was not necessary for the mother of the winning yearling to be present to insure her offspring's victory in a fight with another yearling.

During this period, no yearling of lower rank defeated any yearling of higher rank. In the column, from top to bottom, and in the row, from left to right, the yearlings are listed in the order of their mothers' rank in the hierarchy of adult females. Yearlings defeated yearlings whose mothers ranked lower than their own in the hierarchy of adult females and were defeated by yearlings whose mothers ranked higher than their own.

TABLE 2

FIGHTS BETWEEN YEARLINGS IN 1963

Individuals are ordered in the column and row according to rank of parent (see Table 1 and Fig. 4). Numbers in boxes are fights won by yearling in column from yearling in row. Figure in parentheses indicates a sequence that was possibly not a fight.

WINNER	LOSER				
	1962-Male-1	1962-Male-3	1962-Male-4	1962-Male-5	1962-Male-6
1962-Male-1	—	1	4	(1)	
1962-Male-3		—	1	3	5
1962-Male-4			—	3	2
1962-Male-5				—	5
1962-Male-6					—

Two Year Olds

Fights between the 2-year-old offspring of the females ranked in Table 1 are tabulated in Table 3. The relation between *1961-Male-4* and *1961-Female-5* is doubtful, since I saw no clear fights between them and was unable to arrange any. Two-year-old monkeys defeated those whose mothers ranked lower than their own in the hierarchy of adult females and lost fights to those whose mothers ranked higher than their own.

The mother of *1961-Female-4* died at the end of 1961, while *1961-Female-4* was yet an infant. During 1961, however, *Old-Female-4* ranked above *1956-Female-4* (her own daughter) and below *Old-Female-3* (Table 4). In 1963, *1961-Female-4* ranked between the offspring of those two females; this suggests that dominance relations are formed very early in life and persist. The clear relations between the yearlings support this suggestion. Another possibility is that following her mother's death the orphaned infant associated closely with other older animals who were close to her mother's rank, and that the orphan's rank derived from her new associates. Unfortunately, I saw no fights between *1961-Female-4* and others of her age class in 1961. In 1962 she was defeated by the three 1961 males that ranked above her in 1963, but I did not see her fight with the other 1961 monkeys in 1962.

Three Year Olds

Fights between the 3-year-old offspring of the females ranked in Table 1 are tabulated in Table 5. The top ranking 3 year old is a female. The lowest-ranking monkey rarely fought with others, and I had little success in tricking her into a fight with her age peers. It might be argued that there are too few

TABLE 3

FIGHTS BETWEEN TWO-YEAR-OLD MONKEYS IN 1963

Individuals are ordered in the column and row according to rank of parent (see Table 1 and Fig. 4). The parent of *1961-Female-4* died at the end of 1961 (see text). Numbers in boxes are fights won by monkey in column from monkey in row. Figures in parentheses indicate sequences that were possibly not fights.

WINNER	LOSER						
	1961-Male-1	1961-Male-2	1961-Male-3	1961-Fem.-4	1961-Male-4	1961-Fem.-5	1961-Fem.-6
1961-Male-1	—	4	12	8	3	4	7
1961-Male-2		—	2	5	3	3	1
1961-Male-3			—	4	3	1	2
1961-Fem.-4				—	6	9	2
1961-Male-4					—	(1)	3
1961-Fem.-5					(1)	—	3
1961-Fem.-6							—

monkeys in this age class and too few fights between them to allow definite conclusions. At the least, the statement that rank of offspring corresponds to rank of mother is not contradicted by the data from this age class.

The mother of *1960-Female-4* died at the end of 1961 (see the section on 2 year olds). By that time, *1960-Female-4* had apparently already established her place in the dominance hierarchy of her age class (although I did not see her fight in this class in 1961), for during 1962 she won more fights from these same two monkeys than she lost to them. During 1962 her closest associate was her younger sister, *1961-Female-4,* rather than some older monkey (Sade 1965).

Four Year Olds

Fights between the 4-year-old offspring of the females ranked in Table 1 are tabulated in Table 6. The hierarchy of 4-year-old monkeys is linear. With

TABLE 4

FIGHTS BETWEEN ADULTS (FIVE-YEAR-OLD AND OLDER MONKEYS) IN 1961

Numbers in boxes are fights won by monkey in column from monkey in row. See text for further comments.

WINNER	LOSER										
	Old-Male-A	Old-Fem.-1	1956-Male-1	1956-Fem.-2	1956-Male-B	Old-Fem.-3	Old-Fem.-4	1956-Fem.-4	Old-Fem.-5	1956-Male-5	Old-Fem.-6
Old-Male-A	—	1	7	3		4		1		1	4
Old-Fem.-1		—	1	3	1	4		2	4	2	
1956-Male-1			—	2	3	8	7	11	14	9	3
1956-Fem.-2				—		6	1	7	2	2	1
1956-Male-B					—	1	2	2	1	2	
Old-Fem.-3						—	1	1	7		2
Old-Fem.-4							—	2		1	1
1956-Fem.-4				1		1		—	4		
Old-Fem.-5									—	1	1
1956-Male-5								1		—	1
Old-Fem.-6											—

one exception, rank of 4 year olds corresponds with rank of mother in the hierarchy of adult females. The exception is doubly interesting.

The mother of *1959-Male-4* died at the end of 1961 (see the above two sections), yet in 1963 he ranked higher than the offspring of the female who ranked higher than his own mother while she was alive (Table 4). In 1961 and 1962, *1959-Male-4* lost fights to *1959-Male-3,* but defeated him in 1963; that is, *1959-Male-4* rose in rank. Following the death of his mother, *1959-Male-4* had as his closest associate his older brother, *1957-Male-4.* It is possible that *1959-Male-4*'s rank in 1963 derived from this association. However, in 1963 *Old-Female-3* still won fights from *1957-Male-4,* as she did the previous year (Tables 7 and 8). If *1959-Male-4*'s rank derived only from his association

TABLE 5

FIGHTS BETWEEN THREE-YEAR-OLD MONKEYS IN 1963

Individuals are ordered in the column and row according to rank of parent (see Table 2 and Fig. 4). The parent of *1960-Female-4* died at the end of 1961 (see text). Numbers in boxes are fights won by monkey in column from monkey row. Figure in parentheses indicates sequence that was possibly not a fight.

WINNER	LOSER		
	1960-Fem.-4	1960-Male-4	1960-Fem.-6
1960-Fem.-4	—	9	1
1960-Male-4		—	1
1960-Fem.-6		(1)	—

TABLE 6

FIGHTS BETWEEN FOUR-YEAR-OLD MONKEYS IN 1963

Individuals are ordered in the column and row according to rank of of parent (see Table 2 and Fig. 4). The parent of *1959-Male-4* died at the end of 1961 (see text). Numbers in boxes are fights won by monkey in column from monkey in row.

WINNER	LOSER			
	1959-Male-3	1959-Male-4	1959-Fem.-5	1959-Fem.-6
1959-Male-3	—		3	2
1959-Male-4	1	—	2	2
1959-Fem.-5			—	3
1959-Fem.-6				—

TABLE 7

FIGHTS BETWEEN ADULTS (FIVE-YEAR-OLD AND OLDER MONKEYS) IN 1963

Numbers in boxes are fights won by monkey in column from monkey in row. Figures in parentheses indicate sequences that possibly were not fights. See text for further comments.

WINNER	LOSER											
	Old-Male-A	Old-Fem.-1	1956-Fem.-2	Old-Fem.-3	1957-Male-4	1958-Fem.-3	1956-Male-C	1956-Fem.-4	Old-Fem.-5	Old-Fem.-6	1958-Male-6	1956-Male-5
Old-Male-A	—	2	7	3	4	2	(1)		1		1	
Old-Fem.-1		—	7	3	6	9	(1)	7	4	2		
1956-Fem.-2			—	6	10	7	(1)	13	1	(1)	1	7
Old-Fem.-3			1	—	4	1		5	5	2		
1957-Male-4				(1)	—	2	10	5	7	1	3	9
1958-Fem.-3						—		12	6	1		1
1956-Male-C			(1)				—	2		2	1	1
1956-Fem.-4								—	1	2		1
Old-Fem.-5									—	6		2
Old-Fem.-6										—		
1958-Male-6											—	6
1956-Male-5												—

TABLE 8

FIGHTS BETWEEN ADULTS (FIVE-YEAR-OLD AND OLDER MONKEYS) IN 1962

Numbers in boxes are fights won by monkey in column from monkey in row. See text for further comments.

WINNER	LOSER									
	Old-Male-A	Old-Fem.-1	1956-Male-1	1956-Fem.-2	Old-Fem.-3	1957-Male-4	1956-Fem.-4	Old-Fem.-5	1956-Male-5	Old-Fem.-6
Old-Male-A	—	4	8	2	2	6	3	1	2	3
Old-Fem.-1		—	1	1	18	1	3	2	3	
1956-Male-1			—	4	10	9	8	7	2	2
1956-Fem.-2				—	11	6	2	1	2	4
Old-Fem.-3					—	8	11	4	3	2
1957-Male-4						—	1	2	8	
1956-Fem.-4							—	4		7
Old-Fem.-5					3	1	2	—		5
1956-Male-5									—	
Old-Fem.-6						1				—

with his older brother, he should rank below *1959-Male-3*. In 1963, *1959-Male-4* seemed appreciably larger than *1959-Male-3*. Perhaps differences in size and maturation override former experience and effects of continued association in determining which of a pair wins fights as monkeys near adulthood.

Adults

All fights between adults, male and female, are tabulated in Tables 4, 7, and 8. During each period, some females ranked higher than some males. Since 1961 there have been no changes in the rank of females, save for the removal of *Old-Female-4* from the hierarchy at her death and the addition of *1958-Female-3* when she became adult, at 5 years of age.

In 1961, *1956-Female-4* ranked just below her mother, *Old-Female-4*. Following the death of her mother, *1956-Female-4* maintained her position in the hierarchy of adult females.

When *1958-Female-3* became adult in 1963, she defeated most monkeys who ranked lower than her mother, *Old-Female-3*. The previous year, *1958-Female-3* won more fights from *1957-Male-4* than he won from her, but in 1963 he won in each fight: a gain in rank for *1957-Male-4*.

1956-Male-5, on the other hand, fell from a position just below his mother, *Old-Female-5,* in 1961 to the bottom of the hierarchy of adults in 1963. He was castrated as an adult near the end of 1960; his fall in rank may perhaps be attributed to his altered physiology.

1956-Male-1 left the group near the end of 1962, but in 1961, and in 1962 before he departed, he ranked just below his mother, *Old-Female-1,* and was thus the third-ranking adult (although the second-ranking male) in the group.

The rather doubtful position of *1956-Male-C* in 1963 is due to the fact that he had recently joined the group, was still peripheral to it, and was interacting more frequently with younger monkeys than with adults.

The single victory (in 1962) of *Old-Female-6* over *1957-Male-4* might be considered evidence of a triangular dominance relation, since he did not defeat her during that period. Since he defeated her once in 1963, while she defeated no adults during that period, and since most of the other data indicate that the linear hierarchy is the rule, I prefer to consider that there are too few fights recorded between these two monkeys to constitute a valid exception.

CONCLUSIONS

The following generalizations are warranted for this small group of rhesus monkeys.

1. Within each age class, the hierarchy of individuals ranked according to their wins and losses of fights is linear.

2. Offspring begin to fight as old infants or young yearlings. They defeat their age peers whose mothers rank below their own and are defeated by their age peers whose mothers rank above their own.

3. The hierarchy thus established persists for several years, either because the first few fights set precedents which cannot be easily broken, or because the offspring continue to associate with their mothers or other older monkeys who rank near their mothers, or because of both reasons.

4. As females become adult they come to rank just below their mothers in the hierarchy of adults. This means they defeat not only their age mates but also older females who were adults when they were growing up.

5. As males become adult they tend to rank near their mother in the adult hierarchy, but at puberty or later, if they remain with the group, they may lose or gain rank. I speculate that at about puberty physiological differences between males become more important in fighting and that the differences that derive from past experience and continued association with adults of different rank become less overriding in determining the winners of fights.

ACKNOWLEDGMENTS

This work received major support from Grant B-2385 from the National Institute of Neurological Diseases and Blindness, U.S.P.H.S. The manuscript was prepared while the author was supported by a U.S.P.H.S. Pre-doctoral Fellowship from the National Institute of Mental Health. I thank Carl B. Koford of the Laboratory of Perinatal Physiology, N.I.N.D.B., for permission to return to Cayo Santiago in 1963. I wish to thank James A. Gavan and Earl W. Count for encouraging me to collect these data, and Richard W. Hildreth for his inspiration in the field. The manuscript profited from the constructive comments of William A. Draper, to whom I am grateful.

REFERENCES

Altmann, S. A. 1962. A field study of the sociobiology of rhesus monkeys, *Macaca mulatta*. *Ann. N.Y. Acad. Sci.* 102:338–435.

Chance, M. R. A. 1956. Social structure of a colony of *Macaca mulatta*. *Brit. J. Animal Behaviour* 4:1–13.

Harrison Matthews, L. 1964. Overt fighting in mammals. In *The natural history of aggression,* eds. J. D. Carthy and F. J. Ebling. London: Academic Press.

Hinde, R. G., and Rowell, T. E. 1962. Communication by postures and facial expressions in the rhesus monkey (*Macaca mulatta*). *Proc. Zool. Soc. London* 138:1–21.

Kawai, M. 1958. On the system of social ranks in a natural troop of Japanese monkeys: I. Basic rank and dependent rank. *Primates* 1–2:111–30. Translated in *Japanese monkeys,* ed. S. A. Altmann. Edmonton: The Editor.

Kawamura, S. 1958. Matriarchal social ranks in the Minoo-B troop: a study of the rank system of Japanese monkeys. *Primates* 1–2:149–56. Translated in *Japanese monkeys,* ed. S. A. Altmann. Edmonton: The Editor.

Koford, C. B. 1963. Rank of mothers and sons in bands of rhesus monkeys. *Science* 141:356–57.

Rowell, T. E. 1962. Agonistic noises of the rhesus monkey (*Macaca mulatta*). *Symp. Zool. Soc. London* 8:91–96.

Rowell, T. E., and Hinde, R. A. 1962. Vocal communication by the rhesus monkey (*Macaca mulatta*). *Proc. Zool. Soc. London* 138:279–94.

Sade, D. S. 1965. Some aspects of parent-offspring and sibling relations in a group of rhesus monkeys, with a discussion of grooming. *Am. J. Phys. Anthropology n.s.* 23:1–17.

DISCUSSION OF AGONISTIC BEHAVIOR

DAVID MC K. RIOCH

The excellent chapters in this section on agonistic behavior of nonhuman primates present a wealth of data and a degree of technical sophistication which merit a more detailed, analytical discussion than I can provide. It may be useful, however, to call attention here to certain more general aspects of the problems dealt with and to the bearing which study of the social behavior of nonhuman primates has on our understanding of human groups and societies.

Probably the major difficulty we face in studying behavior is the inevitable tendency to interact with our environment in the stereotyped cultural patterns which direct our activity as social beings. These, of course, include their associated stereotyped patterns of symbolic behavior, overtly expressed in language. In our social system we thus deal with behavior in terms of social conventions, including the conventional objectives or purposes, rather than in terms of the factors controlling the interaction of the organism with its environment. Our language is full of words and phrases which presumably refer to behavior, but which basically refer to conventional postulates *inferred* from the observed behavior—such as feelings, motives, instincts, habits, and so forth. In personal interaction with our fellows, we exchange integrated messages of attitudes, purposes, moods, and so forth, rather than discrete cues, signs, and signals. Together with our vis-à-vis we thus become integral parts of a previously learned social system which follows its own temporal course.

The application of social concepts (probably largely inherited from the fear of bandits in the European forests in the Middle Ages and maintained by hunters for ivory and trophies in the past century) to the jungle still determines the thinking of considerable sections of our professional population. Many students of human behavior are convinced that "the law of the jungle"

David McK. Rioch, Division of Neuropsychiatry, Walter Reed Army Institute of Research, Washington, D.C.

is one of arbitrary, murderous threat and attack and that the "normal" behavior of animals in "the wild" provides an "explanation" for the more distorted aspects of inadequate human mores. Once the chimpanzee and gorilla had been labeled dangerous, it was virtually guaranteed that methods for killing the adults and capturing the young would be greatly improved, but that it would take years to find out that these animals were responding to the form of the human approach and were not driven by some immutable force to attack and destroy.

When one observes human behavior under a series of different circumstances, one is more and more impressed by the "power of the word"—that is, of the mechanism, as it were, which maintains the stability of social communication in the group of allegiance. It was consideration of this phenomenon which led to the Sapir-Whorff hypothesis—namely, that the human sees the world in terms of his language. Although this hypothesis has been largely discarded, a recent series of experiments by Glanzer and Clark (1964) throws new light on the problem. They showed that although subjects accurately perceive forms, shapes, symbols, and so forth presented to them, in "thinking" about the perceived material, it was the *words* or *names* applied to the perceptions which were used and not the raw perceptions. This phenomenon has now come to be known as the "verbal loop" in the analysis of the behavior commonly called thinking. It seems to me that this provides the necessary modification to the Sapir-Whorff hypothesis. Although observers take in chunks of information, as it were, from the environment in relatively undistorted form, their thinking about this information (which directs what they do next) and also their transmission of these data (evoking behavior in their fellows) is in terms of the symbolic behavior derived from previous social experience with their friends and professional colleagues. Depending, then, on the group mores, one may get an extraordinary degree of rigidity in both formulations and theories due to factors quite apart from the observers' actual capacities as recorders of events.

It is quite clear that the terms "dominant" and "subordinate" can be used descriptively for particular patterns of behavior, but when dominance is formalized and then thought of as literally referring to some occult force that directs behavior, it can seriously interfere with careful observation and critical analysis. Symbolic behavior, as all behavior, appears to be hierarchically organized, and certain terms come to refer to, or even to stand for, whole conceptual frames of reference. Considerable attention has been given to these problems in social communication during recent years, but relatively few sociobiological studies have been conducted on the phenomena. An exception is the series of investigations conducted on the symbolic behavior of brain-injured patients (Weinstein and Kahn 1959) and on the influence of social factors on the form of psychotic delusions (Weinstein 1961). In these problems of language, it would be well for us always to bear in mind the comment

Mr. Perceval (Bateson 1961) made during his recovery from a violent psychotic episode: "I thought immediately thus—the spirit speaks poetically, but the man understands it literally."

These considerations place strong emphasis on the need for comparative studies, comparing not only different species under similar circumstances, but the same species under different circumstances (different habitats, different degrees of freedom, different experimental controls of the consequences of behavior, and so forth) and utilizing observers and experimenters with different theoretical and technical histories who consequently use very different symbolic behavior.

The use of the term "agonistic" to classify the patterns of interaction considered in this section of the volume is of interest in that it avoids postulating a purpose or objective, on the one hand, and, on the other, gives freedom to include a considerable variety of activities. By derivation it refers to competition —athletic, musical, poetical, and so forth—with the connotation of strong effort (cf. agony). From certain major Greek athletic contests, it has the extended significance of combat. Most of the authors in this section have described chiefly competitive combative behavior, with emphasis on the complaints of the subordinate member, but they also emphasize the infrequency of disabling or lethal damage. One is reminded of the patterns of ritualistic combat between males in the mating season in phylogenetically lower forms, patterns in which serious damage could hardly occur (cf. Eibl-Eibesfeldt 1963).

There seems to be general agreement among all observers on the importance of this limited agonistic behavior for maintaining the hierarchical structure of the group, though the rigidity-flexibility dimension seems to vary with the species. It is not clear yet what characteristics of the organism-environment interaction favor shift in one or the other direction, though one might guess that greater "intelligence," more abundant food supply, very strong or very weak capacities for aggressive destruction, and so forth might correlate with flexible hierarchies.

All authors also agree on the finding that the dominant animals bear scars from their careers. In this connection, I recall some anecdotes Dr. Ray Carpenter told me about Diablo. Diablo (Carpenter 1942) was the head monkey of Santiago Island as well as of Group II in the early period after the colony was established. He appeared to pay no attention to routine intragroup challenges, but in intergroup fights he frequently killed selected adversaries. He had no scars, but was usually surrounded by a large, devoted harem. Statistically, Diablo was a sport, but theoretically he is very important because he demonstrates that the macaque nervous system is able to mediate patterns of behavior which appear more adequate and which take into account more contingencies and probable consequences than the patterns shown by average dominant males. Parenthetically, it may be noted that these observations illustrate the importance in biology of recognizing the significance of unique situations, in contrast with much current emphasis on statistical research design. I mention this because I believe strongly that although there is no question of the importance of quantitative data, there is still room in this field for anecdotal records of unique events.

In his chapter on tripartite interactions, Dr. Kummer also limits himself to competitive combative behavior. His interest in the ontogenetic developments of this pattern closely parallels a great deal of interest among psychiatrists in tracing the development from earlier origins of symptoms shown by patients. Certainly, being able to relate inexplicable, inadequate symptoms in humans or complex patterns of behavior in older animals to better-known, more primitive, useful patterns in early life is conducive to a sense of understanding of the behavior. There is, however, another aspect of these tripartite relations I would like to note. In studies of comparative behavior, the presence of protective patterns in the mother-infant relationship has been shown in a large number of species. In descriptions of the lower forms, one gets the impression that the interaction is basically dyadic, either mother vs. aggressor or mother-infant complex vs. aggressor. The ability thus to direct behavior in more complex patterns (triadic, with independent participation by the participants) is clearly much greater in primates than in phylogenetically lower forms. It appears that the primate capacity for processing data during a rapidly progressing interaction under stressful conditions is of a higher order of magnitude than that of phylogenetically lower species. One would like to know whether different primates vary in their capacity for handling these more complex patterns of interaction and, if so, whether such differences might be correlated with differences in the relative brain size or structure. Dr. Kummer's baboons apparently integrate information from two other organisms while concurrently sending them separate messages through the course of a rapid interaction under stress.

The precise, quantitative observations reported by Dr. Sade and Dr. Kaufmann demonstrate the value of conducting studies on animals confined to a limited territory, which facilitates continuous detailed observation of particular monkeys engaging in particular patterns of interaction. These studies provide strong support for the concept of the importance of the central group of females for maintaining the over-all stability of the group structure. Other data indicate that this phenomenon is of general significance. Similar concepts are emerging from modern detailed studies of human social groups in spite of the fact that in general these concepts are counter to the popular traditions.

The criteria that Dr. Kaufmann used for including or excluding patterns of interaction under the descriptive heading of agonistic behavior are more inclusive than those employed by most authors. Thus he considered ritualistic mounting—at least under certain circumstances—as falling in this class, whereas most authors would describe it as some form of sexual behavior.

In addition to describing individual patterns of behavior in monkeys from different subdivisions of the group, Dr. Kaufmann has considered the effects of annual rhythms and their influence on the probability of the occurrence of one or another pattern. His interests seem to be in the group organization and in the varieties of interactions which provide the social communication maintaining the stability of the group organization. This directs attention to some important aspects of the general problem of the study of behavior, and of the classification and formulation of the data. Thus, if one looks at the immediate consequences of ritual fighting and of ritual mounting, one will obviously find different behaviors and probably find different physiologically and biologically correlated changes in the organism. However, if one considers a longer time duration, such as that involved in study of group stability under different circumstances, it might well be that these two patterns are equivalent in the sense that, in terms of maintaining the stability of social communication, they are interchangeable.

Although the theoretical frame of reference is not particularly relevant to the problem of observation and description of limited aspects of particular dyadic interaction patterns of primates, there are a number of advantages to using the recently developed concepts of communication theory (and its extension to the interaction between adaptive feedback mechanisms) for studying group organization and functional structure. Behavior is continuous through time, and hence, criteria for selecting the beginning and end of behavioral events are necessary. These may vary considerably for the same event when considered in relation to different factors. Further, since it is not feasible to record or describe all aspects of behavior in its course, it is necessary to develop criteria to determine the relevance of different aspects both to the course of the interaction and to the arrival at the pre-defined end state. No theory will provide ready-made answers to such questions, but communication theory tends to direct attention to them in a manner which the assumption that behavior is determined by instincts, motives, pain and pleasure, and so forth, fails to do.

I should point out that by communication theory I do not mean the limited area of measurement of rate of flow of information to the observer. Shannon's beautiful theoretical development of such measurements is practically applicable at the present time only in situations over which the operator has control and in which the unit of information can be specified. This has been of such importance to the growth of communication systems (telephone, radio, and so forth) that it has occupied the major attention of many investigators. As pointed out by MacKay (1957), the more descriptive aspects of communication theory are of more current significance to biologists. In this sense, communication theory is providing a rigorous system for correlating the *form* of events as they vary over time, independently of the energy required by the system. In contrast, classical energy theory deals with the correlation of events in terms of *force* (as measured by mass, acceleration, and so forth). Thus, information theory is concerned with the effects "of form on form," to use MacKay's expression (1963), whereas energy theory is concerned with the effects "of force on force." I would like to call attention to another aspect of communication which is useful for maintaining a more analytic view of behavior. This is the fact that the "meaning" of a signal is not a property or characteristic of the signal per se (MacKay 1962). Rather, the meaning can be understood as the number of states the recipient may be in (the number of different responses the recipient is capable of making) and the degree to which the signal increases the probability that one rather than some other of these responses will occur. Communication theory, as it has been developed in the past decade, can be useful in directing attention to (1) the situations and circumstances influencing the probability of occurrence of certain patterns of interaction (or behavioral events), (2) the cues or signals which are necessary and sufficient for initiating the events, (3) the factors which maintain and/or direct the course of the events to the signals or cues associated with their termination, and (4) the relationship of different events in the total social organization and in maintaining the group extant. One of the major advantages of communication theory is that one can deal with the whole range of behavior, from the single organism in a laboratory experiment to the social group-environment system, in a unified conceptual frame of reference.

Although a number of investigators in the early part of this century studied the significance of social factors and group organization for the behavior of individual persons, the development of a discipline of social biology has been delayed. At present, one can only speculate on the factors associated with this delay, but one must recognize the acute need for a systemized body of knowledge and the physical and biological factors (including communication and data processing) which control and which limit group organization. The studies on social behavior and group organization of primates reported in this volume and elsewhere during recent years give promise that the need for a social biology will soon be met. One would like to know whether there is a consistent correlation between different functional-anatomical specializations of different species, the size and structures of the groups they form, and the patterns of interorganismal, intragroup social behavior they develop. Thus in the natural habitat where the species have survived for considerable periods, it would be useful to know whether there are reliable correlations between highly specialized functional-anatomical features (such as for escape by running or hiding, for reproduction, for aggression, and so forth), the size and structure of the group, the social behavior within the group, and the ecological conditions. From such data one might begin to derive information on the functions of the total organization which have to be performed for survival, and also information on the degree of specialization in one function and modifications in other functions which are compatible with a total organization which permits survival.

The great advantage of detailed sociobiological studies of primates is that it will be possible to conduct a consistent series of experiments to determine correlations between social behavior and group organization and modification of one or other of the environmental contingencies. Although it is now possible to devise and carry out experiments manipulating separately and quantitatively such factors as the extent of the available territory, the availability of food and water, the size of the group, the rate of increase of the population, and so forth, few studies of group organization and social behavior under such stresses have been carried out. One experiment with which I am acquainted in this area is that of Calhoun (1962) on a colony of rats kept in a limited territory with consequent overpopulation, but with ample food and water. Dr. Calhoun found marked changes in the group organization and in the social behavior of the rats, together with a leveling off of the population considerably below the theoretical number the area would support. Dr. Calhoun's findings are at considerable variance with the present popular attitude toward overpopulation, which centers its attention on the availability of food and water. It is obviously very desirable to get further comparative data on these problems and also to get data on the effects of loads such as overpopulation, inadequate food supply, and so forth on the problem-solving, data-processing capacities of different organisms. Sociobiological information of this nature would be of considerable practical importance at the present time for the rational development of the rapidly growing fields of social and community psychiatry and mental health. (For example, it would appear that the significance of individual rewards and punishments is quite different in different social organization patterns). A cumulative body of knowledge on these very important problems is still lacking, however. It is to be hoped, therefore, that the studies on the social behavior and the group structure of primates will be extended, with attention not only to comparison of species and habitats but also to comparison of different experimental manipulations.

REFERENCES

Bateson, Gregory (ed.). 1961. *Perceval's narrative: a patient's account of his psychosis, 1830–1832.* Stanford: Stanford University Press.

Calhoun, John B. *The ecology and sociology of the Norway rat.* Washington: U.S. Government Printing Office.

Carpenter, C. R. 1942. Characteristics of social behavior in non-human primates. *Trans. N.Y. Acad. Sci. Series II.* 4:248–58.

Eibl-Eibesfeldt, Irenäus. 1963. Aggressive behavior and ritualized fighting in animals. In *Violence and war,* ed. J. H. Masserman, pp. 8–17. New York: Grune & Stratton.

Glanzer, Murray, and Clark, W. H. 1964. The verbal-loop hypothesis: conventional figures. *Am. J. Psychol.* 77:621–26.

MacKay, D. M. 1957. Information theory and human information systems. *Impact of Science on Society* 8:86–101.

———. 1962. Communication and meaning—a functional approach. In *Symposium No. 21, The determination of the philosophy of a culture.* The Wenner-Gren Foundation.

———. 1963. Machines and societies. In *Man and his future,* ed. G. Wolenholme. London: J. & A. Churchill, Ltd.

Weinstein, E. A. 1961. *Cultural aspects of delusion: a psychiatric study of the Virgin Islands.* New York: Free Press of Glencoe.

Weinstein, E. A., and Kahn, R. L. 1959. Symbolic reorganization in brain injuries. In *American handbook of psychiatry,* ed. S. Arieti, pp. 974–81. New York: Basic Books.

Part III

CAUSAL MECHANISMS

8

EXPERIMENTAL APPROACHES TO THE PHYSIOLOGICAL AND BEHAVIORAL CONCOMITANTS OF AFFECTIVE COMMUNICATION IN RHESUS MONKEYS

ROBERT E. MILLER

A stranger in a foreign land is faced with bewildering and frustrating confusion. Unless he can obtain the services of a translator or at least a good bilingual dictionary, he has an extraordinarily difficult time satisfying his most basic needs for food and shelter, to say nothing of understanding the customs and culture of the natives of that land. He may, in time, through careful observation and imitation, acquire a basic vocabulary in the local dialect, and eventually a full and complete exchange of ideas and information may ensue. Until he achieves some mastery of the language, however, he will probably attempt to communicate through a sign language—pointing, gesturing, and acting out. Obviously, the communication problem will prove to be immeasurably more difficult if the natives are hostile and suspicious and flee at the visitor's approach.

The scientist interested in communication processes in animal species finds himself in somewhat the same position as our stranger in a foreign land. Only recently have even rough approximations of bilingual dictionaries been constructed through the patient and meticulous observations of the ethologists. The descriptions of the complex and subtle behaviors which even relatively primitive species employ to communicate with each other regarding territoriality, sexual invitation, aggressive intent, and so forth have revealed for the first time the remarkably specific and intricate nonverbal capabilities of the nonhuman species.

Robert E. Miller, Department of Clinical Science, University of Pittsburgh School of Medicine, Pittsburgh, Pennsylvania.

Most of the data have been collected in natural or seminatural conditions by skilled and patient observers who unobtrusively watched ongoing behavior in their animal subjects. Another approach could be described as analogous to the use of an interpreter who can communicate in two or more languages. This is a laboratory method and, consequently, has the advantages of control and reproducibility and the disadvantages of unnaturalness, restriction, and fractionation. It cannot now, nor can it ever be expected to, explore or define the full range of primate communication, but it can and does isolate segments of behavior for detailed study.

The method depends upon the establishment of a simple artificial language whose definitions are agreed upon by the experimenter and the animal subjects, in our experiments, post-adolescent male *Macaca mulatta* monkeys. We chose to communicate with our subjects by instrumental conditioning with its two general subclasses, instrumental avoidance and instrumental reward. Thus we presented a stimulus followed by a consequence, the nature of which was determined by the response of the subject within a specified time period. In avoidance conditioning, the conditioned stimulus was followed by a noxious stimulus if the subject failed to perform the response which the experimenter had arbitrarily selected as the avoidance response. Likewise, in reward conditioning, the subject secured a gratification only if he performed a certain act upon receipt of the conditioned stimulus.

Instrumental conditioning is rapidly established in most species of animals and is a persistent, reliable, and sensitive phenomenon. It has provided the basic data for several prominent theories of learning (Hull 1952, Spence 1936). However, in order satisfactorily to accommodate most of the experimental data on learning in a conditioning theory, it has been necessary to postulate intervening affective conditions which mediate learning and performance. Hence, we hear a good deal about secondary drives and secondary reinforcement, the acquired drive of fear, anxiety reduction, and so forth (Miller and Dollard 1941, Mowrer 1960). Such theories assume that the conditioned stimuli arouse anticipatory affective responses which, in turn, motivate the animal to perform the appropriate conditioned behavior. Thus, when a monkey presses a bar upon presentation of an avoidance stimulus, the experimenter may conclude (1) that the animal perceived the stimulus, (2) that because of its prior pairing with a noxious stimulus the avoidance stimulus aroused fear, and (3) that the monkey pressed the bar to eliminate the signal, to avoid the shock, and to reduce his fear. A substantial literature supports these interpretations by indicating that autonomic and biochemical responses manifesting emotional discharge do, in fact, accompany the behavioral responses of instrumental conditioning (Mowrer 1960).

From a communications point of view, instrumental conditioning provides the experimenter with a simple language with which he can signal to the animal and receive back a report which reveals something about the affective state of the subject at that time. We have employed such techniques in the study of nonverbal communication of affects in the rhesus monkey.

The monkeys were first trained individually to speak our artificial language; that is, they were given instrumental training. Since we wanted ancillary measures of affective experience besides those of overt behavior (bar pressing), physiological measures, namely heart rates, were obtained throughout training. To obtain these records, we restrained the monkeys in primate chairs so that they could not interfere with the EKG electrodes. After some preliminary adaptation to the restraint of the chair, conditioning was begun. The primate chair was placed in a test room facing an 8 × 6 × 46 inch tunnel. The chair was adjusted so that the center of the tunnel was at eye level for the monkey. The distal end of the tunnel was closed off so that it was completely dark. Four inches from the closed end, a pair of end-lighted lucite plaques were mounted. One of the 3 × 6 inch plaques contained an etched double-pointed arrow, and the other contained a rectangle. The arrow illuminated in yellow was the conditioned stimulus for avoidance, and the rectangle lighted in red was the conditioned stimulus for reward. Two bars were mounted on the primate chair (Plate 8.1). One, with a 6-inch handle, terminated or avoided the shock stimulus when it was pulled three inches toward the animal. The second bar, with a 2-inch handle, activated an automatic pellet dispenser when it was depressed ¼ inch if the appropriate conditioned stimulus (CS) was present. A pellet cup located beside the animal delivered rewards within easy reach of the animal's mouth. Electrocardiogram leads were attached to the monkey's chest over the sternum and on the left chest wall, and a ground lead was attached to the left thigh. Shock electrodes were placed on the left ankle and on the metal seat. When this instrumentation was all attached and tested, the illumination in the test room was dimmed, and the experimenter left the room to begin the test session from a control room located in another area. The automatic test equipment was sufficiently soundproofed and far from the test room so that no audible cues were available to the test subjects. Each session was begun by a 5-minute baseline period during which a heart-rate sample of 1 minute was obtained as well as a measure of the monkey's bar-pressing performance. At the end of this baseline period, automatic tape programers were switched on for the 45-minute conditioning session. Each such session consisted of twenty conditioning trials presented at intervals of from 1 to 3 minutes in a randomized fashion. Throughout conditioning and subsequent communication sessions the CS-US interval was 6 seconds; that is, the animal had 6 seconds from stimulus onset to perform the appropriate response.

The first experiments were conducted on single classes of response; that is, a monkey was required to learn either an instrumental avoidance response or an instrumental reward response, but never both (Miller, Banks, and Ogawa 1962, 1963; Miller, Banks, and Kuwahara 1964). The more recent studies combined both reward and avoidance in a single session. In order to establish this discriminated avoidance-reward conditioning, however, the monkey was first trained to avoid shock, then to obtain reward, and finally both kinds of stimuli were randomly mixed in test sessions until the animals achieved very rigorous criteria of acquisition.

There were six categories of lever pressing to be considered in the criteria of conditioning. There were two kinds of conditioned response, for example, the avoidance bar pulled during avoidance CS presentation or the food bar pressed during reward CS presentation. Likewise, there were two categories of error, for example, the avoidance bar pulled during presentation of a reward CS and vice versa—in our analysis these were called false CR's. Either bar could be operated throughout a session without regard to stimulus presentation. These intertrial or spontaneous responses, if they occurred (at a steady, high frequency rate) could provide a baseline from which many spurious CR's would be expected. In order to insure that a discriminated learning was established, criteria were required for each source of error. The number of observed avoidance CR's in a session were compared statistically with false CR's and with chance expected CR's (derived from spontaneous responses) and, likewise, the same comparisons were made for reward trials. When all four comparisons were statistically significant for three consecutive sessions, the monkey was judged to have acquired discriminated reward-avoidance conditioning. Figure 1 illustrates the heart-rate responses of a group of five monkeys

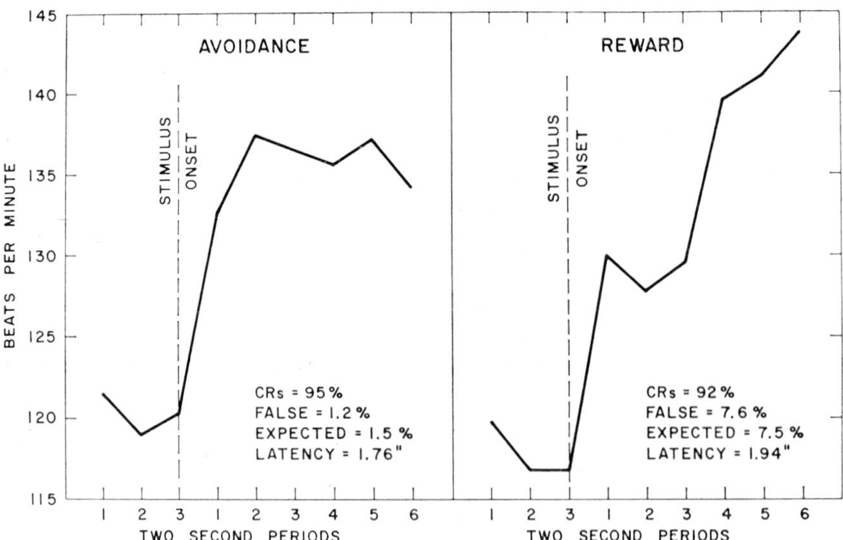

FIG. 1.—Conditioned cardiac responses in five monkeys for the final five conditioning sessions. The heart-rate response to avoidance stimuli differs significantly in trend from the heart-rate response to reward stimuli. Instrumental behavior is summarized in the appropriate panel.

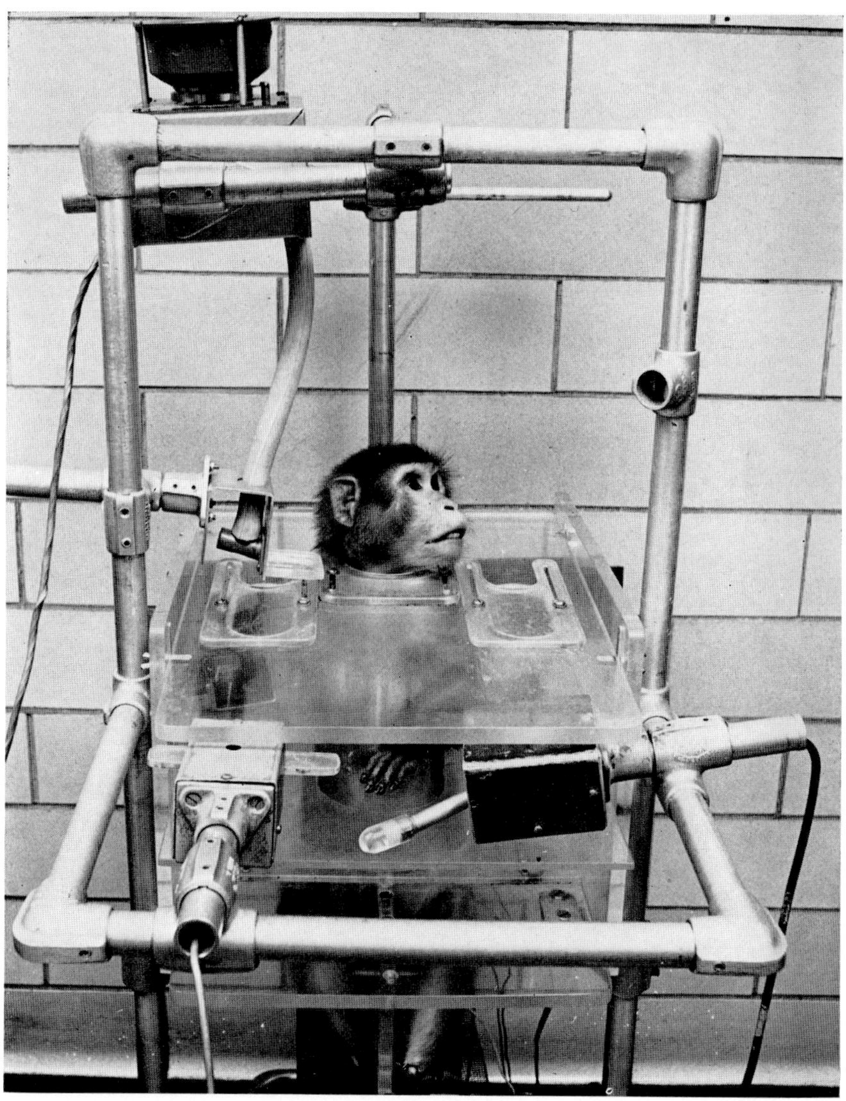

PLATE 8.1.—Front view of an animal prepared to be placed before the tunnel as a "responder." The two bars are attached to the chair and the automatic food vender is in position.

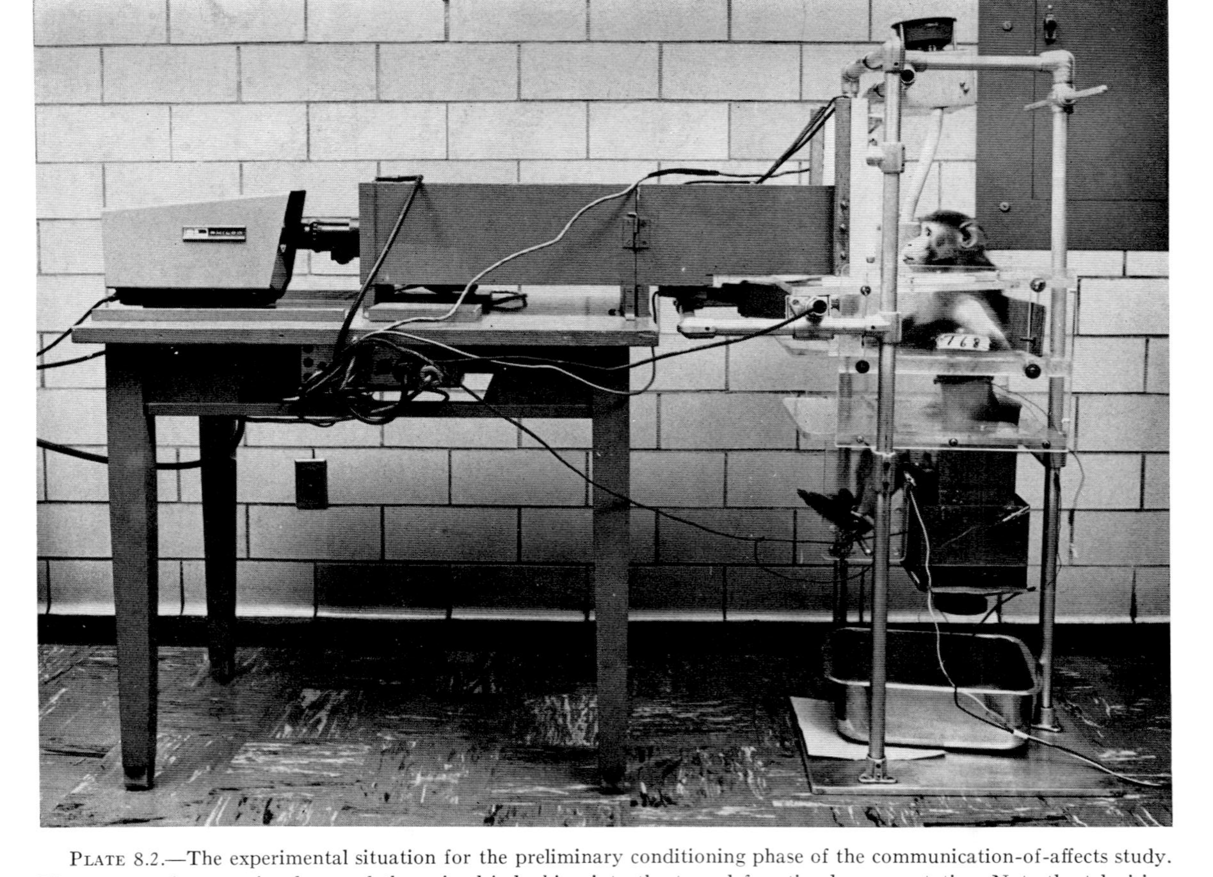

PLATE 8.2.—The experimental situation for the preliminary conditioning phase of the communication-of-affects study. The response bars are in place and the animal is looking into the tunnel for stimulus presentation. Note the television camera in place at the end of the tunnel focused on the head and face of the animal. Electrocardiogram leads are attached to the chest and leg.

PLATE 8.3.—A view from behind a responder looking into the tunnel at a film

PLATE 8.4.—A sequence of consecutive frames of film on an avoidance trial. Sequence begins at top left. This section of film was identified as a reliable trigger for avoidance responses from viewers. Note the trial lamp in the lower left. This lamp was masked from the viewers and centered on a photoelectric cell to activate the appropriate programing equipment located in the control room.

during the last five conditioning sessions. These data reveal that there were, in fact, significant discriminated conditioned autonomic nervous system responses which followed onset of the stimuli.

When the animals had achieved the criteria of conditioning, the communication of affects tests were begun using a paradigm which we have called "cooperative conditioning" (Miller, Banks, and Ogawa 1962). One of the monkeys was selected as a "stimulus animal" and paired with a second trained animal, the "responder." The stimulus monkey was placed in a test room facing the stimulus tunnel. This time the tunnel was equipped with a closed-circuit television camera mounted with the lens directly between the stimulus plaques (Plate 8.2). The forward end of the lens was in front of the stimuli and was shielded so that no reflection, flicker, or change of illumination levels was transmitted via television upon stimulus presentation. The camera was adjusted to televise the face and head of the stimulus monkey as it viewed the conditioning stimuli. The stimulus monkey was prepared as before with EKG electrodes, the automatic feeder and delivery cup, and shock electrodes. However, the levers that permitted the animal to respond appropriately to the stimuli were not attached to the primate chair. The responder monkey was likewise placed before a stimulus tunnel in another test room. The animals were unable to communicate vocally because of the distance between them and the soundproofing. The stimulus plaques had been removed from the tunnel in the responder's room and the distal end of the tunnel had been removed and replaced with a television receiver. Thus, the responder looking into the stimulus tunnel saw only the life-size televised picture of the stimulus monkey's face. The responder was equipped with EKG electrodes, feeder mechanism, and shock leads. This animal also received the two response levers which would permit the pair of animals to avoid shock or receive food rewards. The animals were wired in parallel in the two reinforcement circuits.

When both animals were appropriately placed and instrumented, the television camera was tuned and focused and the experimenters returned to the control room, which was also equipped with a television monitor, and the test session began. As during individual conditioning sessions, a 5-minute preperiod was recorded to obtain baseline lever-pressing and heart-rate measures. The automatic programing equipment was then turned on and ten avoidance and ten reward stimuli were presented to the stimulus animal in a randomly mixed pattern during the 45-minute test session. The animals in such a pair were tested daily for 10 days. Then the roles of the two animals were reversed, the responder becoming the stimulus monkey for an additional ten sessions. All possible pairings of animals within the squad were tested. Since the usual group consisted of three monkeys, there were six such pairings during the communication of affects tests. At the conclusion of the daily sessions the animals were returned to a common housing room where they remained in visual and auditory contact until the following day's session.

The task which confronted a pair of animals linked in the cooperative-conditioning network was primarily one of communication. The stimulus monkey received the conditioned stimuli but lacked the mechanisms to deal with them. The responder had the necessary levers to provide reinforcement but lacked the conditioned stimuli which served as the source of information to tell him which bar was to be pressed and when. Our hypothesis was that the animal confronted with a stimulus would exhibit both autonomic nervous system and behavioral responses which would be manifest in changes in facial expression and/or head movements, and that the responder upon perceiving such affective expressions would react with appropriate heart-rate and instrumental behaviors. If such communication did occur, there should be evidence of autonomic arousal in both subjects and, further, instrumental data which demonstrated a statistically reliable discriminated response.

Earlier experiments had revealed unequivocally that in the cooperative-conditioning technique monkeys were extremely sensitive to facial cues in avoidance and somewhat less sensitive to facial expressions anticipatory of reward when the pair of animals was exposed to only one of the conditioning situations, for example, avoidance only or reward only (Miller, Banks, and Ogawa 1963; Miller, Banks, and Kuwahara 1964). The combined avoidance-reward situation was, of course, much more difficult for the monkeys in terms of both initial individual acquisition and cooperative conditioning.

The results of the communications tests for the twelve pairings obtained from two replications of three animals each indicated that the responder monkeys perceived the affective expressions of their stimulus partners and responded both physiologically and instrumentally to these expressive changes. The communication process was much more effective in the avoidance trials than in the reward trials, confirming the previous experiments on single affects. Only one monkey, which served as responder to the other two members of his squad, failed to show significant instrumental performance on the avoidance bar. This particular monkey set up a high intertrial rate on the avoidance bar, thus obtaining a high probability of chance or spurious CR's. The remaining ten pairs of subjects were each significant with regard to appropriate avoidance behavior beyond the 0.01 level. Likewise, in all twelve pairings there were significant accelerations in heart rate in both stimulus and responder monkeys upon presentation of the avoidance stimulus. Acceleration of cardiac rate in the stimulus monkey preceded the cardiac response of the responder by 2 seconds—a factor which may be representative of the lag in this communication network.

The instrumental data from reward trials were much less impressive than those from avoidance trials. As a group, the responders pressed the proper bar and received food on 39 per cent of the reward trials, but only three of the pairs were found to have achieved statistically reliable discriminations to food in terms of bar-pressing performance. The heart-rate data, however, indicated significant cardiac responses in the viewers in all but three of the twelve pairings.

The cooperative-conditioning paradigm was specifically designed to utilize monkeys as perceivers or judges of facial expressions of other monkeys. As was mentioned earlier, the test situation was analogous in some ways to the use of an interpreter. Thus, by pressing a bar at a certain moment and showing an appropriate cardiac response, the responder "told" the experimenter that something in the facial expression of the stimulus monkey indicated to him, the responder, that a noxious or satisfying event was anticipated. It was our conviction that a monkey was a much more skilled interpreter of facial expression in another monkey than was man. We have deliberately avoided attempts to describe detailed cues which may be the necessary or sufficient expressive changes to convey affective experience from one monkey to another. However, since the experimenters were also watching the stimulus monkey via television, it was impossible not to notice differences in facial response to the avoidance and reward stimuli. There was no question that the avoidance stimulus elicited a more discriminable and intense response than did the reward stimulus. Hence, the finding that communication of fear (anticipation of pain) was more effective than communication of anticipation of reward came as no great surprise.

Since this experiment on concurrent avoidance and reward had demonstrated that behavioral and cardiac responses indicative of communication had been obtained, a more sensitive experiment was designed to factor out specific facial expressions involved in such communicative processes. For this purpose, a monkey trained to discriminate in the reward-avoidance situation was placed before the stimulus tunnel with the stimuli in place and without the response bars. A remote-control 16 mm color camera containing 1,200 feet of film replaced the television camera. Special pilot lamps were placed at the extreme lower corners of the exposure area to indicate presentation of the reward or avoidance stimulus. The movies were taken at 24 fps, a session lasting 33 minutes. Ten reward trials and ten avoidance trials were randomly mixed during a filming session. The experimenter controlled the reinforcement delivery from the control room. One avoidance stimulus, randomly selected, was permitted to continue for the full 6-second CS-US period and was followed by a 0.5-second shock. The other nine avoidance stimuli were terminated between 5 and 6 seconds after stimulus onset. Likewise, one reward stimulus elapsed without pellet delivery, while the reward CS was terminated with pellet delivery between 5 and 6 seconds on the other nine trials. Films were taken of two different stimulus monkeys.

A group of five post-adolescent male rhesus monkeys was then given initial training in discriminated reward-avoidance conditioning. As before, they were first trained in avoidance, then reward, and finally the two kinds of stimuli were randomly mixed in a single session. The animals were trained to the same stringent four-way criterion described above.

Upon completion of training, the stimulus plaques were removed from the tunnel and replaced by a rear-projection screen. The films of the faces of stimulus monkeys were projected on this screen in life size. The stimulus indicator lights on the film were masked from the viewer and projected directly into the lens of two photosensitive cells. Thus, when one of the indicator lamps in the film was illuminated for a trial, the photocell assembly was activated and, in turn, triggered the appropriate timing and control apparatus in the control room. In this way, it was possible to synchronize trial presentations precisely between the filmed session and the replay session.

The trained animals were placed before the tunnel and equipped with bars, reinforcement devices, and heart-rate electrodes. The experimenter left the test room and the usual 5-minute baseline measures were obtained. The projector was then switched on from the control room and the communication-of-affect session was begun (Plate 8.3). As in the televised tests, reinforcements were contingent upon detection and response to the facial expressions of the stimulus animal upon presentation of the conditioned stimuli. Ten sessions were given to each responder for each of the two films.

The heart-rate data revealed highly significant cardiac responses for avoidance trials in each of the two films (Fig. 2). Likewise, statistically significant

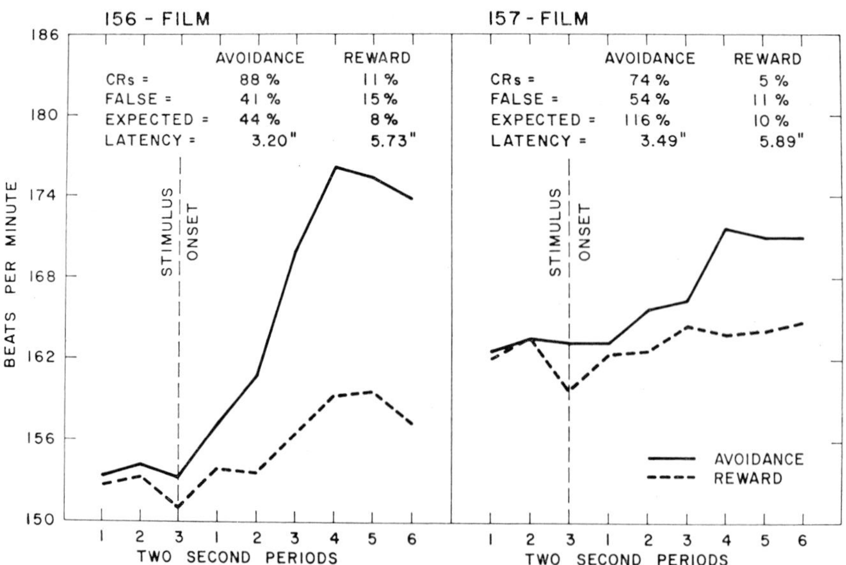

FIG. 2.—Conditioned cardiac response of five monkeys to the films of stimulus monkeys exposed to avoidance and reward stimuli.

cardiac changes were obtained in nine of the ten responder-film combinations for reward trials. The instrumental data, as in the television series, indicated that the avoidance response was discriminated more effectively from facial expression than was anticipation of food.

The big advantage in filming stimulus animals was that a standard stimulus situation could be replayed over and over again in an attempt to isolate the specific facial cues which elicited instrumental and cardiac responses in the viewer. Measurements of response latency and magnitude of autonomic arousal were obtainable from the records. Compilation of these data revealed that there was, in general, inter-subject and intra-subject consistency in speed and magnitude of response to specific trials on the films. Thus it was possible to select the most effective and least effective avoidance trials for comparison and, likewise, for reward trials. These specific sections of film were then inspected and analyzed to determine what specific differences were apparent. Some differences between behaviors on effective reward and effective avoidance trials have been observed. The most communicative fearful behaviors seem to be associated with many brief glances at the stimulus plaque followed by head and eye movements away from the stimulus (Plate 8.4). This repetitive peeking at the avoidance stimulus was most notable for the most effective avoidance trials and less frequent on noneffective trials. The behaviors upon presentation of the reward stimulus were markedly different. The stimulus monkey reduced his ongoing activity upon stimulus presentation and fixed his eyes directly on the stimulus with only a few eye and head movements away from the tunnel. The eyes were wide open and the facial musculature relatively slack.

It became apparent in analysis of these data that further refinements of data collection and processing would be desirable. All heart-rate data were recorded as sums of R-peaks over 2-second periods. This is too gross a measure for the precise localization of cues in the film. There were two reaction times to be measured—that of the stimulus animal perceiving and reacting to the conditioning stimulus and a second lag in the perception and response of the viewer to the facial changes of the stimulus monkey. A beat-to-beat measure of heart rate from a cardiotachometer would enable one to determine within a few frames of film when a communication of affect had occurred. Future studies will incorporate cardiotachometry.

As was mentioned previously, we have no illusions about the generality of our present techniques. The communicative capabilities of nonhuman primates are exceedingly more subtle, complex, and wide ranging than the relatively simple avoidance-reward behaviors which have been described here. Yet, the cooperative-conditioning method does provide very detailed analyses of the necessary and sufficient expressive cues which convey affect from one animal to another and, in addition, permits the measurement of autonomic nervous system activities which accompany such communication of affects. It is perhaps no accident that the facial expressions connotive of apprehension of shock, the quick peeking at stimuli, resemble the behaviors of submissive monkeys in dominance encounters where direct eye contact is usually avoided. The use of laboratory experimental techniques such as the one outlined here may provide additional information regarding the nonverbal communication capacities and cues of the nonhuman primates and may be extended to investigations of relationships between communication and such phenomena as social dominance status and effects of social deprivation.

These investigations were supported by a grant (M-487) from the National Institute of Mental Health, National Institutes of Health, U.S. Public Health Service.

REFERENCES

Hull, C. L. 1952. A behavior system. New Haven: Yale.

Miller, N. E., and Dollard, J. 1941. Social learning and imitation. New Haven: Yale.

Miller, R. E., Banks, J., and Kuwahara, H. 1966. The communication of affects in monkeys: cooperative reward conditioning. *J. Genet. Psychol.* 108:121–34.

Miller, R. E., Banks, J., and Ogawa, N. 1962. Communication of affect in "cooperative conditioning" of rhesus monkeys. *J. Abnorm. Soc. Psychol.* 64:343–48.

———. 1963. The role of facial expression in "cooperative-avoidance" conditioning in monkeys. *Ibid.* 67:24–30.

Mowrer, O. H. 1960. Learning theory and behavior. New York: Wiley.

Spence, K. W. 1936. The nature of discrimination learning in animals. *Psychol. Rev.* 43:427–49.

9

NEUROLOGICAL ASPECTS OF EVOKED VOCALIZATIONS

BRYAN W. ROBINSON

Naturalistic techniques are valuable and essential but, as with every scientific technique, are inevitably limited in some fashion. There are potential advantages in supplementing field observations with direct behavioral manipulation via brain tele-stimulation. These will be pointed out and some recent technical advances in this area will be described. Following this, an abbreviated review of the localization and organization of electrically evoked vocalizations will be presented. A complete technical report is in preparation (Robinson and Mishkin 1966).

There is good behavioral evidence, confirmed amply by other contributors to this book, that vocal responses are of prime importance in primate communication and social interaction. In the second part of this chapter, the study of vocal responses is shown to be helpful in gaining a deeper insight into the broad and enduring questions of cerebral organization.

TELE-STIMULATION

The use of brain tele-stimulation in primate behavioral studies is justified since there are inherent limitations in purely observational techniques that can be corrected with tele-stimulation.

The value of naturalistic studies derives from the assumption that present environmental conditions are not significantly different from those existing during the epoch of evolution of the behaviors under study. Unless this assumption is considered valid, there can be no assurance that the present environment is any better as a context for studying the behavior than other possible environments, including those in a large outdoor compound or even in the

Bryan W. Robinson, Laboratory of Neurophysiology, Yerkes Regional Primate Research Center, Emory University, Atlanta, Georgia.

laboratory. We cannot know with certainty what food, cover, pressure from other species, and other conditions existed at various times in the phylogenetic history of the various types of primates. It therefore seems justifiable to view "natural" conditions as "contemporary" conditions and not to endow these with semi-mystical properties for nurturing behavior. The behavioral mechanics used by particular primate species and troops to survive and to prosper under contemporary conditions are valuable but can hardly be considered an exhaustive accounting of the species' full adaptive capacities. The illusive nature of natural conditions is highlighted when we consider man as a primate species. Man's natural habitat is so diverse that it defies definition. The important point is not the mere complexity and variability of man's environment, however, but rather the realization that whatever these conditions are—feast or famine, youth or age, peace or war, solitariness or crowding, terrestrial or extra-terrestrial—each of them supplies a context that permits the observer to learn more about the nature, meaning, and effectiveness of the various human behaviors.

Second, the presence of a human observer immediately introduces some artificiality into naturalistic studies, particularly when the observer makes no effort to camouflage himself, a technique prominent in several recent studies. The effects of such artificiality may appear to be small; nonetheless, their influence cannot as yet be assessed. Animals need not constantly attend to an intruder to be constantly aware of his presence, and it is likely that his presence evokes subthreshold anxiety or aggressiveness that will influence overt behavior from time to time.

In addition to these considerations, observational techniques may not be the most desirable procedure in special conditions. If, for example, a behavior is rarely seen in the field, even prolonged observation is insufficient to study it. Such an impasse may occur with behavior patterns of obvious importance. Schaller (1963) observed copulation in the mountain gorilla only twice. From this observation the experimenter cannot say whether this lack of copulation resulted from the mild disturbance created by the presence of the observer, or from a natural celibacy, or from a preference for nocturnal copulation. The end result was that after a prolonged and careful period of observation under natural conditions a process as fundamental as copulation was not seen often enough to be analyzed. Another special condition occurs when the stimuli necessary to evoke a behavior are not prominent. For example, in many primate groups, including the gorilla, dominance is established, maintained, and transferred so smoothly that the dynamics of the process are anything but evident. A third special condition is one in which specific behaviors, evolved under conditions differing from those now existing, may be imperfectly understood or even incomprehensible when observed in the field. Displacement activities represent one example of this; another, more specific, may be represented by intertroop encounters among Ceylon gray langurs reported by Ripley in this volume.

Since there is no unique value to natural conditions, since naturalistic field studies inevitably introduce some artificial elements into the animal's environment, and since any one environment may result in certain behaviors occurring rarely, or with muted dynamics, or without obvious meaning, some manipulation of the animal's behavior may be necessary to arrive at a full understanding of primate behavior patterns. Brain tele-stimulation is behavioral manipulation technique that has rapidly developed in the past five years. By means of electrical stimulation of the brain through chronic, implanted electrodes, physiologists have long been able to evoke behaviors such as eating, drinking, aggression, flight, penile erection, alerting, sleep, vocalization, and motivational responses, that is, self-stimulation and escape-from-stimulation. These responses and others have now been localized with good accuracy in many regions of the brain. If the electrode is well placed and if the stimulating current is not permitted far to exceed threshold the evoked behavior can resemble spontaneous behavior to a remarkable degree and at times cannot be distinguished from it. Such stimulation, if carried out carefully, can be used intermittently for a period of months without appreciable brain damage (Delgado 1959).

Robinson, Warner, and Rosvold (1965) have recently developed a system designed expressly for primate behavioral studies in either the field or the laboratory. The technical features need not be described here; however, the main design concepts that are of particular relevance for behavioral work should be mentioned. First, in order to eliminate battery changes with the disrupting effects of capturing the animals under study, the stimulator unit, its upper surface covered with an array of solar cells, is mounted directly on the animal's head. These solar cells extract enough energy from either artificial or natural light to power the unit indefinitely. Thus, once the electrodes and stimulator are in place, the animals can be returned to their natural environments and followed just as one would do in any field study. Second, in order to permit several responses rather than only a single one from each implanted animal, a small 12-position remote-control stepping switch is incorporated. By means of this switch, any of eleven electrodes may be selected as often as desired for stimulation; the twelfth contact is left blank, and the stepper is returned to this position during all rest periods to insure against unwanted stimulation from extraneous electrical disturbances. Third, to permit more than one member of a primate group to be manipulated and to permit such multiple manipulations to be carried out independently and, if need be, simultaneously, all circuits are crystal stabilized. By switching the transmitter to slightly different carrier frequencies and employing tele-stimulators similarly tuned, independent but sequential stimulation of several animals may be carried out. If more than one transmitter is used, these stimulations can be performed simultaneously.

This system and similar ones will permit many otherwise impracticable studies on primate behavior. Particular behaviors in specific animals will fall under the direct influence of the experimenter, and he will be able to modulate the intensity and timing of these behaviors while the animals are roaming freely. Some suggested experimental uses of tele-stimulation in behavioral work have been discussed by Robinson (1964). For a few specific examples of ways in which tele-stimulation can assist primate behavioral studies, let us return to the three previously mentioned special conditions. Copulation in an implanted primate could be evoked and analyzed in detail. The dynamics of dominance could be studied by a number of approaches: by evoking fear in the dominant male; or by punishing the dominant male when he successfully asserts his position; or by evoking challenges from subordinate animals; or by rewarding subordinate animals when they challenge the dominant male. Displacement activity could be studied by noting its appearance during brain stimulation and correlating it with the functional and anatomical systems that were activated by the stimuli.

In summary, brain tele-stimulation is a technique that can be employed in either natural or modified environments to induce changes in the behavior both of individual animals and of groups of animals. It will not replace pure observation but offers the opportunity, when judiciously used, to supplement it by giving the experimenter control over the time of appearance and duration of selected behaviors.

VOCALIZATION

Let us now turn to a topic which is quite different from the preceding one yet which is relevant to it. If we are to manipulate behavior by stimulating various parts of the brain, meaningful results will depend on our knowledge of the organization of the neural substrate underlying these behaviors. "Organization" here means not merely the localization of a response but also the relation of the response to others. Despite the extensive work with electrical stimulation, principles of neural organization are known very imperfectly and so far have been of little use in making behavior more meaningful.

In a previous paper (Robinson 1964), I outlined three organizational principles that summarize some of the invariants governing intersystem and intrasystem relationships. The first of these is that *functions are structurally distributed*—that is, all evoked behavioral responses so far reported are extensively distributed in noncontiguous regions in the forebrain and other areas. This distribution is not unimportant, for quantitative analysis indicates that many of these structures contribute substantially to the total neural representation. As an example, evoked food intake is traditionally associated with the lateral hypothalamus; yet, in a quantitative study, 85 per cent of the occurrences of this response were found to be evoked from outside the lateral hypothalamus. It must be assumed that these extrahypothalamic structures have an important role in organizing and integrating food intake with other adaptive behaviors.

The second principle is that *structures are functionally distributed*—that is, from most structures that are readily studied by electrical stimulation, several different responses can be obtained. It can be shown that this functional distribution is not dependent on the more or less arbitrary nature of defining boundaries for the structures or on spread of current. Functional distribution indicates that in some structures there is an intimate and obligatory intermingling of functional systems.

In many places in the nervous system, intermingling of neural elements implies, and is a means of, functional interaction of these elements. Of the many ways of possible intermingling, three situations can be evaluated quantitatively. First, systems may be maximally associated spatially. If this is true, the probability of evoking the responses at the same time will be high and will be limited only by the probability of evoking the least frequent of the responses. Second, systems may be minimally associated spatially. If this is true, the probability of evoking the responses simultaneously will therefore be low and will approach zero. Third, a possibility that has many fascinating genetic and behavioral implications, systems may be randomly associated spatially. If this is true, the probability of evoking the responses simultaneously will equal the product of the individual probabilities of evoking them without regard to each other. This type of stochastic analysis has been carried out with several behaviors. It was surprising to find that the usual mode of intermingling was not ordered but was random and that neural systems maintained, within limits, spatial *stochastic independence*. This is the third principle of organization. It should be noted that this independence is anatomical and not functional, for we know the behavior of primates consists of highly ordered sequences that are adaptive. We will speculate later about the possible utility of a system that is partly random.

Data were gathered on 5,885 points distributed along 205 electrode tracks in fifteen 8- to 10-lb male *Macaca mulatta*. Roving-monopolar electrodes were used in conjunction with a head-mounted stereotactic electrode-guidance platform made of dental acrylic. This platform allowed the insertion of individual electrodes in unanesthetized animals, exploration of a column of brain tissue, and fixation of the electrode for chronic use in any desired point along the track. The number of electrodes per animal varied from four to thirty-two, averaging thirteen. The stimulating electrodes consisted of the terminal 0.5 mm of a 0.005-inch teflon-insulated stainless steel wire mounted in a 28-guage stainless steel hypodermic tube. The return electrode was the head of an inverted stainless steel screw in the skull, which also served as one of the platform mounting screws. The surface area of the return electrode exceeded that of the 0.5-mm-tip stimulating electrode by a factor of 300. The usual stimulation parameters were a monophasic 1.0-msec d.c. cathodal pulse at a rate of 50 pulses per second. Current per pulse varied but was not allowed to exceed 1.0 ma. Length of application of the pulse train and the rest interval between successive applications of pulse trains varied as the evoked behavior varied. The animals were kept in a restraining chair for all stimulations.

Following completion of the experiments, animals were sacrificed with Nembutal followed by perfusion of the vascular system with saline and 10 per cent neutral formalin. All electrode tracks were identified histologically with thionine and Weil stains, and photographs of the Weil-stained sections were used for reconstruction. By a combination of stereotactic, behavioral, and impedance (Robinson 1962) clues, the error of reconstruction along the entire track was 1.0 mm or less.

To facilitate rapid and accurate retrieval and analysis, data were kept in both analog and digital form on special cards. The anacards (analog cards) were 8½ × 11-inch cards, one for each point stimulated. These cards were given an identifying "point number," from 0 to 5884 in accordance with the chronological sequence in which points were stimulated, and they were filed in numerical order. On these cards were entered all the data gathered in the experiment: each card contained information concerning the animal, position and number of the electrode track, position of the electrode, stimulation parameters, date, experimental situation, evoked or spontaneous changes, and any other facts of interest. A photograph of the histological section containing the electrode track was permanently attached to each card. The locus stimulated was identified and its name entered on the card. Anacards served as a permanent repository of all the data and permitted data retrieval in accordance with animal or electrode track. Data retrieval in accordance with behavioral or anatomical characteristics was very difficult, however, because of the large number of cards.

To complement the analog system, 250 digicards (digital cards) were employed. These were plastic cards, approximately 8½ × 11 inches, each with a capacity of 10,000 bits, into which information was entered by drilling holes. Hole positions were identified by a 4-digit number (0000 to 9999): the first two digits represented the y-coordinate (00 to 99); the second two, the x-coordinate (00 to 99). All desired biographical, technical, behavioral, and anatomical aspects of the data on the anacards were digitized, utilizing devised numerical codes and entered in the digicards in a manner inverse to that used in the anacards. (By means of digicard combinations, the capacity of the digicard set was much greater than 250 × 10,000.) Each digicard was assigned one category, for example, food intake, vocalization, lateral hypothalamus. The point numbers of those characteristics present were drilled into the corresponding hole positions. Altogether, 2.8×10^6 bits were entered in this manner.

Readout was accomplished by placing the card over a light box. In summary, each anacard represented a point of stimulation identified by its behavioral, anatomical, and similar characteristics; each digicard represented a characteristic identified by its point numbers. It was unnecessary to maintain the digicards in any sequence, since they were identified by a combination of color and numbers.

Logical sums and products of any number of cards are readily performed; therefore, any question that can be expressed logically can be asked in this system. Many analyses are satisfactorily made by simple counts of lighted holes. Access to the anacards for recapturing any information lost or not included in the analog–digital coding is easily performed by superimposing a plastic grid over the digicards and noting the point numbers of lighted holes.

Accurate and meaningful analyses of the results required a quantitative assessment of the adequacy of sampling by the exploring electrodes and correction for sampling heterogeneities. Adequacy of sampling can be specified by two numbers: (*1*) the per cent of the volume of a structure included in the sample, and (*2*) the number of stimulations per cubic millimeter within the portion sampled. An over-all "figure of merit" of sampling can be obtained by multiplying these two numbers. To determine these it was necessary to measure the volumes of various brain structures. A normal brain was embedded in celloidin, cut in the frontal stereotactic plane, and matched to the Olzewski atlas for *M. mulatta*. By extrapolation, a set of sections from A33.0 to P2.0 was obtained. Enlarged line drawings were made at millimeter intervals, and the cross-sectional areas of each structure were measured mechanically by a polar planimeter at these intervals. The volume in arbitrary units was then obtained by simply adding these areas together. Conversion to cubic millimeters was made by comparing precisely defined areas of this atlas to the Olzewski atlas and deriving a conversion factor, which was in turn corrected for shrinkage in accordance with data provided in the introduction to the Olzewski atlas.

Figure 1 illustrates at three representative levels of the forebrain those areas from which electrical stimulation evokes vocalization. The sounds can be categorized into several recognizable types and would appear either during stimulation (Vocalization) or immediately following stimulation (After-vocalization). The points are concentrated in specific areas. At the same time, they are not confined to any one anatomical structure but are distributed among several. A quantitative aspect of this structural distribution is given in Table 1 as Per Cent of Corrected Total. It was found that 87 per cent of all vocalization points lie in neural systems coursing in the eleven structures listed.

The values were derived in two steps. First, if a group of points appeared to be centered on a structure, all were counted as lying in that structure even if some actually were just outside. This gathering of points is necessary and is justified since boundaries in the central nervous system cannot be precisely defined and since studies of degenerating axons indicate a degree of anatomical diffusion of fibers beyond the accepted limits of most structures. For example, all the points in Figure 1 at +31 are considered to be in the anterior cingulate gyrus. Second, the subtotals of these points in each structure were corrected for sampling inequalities to a standard criterion of one stimulation/cubic millimeter, and the contribution of each structure was expressed as a fraction of the corrected total. The possible error introduced by this correction will vary with the inadequacy of the original sampling and can be judged from the Sampling Factor, which gives the actual number of stimulations per cubic

TABLE 1

VOLUMES, NUMBER OF TIMES STIMULATED, AND SAMPLING DATA FOR VARIOUS STRUCTURES

Structure	Volume (mm^3)	Exploration Fraction	Times Stimulated	Sampling ($stim/mm^3$)	Figure of Merit	Per Cent of Corrected Total	Rate of Evoked Vocalization
nuc. acc...	110	1.00	52	0.47	0.47	0.08	0.23
gca......	693	0.82	309	0.54	0.44	.08	.05
DB......	171	0.97	130	0.78	0.77	.10	.34
POR.....	30	1.00	48	1.60	1.60	.03	.64
LH......	54	1.00	76	1.41	1.41	.03	.11
DH......	70	1.00	110	1.58	1.58	.01	.09
PH......	57	1.00	60	1.05	1.05	.07	.25
MH......	52	1.00	78	1.51	1.51	.03	.17
teg.......	225	0.88	105	0.53	0.48	.22	.31
amyg.....	395	0.78	90	0.38	0.30	.16	.20
IC.......	838	0.33	128	0.47	0.16	0.06	0.16

LIST OF ABBREVIATIONS

nuc. acc. ....nucleus accumbens
gca. ....anterior cingulate gyrus
DB ....diagonal band of Broca
POR ....pre-optic region, medial and lateral
LH ....lateral hypothalamus
DH ....dorsal hypothalamus
PH ....posterior hypothalamus
MH ....medial hypothalamus
teg. ....tegmentum
amyg. ....amygdala
IC ....internal capsule
ST ....stria terminalis

millimeter. The over-all figure of merit for each structure emphasizes the deficiencies in sampling and is useful as a conservative indicator of the confidence that may be placed in the analysis.

The results show clearly that, although the eleven structures contain 87 per cent of the corrected total of points, no one structure dominates or can be termed a center for vocalization. This distribution is irreducible, and reflects the involvement of several definite neural systems in vocalization.

Not every stimulation in the structures concerned with vocalization is successful in evoking sound. Most of the time a new placement will fail to elicit a vocal response. The probability of evoking a desired response may be measured by considering only those placements within each structure. These results are given in Table 1 in Rate of Evoked Vocalization. These values show clearly

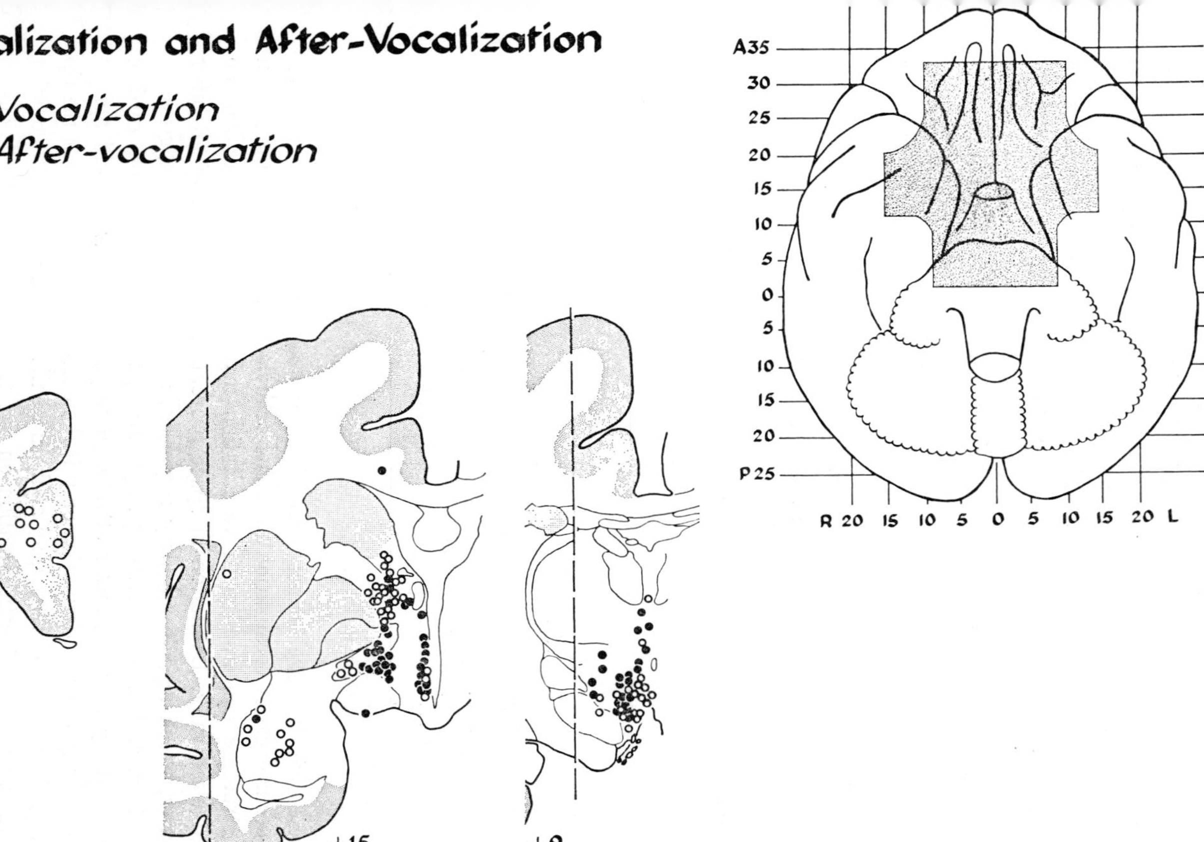

FIG. 1.—Map of electrically evoked vocalizations at three representative levels. "Vocalization" indicates that the response occurred during stimulation; "after-vocalization," that it occurred immediately following the stimulus. The shaded area on the base view shows the extent of the electrode exploration. At each of the three levels shown, the dashed line indicates the extent of exploration. All areas medial to the dashed line were stimulated.

that the probability of evoking vocalization in different structures varies greatly. A dilemma from the organizational point of view is that structures with high rates of evoked vocalization may contribute rather insignificantly to the total system of vocalization. For example, the pre-optic region, with a high sampling figure of merit of 1.60, has the highest rate of evoked vocalization (0.64) yet contributes only 3 per cent of the corrected total; the anterior cingulate area, with a much lower rate (0.05), nonetheless contributes a higher amount (8 per cent) to the corrected total. Such divergent indexes of the possible importance of the anatomical substrate of specific behaviors underscores the inherent difficulties in organizational schemes built on the concept of neural centers or derivatives of it.

TABLE 2

STOCHASTIC ANALYSIS: VOCALIZATION VS. AGGRESSION

In this and succeeding tables the figures in parentheses following structures are respectively the number of times stimulated and the rates of evoking the behaviors in the caption. The column p is the product of these rates.

STRUCTURE	p	CONJOINT OCCURRENCES		
		Random	Actual	Maximum
PH (60, .25, .25)........	0.06	4	6	15
DH (110, .09, .15)......	.01	1	1	10
LH (76, .11, .13)........	.01	1	1	8
DB (130, .34, .11)......	.04	5	10	14
ST (86, .30, .16)........	0.05	4	10	14
Mean..............		3.0	5.6	12.2

t-test: actual vs. random = N.S.
actual vs. maximum <.01

These difficulties are further emphasized by stochastic analyses of the conjoint occurrences of vocalization and other behaviors during stimulation. From each structure concerned with vocalization many other responses can be evoked electrically. Limitation of space prevents a description of all other responses, which include pupillary responses, alerting, searching, aggression, biting, fear, eating, drinking, sleep, defecation, urination, salivation, cardiac and respiratory changes, and motivational responses. Behaviors will appear simultaneously during stimulation only if their neural systems lie near the tip of the probing electrode; thus, an analysis of the frequency of such conjoint appearances will provide information on the tendency of neural systems to intermingle. Tables 2 through 6 illustrate examples of this type of analysis for evoked vocalization and for some behaviors commonly associated with it. Table 2 examines the association of vocalization and aggression in each of five structures. It can be seen, when the differences are evaluated with the t-test for correlated means, that the neural systems underlying these two responses are randomly intermixed. Table 3 shows a similar result for vocalization and lip-smacking. Table 4 gives the same result for the triad of vocalization, agitation, and smacking; Table 5 shows the correlation for the tetrad of vocalization, agitation, smacking, and aggression. Table 6 gives it for the pentad of the tetrad plus biting. Many other combinations have been analyzed with identical

TABLE 3

STOCHASTIC ANALYSIS: VOCALIZATION VS. SMACKING

STRUCTURE	p	CONJOINT OCCURRENCES		
		Random	Actual	Maximum
PH (60, .25, .32)........	0.08	5	8	15
DH (110, .09, .35)......	.03	3	5	10
LH (76, .11, .41)........	.05	4	6	8
DB (130, .34, .42)......	.14	20	20	44
ST (86, .30, .44)........	0.13	11	15	26
Mean..............		8.6	10.8	20.6

t-test: actual vs. random = N.S.
actual vs. maximum <.01

TABLE 4

STOCHASTIC ANALYSIS: VOCALIZATION VS. AGITATION VS. SMACKING

STRUCTURE	p	CONJOINT OCCURRENCES		
		Random	Actual	Maximum
PH (60, .25, .37, .32)....	0.03	2	3	15
DH (110, .09, .25, .35)...	.01	1	1	10
LH (76, .11, .36, .41)....	.02	2	3	8
DB (130, .34, .32, .42)..	.05	6	8	42
ST (86, .30, .13, .44).....	0.02	2	2	11
Mean...............		2.6	3.4	17.2

t-test: actual vs. random = N.S.
actual vs. maximum <.001

results. There appears, from these data, to be a consistent lack of order in the spatial organization of neural systems within restricted anatomical areas. The results are remarkable in some ways. For example, Table 5 shows that four of the most common responses evoked from the posterior hypothalamus never occur together, despite the fact that the probability of evoking each is at least 0.25.

The dilemma in organization is now sharply etched: how is orderly, adaptive behavior generated from neural systems which are in part inherently unordered? What can be the utility of such a method of constructing a brain? I feel that the ultimate answer will be related to the long-standing problem of explaining individual variability between brains of the same species. When neural systems are in close spatial proximity, an opportunity for interaction is provided by short-axon synapses, particularly in cell fields rich in Golgi

TABLE 5

STOCHASTIC ANALYSIS: VOCALIZATION VS. AGITATION VS. SMACKING VS. AGGRESSION

Structure	p	Conjoint Occurrences		
		Random	Actual	Maximum
PH (60, .25, .37, .32, .25)	0.01	1	0	15
DH (110, .09, .25, .35, .15)	–	0	0	10
LH (76, .11, .36, .41, .13)	–	0	2	8
DB (130, .34, .32, .42, .11)	0.01	1	0	14
ST (86, .30, .13, .44, .16)	–	0	2	11
Mean		0.4	0.8	11.6

t-test: actual vs.random = N.S.
actual vs. maximum <.001

TABLE 6

STOCHASTIC ANALYSIS: VOCALIZATION VS. AGITATION VS. SMACKING VS. AGGRESSION VS. BITING

Structure	p	Conjoint Occurrences		
		Random	Actual	Maximum
PH (60, .25, .37, .32, .25, .43)	–	0	0	15
DH (110, .09, .25, .35, .15, .23)	–	0	0	10
LH (76, .11, .36, .41, .13, .16)	–	0	0	8
DB (130, .34, .32, .42, .11, .07)	–	0	0	9
ST (86, .30, .13, .44, .16, .10)	–	0	0	9
Mean		0	0	10.2

t-test: actual vs. random = N.S.
actual vs. maximum <.001.

Type 2 cells, such as the various limbic areas. This spatial proximity is highly stereotyped in brains of the same species. If the neural elements of these systems intermix randomly to a degree, then the resulting short-axon synaptic connections will also be to an extent unspecified, and we can visualize an interplay of specificity and randomness that together account for both the behavioral stability that characterizes the species and the behavioral variations that characterize the individuals of that species. This concept of an interplay between randomness and specificity in neural organization may be termed *a theory of random specificity,* a name paradoxical enough to be remembered.

The author would like to acknowledge the participation of H. E. Rosvold and M. Mishkin in all phases of this experiment.

REFERENCES

Delgado, J. M. R. 1959. Prolonged stimulation of brain in awake monkeys. *J. Neurophysiol.* 22:458–75.

Robinson, B. W. 1962. The impedance method of localizing intracerebral electrodes. *Expertl. Neurol.* 6:201–23.

———. 1964. Forebrain alimentary responses: some organizational principles. In *Thirst: First International Symposium on Thirst in the Regulation of Body Water.* New York: Pergamon Press.

———. Some new horizons in experimental neurology. Commemorative volume: P. M. Sarajishvili, Moscow, 1964 (in press).

Robinson, B. W., and Mishkin, M. 1966. Evoked vocalization in *M. mulatta.* In preparation.

Robinson, B. W., Warner, H., and Rosvold, H. E. 1965. A brain tele-stimulator with solar cell power supply. *Science* 148:1111–13.

Schaller, G. 1963. *The mountain gorilla: ecology and behavior.* Chicago: University of Chicago Press.

10

THE BEHAVIOR OF SQUIRREL MONKEYS (*Saimiri sciureus*) AS REVEALED BY SOCIOMETRY, BIOACOUSTICS, AND BRAIN STIMULATION

DETLEV W. PLOOG

This chapter is the result of the combined efforts of my co-workers, Rolf Castell, Sigrid Hopf, Manfred Maurus, J. Mitra, Peter Winter, and myself, the general theme of our observations of the last 3 years being the social behavior of the squirrel monkey. As the title of this chapter suggests, our study is divided into three parts. The first part is an exposition of group behavior analyzed with the aid of a sociometric method. Initial observations were made on the ontogeny of social behavior with special consideration of genital display, a characteristic mode of behavior for squirrel monkeys. The second part presents a survey of the extensive sound repertoire of these animals and the meaning of the sounds with reference to social behavior. In the third part we offer the results of experiments in brain stimulation suggesting that an essential component of the social behavior under investigation has been found represented in certain subcortical structures of the brain. As a whole, this chapter is meant to show how innate behavior, on the one hand, affects group structure and how the group influences the modification of innate behavior, on the other.

GROUP BEHAVIOR

SUBJECTS (see Plate 10.1)

The groups studied consisted of four to six members, their spontaneous behavior having been observed over longer periods of time (weeks to years) ac-

Detlev W. Ploog, Max-Planck-Institute of Psychiatry, Munich.

cording to definite rules and with all external influences of a disturbing nature eliminated whenever possible. The presence of observers in no way influenced the usual behavior of the animals.

Colony I was made up of four males (B, C, D, E) and two females (F and G); D and F were immature at the time.

Colony II consisted of two males (C and I) and four females (J, K, L, M), all mature. C had been introduced into this group from Colony I.

The observations of these colonies date back to the years 1958 to 1960 (Ploog 1963; Ploog, Blitz, and Ploog 1963).

Colony III, at the beginning, was made up of two males (N and O) and two females (P and Q). However, after 2½ years, a female infant (Ri) was born, and its birth as well as its development within the group was also studied (Bowden, Winter, and Ploog 1966; Hopf and Ploog; Ploog *et al.* in preparation).

Colony IV started out with two mature males (R and S), two mature females (T and V), one immature female (U), and W, the son of V, who was 4 months of age at the beginning of the observations and 12 months old at the end. The death of R ended the first series of observations, so that the second series was carried out with the remaining five animals.

Colony V consisted of S, U, V, W (from colony IV), and Y with her male infant X who was 3 months old when both, mother and infant, were introduced into the group.

METHODS

During each period of observation, lasting usually 30 minutes, the observer noted the various modes of behavior of only one animal in the group. Each member of the group was observed an equal number of times over equal lengths of time but in random order during an observation series. In accordance with the plan "who does what to whom," three categories were set up: (1) behavior directed toward a partner, called "active"; (2) behavior directed toward the observed animal by some member of the group, called "passive"; (3) behavior of the observed animal directed toward itself, called "self-directed."

This method has already been described (Ploog, Blitz, and Ploog 1963) and is similar to that of Altmann (1962). Later, recording was technically improved in such manner that the sequence of events as well as their lengths could be noted. A polygraph with a time printer was developed (G. Peiseler Co.), so that it was possible to record short and long lasting modes of behavior occurring simultaneously (for example, carrying an infant on the back and threatening), and this for every chosen member of the group concurrently with others of the same group.

By recording every event observed in this manner, we could place these occurrences on IBM punched cards to ascertain frequency distributions of

PLATE 10.1.—Squirrel monkey mother with 74-day-old male infant

PLATE 10.2.—108-day-old squirrel monkey displaying to his mirror image

PLATE 10.3.—49-day-old squirrel monkey displaying with vocalization to a human. Fur on the ventral surface of the trunk, extremities and tail is still very sparse.

PLATE 10.4.—Black- and yellow-headed groups fighting by means of aggressive vocalization

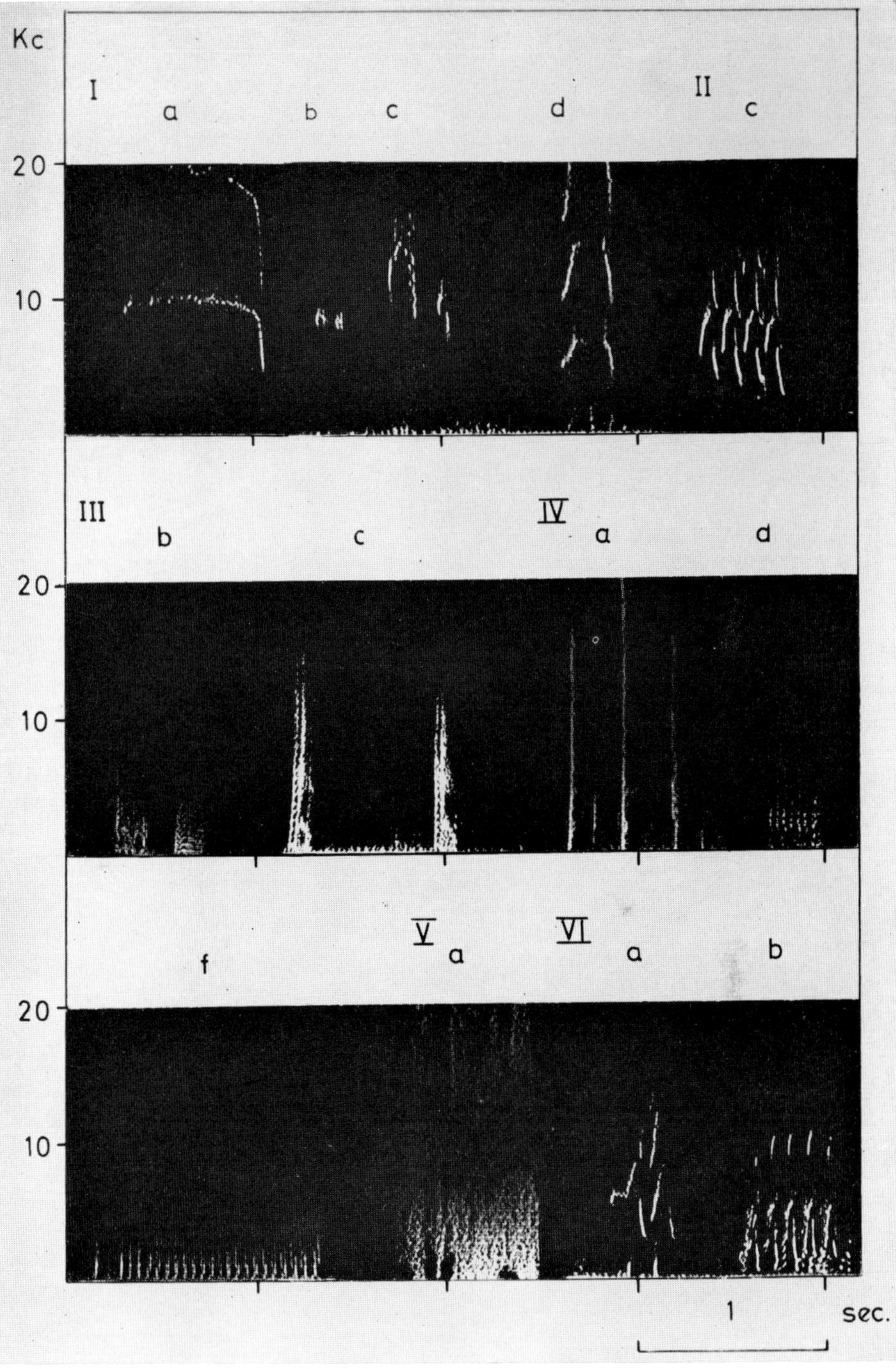

Plate 10.5.—Spectrograms of the listed sound types

List of vocalizations:

I a—isolation squeak; b—peep; c—alarm squeak; d—play squeak.

II c—twittering.

III b—cackling; c—yapping.

IV a—spitting; d—churr; f—purr.

V a—shriek.

VI a—chirp; b—clicking.

modes of behavior, correlation calculations, sequences of behavior, and so forth.

Statistical analysis was based on the null hypothesis that for any given set of observations the animals did not differ. To this end, significance calculations were based on the Chi^2 test when there were distributions for more than two classes or on the Chi^2 test as well as the binomial test (listed values) when there were distributions for only two classes.

List of Behavior Elements

The list of behavior elements has been changed and supplemented during the years in accordance with the task at hand and with an increase in experience.

Colonies I, II, III, and IV are the groups mainly dealt with in this chapter. The following behavior elements were noted and recorded *jointly* for these colonies:

Erections (undirected)
Genital display
Grooming
Huddling
Inspection
Licking/oral activity
Pushing
Scratching
Sniffing/genital inspection

The following were noted only in Colonies I and II:

Chafing
Embracing
Gnawing
Pulling
Rubbing
Tugging

The following were noted only in Colony III:

Chasing
Eating
Defense posture
Looking at object or cage mate
Masturbation
Playful biting
Resting
Running after
Pilfering objects
Handling objects

The following were noted only in Colonies III and IV:

Avoiding
Carrying (baby on back)
Contact
Copulation
Courting
Crowding
Fighting
Marking
Pilfering (food)
Playing
Suckling
Threatening

The behavior repertoire of squirrel monkeys is by no means exhausted with this list. We increased the list for three other colonies.

RESULTS

Types of Genital Display

In order that we might present the principle upon which our investigations are based and, at the same time, an important part of the results obtained, we have chosen from the above list a particularly characteristic mode of behavior in the social life of squirrel monkeys, namely, genital display (Ploog 1963; Ploog, Blitz, and Ploog 1963; Ploog and MacLean 1963).

As shown in Figures 1*a* and 1*b*, genital display is always directed toward one partner. In its fully developed form, it consists of several components: laterally positioned leg with hip and knee bent and marked supination of the foot, abduction of the big toe as well as an erection of the penis. In females, there is an enlargement of the clitoris instead of an erection. The display is frequently accompanied by vocalizations and occasionally a few spurts of urine.

Various types of genital display have been observed: (1) the open position, whereby the partners maintain a distance of approximately 10 cm to 3 or 4 m from one another (Fig. 1*a*). Here, the position of the displaying animal is more erect than in (2), the closed position, where the animals touch each other. In the closed position, the displaying monkey bends over its partner, seeming to jab the partner frontally with its penis (Fig. 1*b*). In both aforementioned positions (3), the counter-position, may also be observed from time to time, where both partners display to each other (Fig. 1*b*); (4) the monkey displays to its own image in a mirror (Plate 10.2); (5) the monkey displays to humans (Plate 10.3).

FIG. 1*a*.—Genital display in open position. Displaying partner looks at the passive partner. (Drawings by Hermann Kacher.)

The erection of the penis or enlargement of the clitoris (which is rather difficult to observe) may be absent entirely in all positions if display is of short duration. The ceremony usually lasts from 3 to 10 seconds, but there have been special situations where genital display was observed for as long as several minutes. From time to time an erection may precede a full display.

The following partner relationships have been observed to date (Table 1). The two exceptions to all of the imaginable possibilities appear to us to be of importance. Never to date have we observed an adult female displaying toward a male in the closed position nor have we seen a male answering the display of an adult female with a counter-display.

TABLE 1

PARTNER RELATIONS IN DISPLAY BEHAVIOR

Open Position	Closed Position	Counter-Position
♂→♂	♂→♂	♂↔♂
♂→♀	♂→♀	♂↔♀
♀→♂	—	—
♀→♀	♀→♀	♀↔♀

FIG. 1*b*.—Genital display in closed counter-position. Partners touch each other but do not look at each other.

Distributional Patterns of Genital Display

In connection with all modes of behavior having a social function, a large part is played by the frequency with which such meaningful information is transmitted, who does the transmitting of this information, and to whom it is addressed. By analyzing the distributional patterns of information we are able to obtain an insight into the group structure. Every observed mode of behavior can, in principle, be examined for its distributional pattern. The one chosen for this purpose here is genital display.

Table 2 shows the distribution of genital display in Colony I. C, the alpha animal of this group, directs his display fifty-eight times toward his rival B, who is approximately the same weight as C and also an adult male. E, the scapegoat of this group, who is also an adult but somewhat lighter male, follows. The other animals, D, an immature male, and two females, F and G, are ignored by C to a large extent. Nevertheless, C did display toward every animal of his group sometime during the course of a year. On the other hand, no animal of the group ever displayed toward C. The immature male, D, directed his display almost exclusively toward the likewise immature female, F. Scapegoat, E, displayed toward none of the group members.

In view of these observations, we might be tempted to conclude that the boss of the group is the one who displays more often than any other member. That this assumption is false is shown in Colony II (see Table 2). This colony was made up of a different combination of the sexes (see p. 150), C having been removed from Colony I and placed in this group in order to see what his behavior would be in relation to a stronger, somewhat heavier rival.

The analysis of *all* recorded modes of behavior showed that 88 per cent of C's total behavior was self-directed (see Table 3). Ten per cent was directed toward the group, although only 2 per cent was directed toward him by his companions. In observation series 1 of Colony II, the picture is somewhat different, however. Here, C is much less busy with himself, directing a third of his recorded behavior toward the group. Series 2 was begun 2 weeks after the end of series 1. During the first series, C had several serious fights with I in which C was the victor and during which time he unquestionably became the alpha animal of the new group. C's behavior patterns in series 2 again approached those he exhibited while in Colony I. Now having shown emphatically that he was again the boss, he slipped back into his previous modes of behavior.

With reference to displaying, Table 2 shows that C directs most of his display toward his main rival, I, (sixteen times) as he did when he was a member of Colony I. But he is far outdistanced in total displaying by I. I directs his display mainly toward M, a very active female striving for dominance (see M in Table 3, series 1). In series 2, C and I unite in their display toward M, whereupon M's total behavior pattern changes markedly (see

TABLE 2

Distribution of Genital Display
(Total Observation Time of 12 Hours per Male per Colony)

COLONY I

Passive Participants

Active Participants / Name	B	C	D	E	F	G	Total	Signif.
♂ B		0	2	2	4	4	12	n.s.
♂ C	58		7	33	1	1	100	$p < 0.01$
♂ D	6	0		3	32	4	45	$p < 0.01$
♂ E	0	0	0		0	0	0	—
♀ F	0	0	0	0		0	0	—
♀ G	0	0	0	0	0		0	—
Total	64	0	9	38	37	9	157	
Signif.	$p < 0.01$	—	n.s.	$p < 0.01$	$p < 0.01$	n.s.		

COLONY II, SERIES 1

Passive Participants

Active Participants / Name	C	I	J	K	L	M	Total	Signif.
♂ C		16	1	4	1	20	42	$p < 0.01$
♂ I	3		4	19	6	42	74	$p < 0.01$
♀ J	0	0		0	0	0	0	—
♀ K	0	0	0		0	0	0	—
♀ L	0	0	0	0		0	0	—
♀ M	2	3	0	0	0		5	n.s.
Total	5	19	5	23	7	62	121	
Signif.	n.s.	$p < 0.01$	n.s.	$p < 0.01$	n.s.	$p < 0.01$		

COLONY II, SERIES 2

Passive Participants

Active Participants / Name	C	I	J	K	L	M	Total	Signif.
♂ C		11	4	4	0	19	36	$p < 0.01$
♂ I	2		0	7	2	32	43	$p < 0.01$
♀ J	0	0		0	0	0	0	—
♀ K	0	1	0		0	0	1	n.s.
♀ L	0	0	0	0		0	0	—
♀ M	5	2	0	0	0		7	n.s.
Total	7	14	4	11	2	49	87	
Signif.	n.s.	n.s.	n.s.	n.s.	n.s.	$p < 0.01$		

Tables 2 and 3, series 1 and 2). M's active participation drops from 44 to 29 per cent.

When making a comparison with Colony I, note that the most active animals of Colony II, I and M, displayed toward the boss from time to time (I, three and two times; M, two and five times).

Comparing the statistics of series 1 and 2 of Table 2, we see the following:

C changes neither the total frequency of his display nor the distribution toward the members of the group.

I reduced his total display in series 2, but the reduction in display toward M was very small. His display toward K was decreased by an amount that was no more than to be expected when considering total reduction. K and M received considerably more display from I than from C in series 1, while in series 2 only M received more from I than from C.

Viewing this result, we might deduce that, although the alpha animal of a group concentrates its display toward the rivals in that group, it is the beta

TABLE 3A

TOTAL BEHAVIOR EXPRESSED IN PERCENTAGES (COLONY I)

	C	D	B	E	F	G
Self-directed.	88	57	60	77	67	75
Active......	10	24	26	3	13	2
Passive......	2	19	14	20	20	23

TABLE 3B

TOTAL BEHAVIOR EXPRESSED IN PERCENTAGES (COLONY II)

Series 1	C	I	J	K	L	M
Self-directed.	57	36	41	30	19	26
Active......	31	39	31	9	49	44
Passive......	12	25	28	61	32	30

Series 2	C	I	J	K	L	M
Self-directed.	72	47	35	29	27	26
Active......	22	36	30	15	41	29
Passive......	6	17	35	56	32	45

animals, second in the hierarchy, who display most frequently, albeit only seldom toward the boss.

Colony IV shows that this hypothesis also cannot be generalized. The composition of this group was described on page 150. The primary difference between it and previously described groups is that a male infant (W) was included in it. W was 19–38 weeks of age during series 1 of Table 4; during series 2, he was 39–52 weeks old.

TABLE 4

Distribution of Genital Display
(Total Observation Time for Series 1 = 8 Hours per Animal;
for Series 2 = $8\frac{1}{2}$ Hours per Animal)

COLONY IV, SERIES 1

Passive Participants

Active Participants

Name	R	S	T	U	V	W	Total	Signif.
♂ R		1	5	8	9	0	23	$p < 0.01$
♂ S	29		14	46	4	0	93	$p < 0.01$
♀ T	0	0		0	0	0	0	—
♀ U	2	3	1		38	2	46	$p < 0.01$
♀ V	0	0	0	0		1	1	n.s.
♂ W	9	25	4	6	13		57	$p < 0.01$
Total	40	29	24	60	64	3	220	
Signif.	$p < 0.01$	$p < 0.01$	$p < 0.01$	$p < 0.01$	$p < 0.01$	n.s.		

COLONY IV, SERIES 2

Passive Participants

Active Participants

Name	—	S	T	U	V	W	Total	Signif.
—		—	—	—	—	—	—	—
♂ S	—		3	2	3	2	10	n.s.
♀ T	—	0		0	0	0	0	—
♀ U	—	0	4		83	0	87	$p < 0.01$
♀ V	—	0	0	3		2	5	n.s.
♂ W	—	42	7	16	28		93	$p < 0.01$
Total	—	42	14	21	114	4	195	
Signif.	—	$p < 0.01$	n.s.	$p < 0.01$	$p < 0.01$	n.s.		

Series 1 of Table 4 shows that W displayed toward all members of the group and particularly toward the boss, S, and his mother, V—such behavior increasing with advancing age (series 2).

Table 4 reveals further novelties. S, as well as his rival, R, do not display toward the infant W in series 1 but they do display toward the immature female, U, and W's mother, V—although seldom. Incidentally, the boss behaved according to expectations. He displayed, as did C, toward his rival (twenty-nine times in series 1) but much more toward the immature female, U. Further quantitative analysis of other modes of behavior shows that displaying not only indicates a striving toward dominance but is also part of courtship.

U, on her part, displayed toward all animals of the group, as did W, but very markedly concentrated this behavior toward V (thirty-eight times in series 1 and eighty-three times in series 2) while V answered very seldom or not at all. Further behavior analyses show that this display by U toward V took place in an importunately aggressive context, this tendency increasing in series 2. V, on the other hand, increasingly limited all contact with U. U's behavior can be understood as a result of frustration by V.

Series 2 of Table 4 shows particularly that marked changes in the behavior of S took place. A comparison of the figures for series 1 and 2 shows the following:

After the death of his rival, R, through illness, displaying by S decreased very greatly, the remaining display being uniformly distributed. Even the preference for U disappeared.

U increased her display activity greatly and concentrated it toward V even more markedly in series 2 than in series 1. W, too, increased his displaying greatly during series 2. However, the distribution pattern remained the same.

Ontogeny of Genital Display

As a framework for the following, Table 5 gives a survey on the main ontogenetic phases, especially of the social behavior in squirrel monkeys raised in laboratory groups (Ploog, Hopf, Winter, in preparation). By observing the various types of display and from sociometric data, we have concluded that genital display is a social signal stimulus that contributes decisively to the formation of group structure. It has different meaning in various situations and is used in various ways in accordance with the standing of the animal in the group. An infant in particular employs this signal differently from the way it is employed by adolescent and adult animals. Adult monkeys also react differently to the display of an infant than to the display of an adult.

The question arose, therefore, as to when and under which circumstances genital display takes place for the first time in the life of a squirrel monkey, how this behavior develops, and in what manner does the group react to this behavior at various ages of the infant.

TABLE 5

SURVEY OF THE MAIN PHASES OF ONTOGENY IN THE SQUIRREL MONKEY

	Infant (*I*) and Mother (*M*)	Infant (*I*) and Group
First week	Progressing *M*-*I*-adaptation by means of tactile stimuli. Stabilization of *I*'s orientation on *M*'s body, depending on its state of maturation. *M* has nest function for *I*.	Social interactions with some females (mature or adolescent), which are attracted by the visual, vocal, and olfactory stimuli from *I:* Vocalizations, looking at, genital display toward, turning toward, and turning away from partners. Males defend *I* but have little or no interaction with *I*.
Second week	Training of motor coordination by climbing on *M*'s body.	Manual contacts with partner broaden *I*'s social behavior.
Third to fourth week	*I* first leaves *M* for short intervals under her surveillance (enlargement of *I*'s territory as a result of motor training).	A female takes a special interest in *I* (so-called aunt-*I*-relationship) and subsequently forms a dyad with *M* that may outlast both *M*-*I*- and aunt-*I*-relationships.
Fifth to seventh week	Amount of daytime spent on own feet increases rapidly. *M*-*I*-behavior is mainly composed of four characteristics: (1) *I* striving away from *M* (2) Strong *I*-*M*-ties (3) Onset of weaning (4) Strong protection by *M* Weaning: *I* runs toward *M*, *M* avoids *I*; this repeated many times until *M* allows sucking. It seems that the four characteristics repeatedly occur without the influence of external stimuli. Visual interaction increases, *M*-*I*-relationship becoming more varied and "personal." First genital display toward *M*. This behavior is seen frequently throughout weaning. Persistence of nest function. Fur on ventral side appears.	
Eighth week to fourth month	*M*-*I*-relationship remains stable. The four characteristics are balanced and are employed adequately: (1) Interest in environment (2) Loosening of *M*-*I*-ties (3) Continuation of weaning (4) Protection when necessary Feeding: both sucking and eating solid food.	Play with peers develops; aunt-*I*-relationship still present. *I* displays toward adult males without further interaction and toward unfamiliar group mates and humans.

TABLE 5—*Continued*

	Infant (*I*) and Mother (*M*)	Infant (*I*) and Group
Fifth to eighth month	*I* spends less time with *M*. *I* more independent. Characteristics (1) and (3) predominant: (1) *I* occupied with peers or objects (3) *M* avoids *I* for long periods	Intense play with peers including first copulatory behavior, which starts during fifth to sixth or ninth to tenth months. In both periods of age *I* stops gaining weight. Male copulatory behavior appears earlier than female. End of aunt-*I*-relationship. Frequent genital display toward the alpha animal.
Ninth to fourteenth month	Intense or sometimes aggressive weaning. End of *M*'s positive actions directed to *I*, except for her carrying *I* on back during nighttime; rare nursing and protection in violent danger.	
Second year	Juvenile-adult-interaction replaces former *M*-*I*-interaction. Long periods without any interaction.	Play with peers continues, but the now juvenile animal less frequently instigates play. More interaction with other group mates, especially the alpha animal: role of scapegoat, and play-fighting are exercised; alpha animal suppresses juvenile male. Frequency of genital display toward alpha animal decreases, threats toward juvenile increase in frequency and intensity. Juvenile male shows dominance toward females and exerts both sexual roles.
Third year		Stop in gaining weight. Sexual maturity in males at approximately 2 years, 9 months. Play decreases and is dependent on initiative of partner. Role of scapegoat continues. Intense suppression of juvenile male by alpha animal; fighting may result in injuries.

Our comments are based on the well-studied lives of three infants W (male), X (male), and Ri (female). W was a member of Colony IV until he was placed in Colony V where he grew up together with X, who was 9 months younger than he. Ri was born and developed as a member of Colony III. As an example for statistical analysis results are presented from W only, who was observed for the longest period of life. Data collected on X, Ri, and four more infants (two male and two female) will be published elsewhere (Hopf and Ploog, to be published; Ploog, Hopf, and Winter, to be published). Genital display first occurs within the first four weeks of life but is only rarely seen during this period. Figure 2 shows Ri on the day after her birth as she displays toward a member of the group from her mother's neck while simultaneously vocalizing in typical fashion. The picture shows the bent right leg and the supinated foot as well as the abducted big toe. Head and glance are turned toward the intended partner while the mouth is open for vocalization. This drawing was taken from a motion picture film that clearly showed the typical leg movement starting with the clamped position around the mother's belly up to thigh spreading as well as the return movement of the leg to the initial position.

Plate 20 shows X as he displays at the age of 7 weeks toward a human with typical vocalization. X spent the first few weeks of his life alone with his mother, Y. Displaying was first observed in his case 28 days after birth. Having run after his mother for a long time to nurse at her breast with no result, he displayed toward her in his frustration.

W grew up alone with his mother at first. Only at a later date did she keep him from nursing regularly. W displayed toward his mother the first time, namely on the thirty-eighth day after his birth. Plate 10.2 shows W at the age of 15 weeks as he displays toward his own image in a mirror.

Table 6 provides an example of how the distribution pattern of infantile genital display fits in with the distribution pattern of other behavior important for the infant. The upper portion of the table (active) shows four modes of behavior, other than displaying, that are dominant in this stage of the infant's life. The lower part shows a selected number of behavioral aspects in which the behavior of the group members toward the infant is best reflected.

In general, neither the behavioral pattern of W toward the group nor the behavior of the group toward W indicates a random distribution. Aside from this, it is quite evident that W is considerably more active with respect to the group (total 860) than the group is toward him (total 422).

The behavior of W toward the boss, S, is clearly distinguishable from that behavior shown toward the remainder of the group. W displays most frequently toward S, although he has the least interactions with him in all other categories. The same holds true in reverse, with the boss paying the least attention to W. One exception to this, however, is the threat behavior. Here, the boss and, even more so, his rival, R, threaten W frequently; the females are far less active in this connection. Without going into sequential behavior in detail, suffice it to say that more frequent threatening of R is a natural consequence of W's much greater "molestation" of R as compared to his "molestation" of S (for example, ninety-one contacts of W toward R). Within the group, R holds second place with one hundred eighty-four interactions and holds the same place with his one hundred two actions toward W.

The behavior of W toward his mother is clearly distinguishable from his behavior toward all other members of the group. With the exception of the boss, W displays most frequently toward his mother. He runs after her most frequently and seeks contact with her to the greatest extent. His mother, for her part, continually seeks to avoid W and to get rid of him, while T, a mature female, most often seeks contact with W, never avoiding him and seldom trying to get rid of him. This pattern of behavior between mother and

FIG. 2.—1½-day old squirrel monkey positioned at its mother's neck displays to a cagemate who looks at it and subsequently contacts it. Infant's mouth is slightly open for vocalization (as revealed by the original motion picture frames).

infant is typical for this stage of life wherein the infant becomes independent quickly and there is an increase of its activity.

After having shown that the infant behaves differently from the mature monkey in the matter of genital display, we must ask exactly how long this difference continues, whether there are transitional phases, and, if so, under which circumstances do these transitional phases unfold. We limit ourselves here to a description of the infant's development and shall publish the supporting statistical data elsewhere (Hopf and Ploog; Ploog, Hopf, and Winter).

The transition from the infantile distribution pattern for displaying to that of adulthood is long and completed very gradually, and occurs under the influence of certain group members who slowly increase their threats toward the young animal. There are various forms and intensity levels of threatening, the choice of means and frequency of threatening depending upon the age of the infant or juvenile addressed.

TABLE 6

THIRTY-MINUTE OBSERVATIONS FOUR TIMES A WEEK

OBSERVATIONS ON WASTL: 19-27 WEEKS OF AGE

		R	S	T	U	V	total	signif.
active	display	29	49	8	13	40	139	$p < 0.01$
	running after	2	2	0	0	53	57	$p < 0.01$
	contact	91	18	61	83	120	373	$p < 0.01$
	fondling	41	6	47	10	50	154	$p < 0.01$
	pilfering	21	8	15	54	39	137	< 0.01
	total	184	83	131	160	302	860	$p < 0.01$
passive	display	0	2	0	2	1	5	n.s.
	threatening	48	21	6	4	1	80	$p < 0.01$
	avoiding	8	4	0	2	40	54	$p < 0.01$
	pushing	7	0	8	14	66	95	$p < 0.01$
	contact	24	1	60	21	23	129	$p < 0.01$
	fondling	1	0	7	10	1	19	$p < 0.01$
	pilfering	14	2	8	9	7	40	n.s.
	total	102	30	89	62	139	422	$p < 0.01$

A 22-month old animal, for example, may, in some contact situation, be more harshly attacked by the boss and by other adult males as well, while a 12-month old male is only threatened. A mother threatens her infant more frequently and more intensely at about 1 year of age, this point of time coinciding with final weaning. Up to about 1 year and some weeks beyond, the infant still nurses although ever less often.

An interesting variation of displaying by the infant was observed at about the age of 2. Upon displaying, the young animal turned its back toward the intended partner (especially boss) while its glance was still directed toward him. Sometimes the glance, too, was turned away from the one addressed. In this way the juvenile seemed to escape further threatening without completely giving up displaying.

An exception to these above rules was seen during play, especially where the boss was included. In this situation, the young animal was not intensely threatened. There was mutual displaying by the playmates, this not to be mistaken for the counter-position. This play resembled scuffling, whereby both, or several participants, had erections very often.

As far as we can state now from observations in Ri and in two more female infants there are no fundamental differences between them and male infants concerning the social behavior during the first 2 years.

Genital Display in Larger Colonies

Further aspects of genital display were recognized in the course of observing larger colonies containing up to twenty-seven animals. Such observations were carried on during the months of June to September, 1964, while the animals were in spacious outdoor cages. One colony consisted of five males and six females, the so-called black heads (The hairs on the tops of their heads having been dyed black for easier identification.), while the other consisted of three males and thirteen females, the so-called yellow heads. After a period of 50 days, in which each colony was thoroughly examined sociometrically, the animals were permitted contact with one another by opening a small door between cages. Almost immediately an intensive and continuing shrieking battle began (see Plate 10.4) with the male leaders displaying toward each other for a long time. It was only at a later time that large-scale fighting broke out in which the females were also active participants. The result was that the yellow heads drove the black heads from their sunnier territory. The subsequent fusion process of the two colonies was followed sociometrically for a number of weeks, and these further observations provided interesting insights into the group dynamics of unstable colonies (Castell and Ploog).

After the alpha male of the yellow heads had become boss of the entire group, the former alpha black head and the alpha yellow head together attacked the challenging beta animal of the black heads. The latter was severely wounded several times, kept itself completely isolated, and never again appeared as a rival of the former alpha black head.

The types of display already described (see p. 151) were also observed in these large groups.

Before the fusion of both groups, the black head males displayed much more frequently than did their females. Their displays were directed mostly toward the females (eighty-eight times) and remained constant during the period of observation. Displays of male toward male took second place (thirty-seven times), and this decreased after a period of time.

TABLE 7

GENITAL DISPLAY OF THE BLACK HEADS

	5♂♂	6♀♀	Total	Display/Animal (*m*)
5♂♂	37	88	125	25
6♀♀	8	24	32	5

Total observation time of 31 hours.
m = Average value.

TABLE 8

GENITAL DISPLAY OF THE YELLOW HEADS

	3♂♂	13♀♀	Total	Display/Animal (*m*)
3♂♂	13	147	160	53
13♀♀	63	173	236	18

Total observation time of 31 hours.
m = Average value.

The female yellow heads were more active in displaying than the black head females. Their displays, on the other hand, were mainly directed toward females (one hundred seventy-three times) and remained constant during the period of observation. The marked displaying of the males toward the females (one hundred forty-seven times) decreased during the period the group was together.

After the fusion of both groups (eight males and nineteen females), the males displayed most frequently toward the females (twice as frequently toward females of the opposite group as toward their own females). Displays of males toward males (sixty-four times) without distinction of group took second place as did displays of females toward females of their own group (seventy-eight times).

In all of the observation series that have been carried out, males displayed much more frequently than females. These displays were mostly directed toward the females. With the sex ratio of the black heads at about 1 ♂ :1 ♀, the average display activity per male and female was similar to the display activity of the total group after fusion with a sex ratio of around 1 ♂ : 2 ♀♀. On the other hand, in the yellow head group, with a sex ratio of around 1 ♂ : 4 ♀♀, display activity per animal of both sexes was noticeably greater. It would seem, therefore, that the inclination of all animals to display is increased owing to the increased number of females per male. This result was already indicated by a comparison of colonies I and II, the one having a reverse sex ratio from the other (4 ♂♂ : 2 ♀♀ and 2 ♂♂ : 4 ♀♀, respectively).

TABLE 9

GENITAL DISPLAY OF THE COMBINED GROUP OF YELLOW AND BLACK HEADS

	8♂♂	19♀♀	Total	Display/Animal (*m*)
8♂♂	64	187	251	31
19♀♀	22	78	100	5

Total observation time of 29 hours.
m = Average value.

DISCUSSION OF GROUP BEHAVIOR

The results show how information about group structure and dynamics of monkey colonies can be obtained with the aid of sociometric observations. The number and distribution of the behavioral elements for each animal varies significantly within a group so that an individual behavior pattern may be seen for each animal. By means of these quantitatively obtainable differences it is thus possible to ascertain the nature of comparable group structures.

Behavior patterns of individuals in stable groups change very little. In unstable groups, however, individual behavior patterns change until each member of the group has found the role he is to play within the structure. The degree of group stability is dependent upon the stage of rivalry between the males, the ratio of the sexes, age distribution, the time together in the group, and probably the individual characteristics of group members as well.

In the case of squirrel monkeys, the alpha animal of a group always seems to be a male. It isolates itself from the group to a greater degree than do the other animals and is mostly occupied with itself. It uses few, but effective, means of dominating the group (weak threatening, displaying, vigorous threatening, biting). On the other hand, squirrel monkeys have no rigid social order (Ploog, Blitz, and Ploog 1963; Ploog and MacLean 1963). A multi-directional social structure replaces the concept of a linear hierarchy whereby each animal shows different capacities in different aspects of behavior (Ploog 1963; Ploog, Blitz, and Ploog 1963).

Displaying, specially analyzed in this study, is very characteristic of the social behavior of squirrel monkeys. Genital display (see Figs. 1 and 2 and Plates 10.2–10.3) is a ritual derived from sexual behavior but employed as a social signal stimulus (Ploog, Blitz, and Ploog 1963). This means of communication was observed as early as the day after birth in one of the animals born to the group. If newly born infants are isolated from other animals and grow up alone with their mothers, displaying seems to make its appearance only several weeks after birth at a point of time characteristic for its development, that is, when the mother removes the infant from her back and, for a given time, does not permit it to nurse at her breast. The infant displays toward its mother for the first time in this situation. We therefore believe displaying to be an innate social signal stimulus, its initial appearance being dependent upon the social situation. It has a different meaning under different circumstances and with different ages.

Generally, the alpha animal displays toward all members of the group except infants under 4 months of age, displaying preferably toward male rivals and females.

Displaying is seldom directed toward the alpha animal but when it is, it comes from beta males who are rivals in unstable group situations and from females toward whom the alpha animal has displayed very often.[1] Infants are the outstanding exception, displaying very frequently toward the boss.

Beta males display often in unstable groups, especially when the group is first assembled or when they are new members of an already existing group. In this situation, displaying is directed exclusively toward the boss.

During the period of time when they are constantly carried by their mothers, infants display toward those group companions who occupy themselves with the infants to the greatest extent (see Table 5).

During the period when infants mostly play, they display toward their mothers when they are not permitted to nurse immediately, toward the boss and other grown animals when they are threatened after having made an attempt at contact or to play with them, toward their playmates while playing, and toward the boss when a playmate has been playing too roughly—almost as if it were "letting off steam" to the boss about the playmate (in these instances, the infant often flees onto the back of its mother). A playmate can display in similar fashion toward the mother of another playmate if the latter spoils the "game."

During the first 4 months the infant displays only in the open position. Later, it tries the closed position with animals playing a lesser role in the group. Displaying after having been threatened always occurs in the open position at all ages.

[1] There is another variable not reported in this chapter which influences the frequency and distribution of genital display: This is the estrus cycle of the females. Our data analysis in this respect is not as yet sufficient for a more definite statement.

Keeping in mind all situations in which displaying occurs, we may assume the following meanings: demanding, self-assertion, courting and desiring closer contact. Dominance is accentuated in the closed position, the aggressive component appearing most strongly. In young animals and females the open position may indicate frustration and defense.

All of this points to displaying in its various forms as being a means of transmitting specific information to definite partners while simultaneously expressing the mood in which the displaying animal finds itself. The exact meaning of the information may be ascertained from the context in which it is transmitted. Displaying influences the behavior of the recipient and, under certain circumstances, elicits an observable behavior response, this latter influencing the behavior of the sender in turn. In order for it to be effective, information must be exchanged frequently. As a consequence of these exchanges, the group members modify their behavior and find their roles in the social structure. Displaying, of course, is not the only means by which group life is governed and regulated. Group dynamics, however, can be shown especially well using this mode of behavior as an example.

In the next section we shall show what role the vocalization repertoire of these animals plays in their social behavior.

VOCALIZATIONS OF SQUIRREL MONKEYS

In the course of observing the behavior of these animals over a long period of years we were repeatedly made aware of the large repertoire of sounds brought forth by them, permitting us, in some instances, to predict certain modes of behavior. From other observations of various species, especially those on social behavior, it has been ascertained that auditory communication plays an outstanding role in controlling the social conduct of animals living in groups (Altmann 1962; Andrew 1963; Carpenter 1934, 1958; Chance 1956; Hinde and Rowell 1962; Itani 1963). Since observations made to date of vocalization have been carried out in an unsystematic manner, we made a list of vocalizations of our animals in our study and, as far as possible, arranged the individual sounds in accordance with functional areas of behavior (Winter, Ploog, and Latta).

MATERIAL AND METHODS

Observations were made throughout the year 1964, during which more than 250 hours of tape recordings alone were made. All animals belonging to the department, fifty or sixty of them, were included in the program, and were divided into various groups and placed in separate rooms. Their surroundings varied from cages in which the animals were kept singly or in pairs for experimental reasons, from observation cages measuring 0.35 to 15 m^3 containing five to six animals to large collective cages (20 m^3) with an outdoor enclosure

($56\ m^3$) containing eleven to sixteen animals (see Group Behavior, above). The greatest number of observations were made of Colony III (see p. 150). To obtain high-grade tape recordings suitable for spectrographic analysis, the animals were caged individually or in groups in an echoless and soundproof room. A capacitor microphone (20–20,000 cps linear) and a stereo tape recorder (20–18,000 cps $\pm$ 3db) were available for the recordings. Comments on behavior were spoken simultaneously on the second track. Spectrographic analysis was executed in accordance with the method developed by W. Schleidt (1964) and modified by us.

Figure 3 is a block diagram showing the layout of the equipment. The endless tape carries the signal to be analyzed and is read continuously from 0 to 20 kc by a frequency analyzer. The amplified output of the analyzer controls the intensity modulation of an oscilloscope. Triggering of the signal and the time sweep of the oscilloscope occurs by means of a light barrier that is activated by the rotating endless tape. By tilting a camera around the horizontal axis of the oscilloscope screen, we recorded many lines (about 300) over one another, thus recording the values analyzed in one picture. Characteristic data for each type of sound, such as frequency and length, number of overtones, and pattern, were ascertained and averaged.

A total of 275 spectrographic analyses were made for the entire list of 26 types of vocalization, the values given in the following 13 examples being supported by measurements of 166 individual sounds.

RESULTS

Purely formal points of view sufficed initially for a systematic arrangement of the list of sounds, and sound designations were chosen accordingly. The limits of this chapter do not permit us to display the entire vocal repertoire (Winter, Ploog, and Latta). We shall, therefore, attempt to give an impression of the variety and meaning of the vocalizations, as well as an insight into our methods of observation by presenting several characteristic examples.

A schematic presentation of the examples given may be seen in Figure 4. Quantitative data are summarized in Table 10. In the original paper (Winter, Ploog, Latta) the designation of two sounds is changed: "peep" instead of "squeak," "keckering" instead of "clicking."

Group of Squeaks

These calls are characterized by a uniform frequency pattern with the change of frequency being less than one octave and the fundamental frequency for the majority of sounds being above 5 kc.

(See Fig. 4Ia and Plate 10.5Ia.) The high and long isolation squeaks (up to 1 sec) are always emitted by the animal as soon as it loses visual contact with, or is separated by a long distance from, the group. Hearing this sound, members of the group will answer in the same manner, the result being an exchange of these squeaks that can continue for minutes, ceasing only when the animal is returned to the midst of the group. This type of reaction can be easily and continuously brought forth by merely playing these sounds on a tape recorder.

(See Fig. 4Ib and Plate 10.5Ib.) In general, peeps vary greatly in frequency and length as well as in occurrence. They can be heard mostly under two conditions: as soon as an animal is separated from the group while exploring unknown surroundings and as soon as a stimulus excites the attention of an animal.

(See Fig. 4Ic and Plate 10.5Ic.) If an animal suddenly becomes aware of a fast-moving object, it reacts to it in a stereotyped manner by emitting alarm squeaks. This very short and high-pitched sound (> 13 kc) immediately results in the entire group fleeing to the highest point of the cage and remaining completely silent there for some time. We can obtain this reaction very easily by either using objects made to appear similar to those found in nature (dummies) or by playing alarm squeaks on a tape recorder.

(See Fig. 4Id and Plate 10.5Id.) Play-squeaking always accompanies playing, chasing, grasping, and playful nipping as well as preparatory sex play. Since we never observed that playing ever changed to fighting, in spite of the

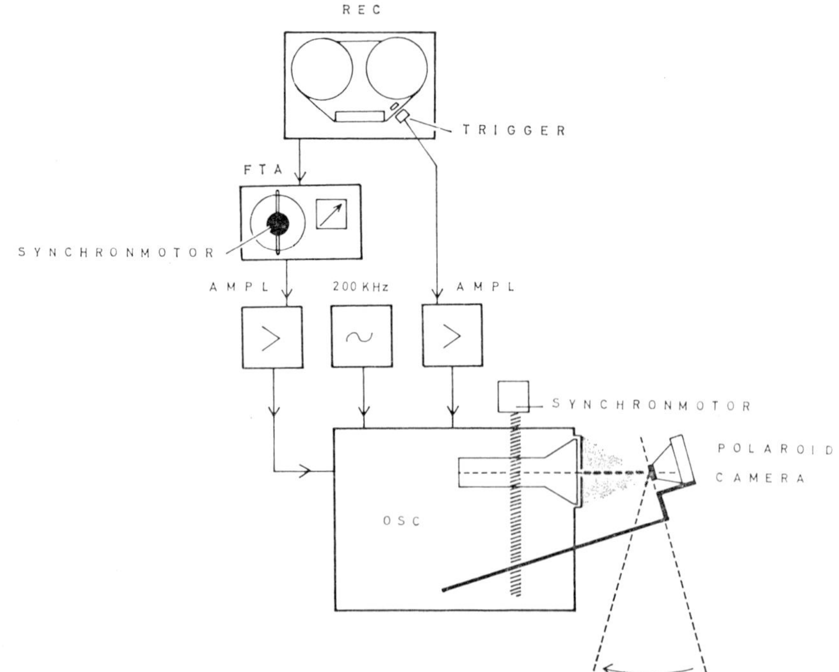

FIG. 3.—Block diagram of the apparatus employed for spectrographic analysis of pitch frequencies.

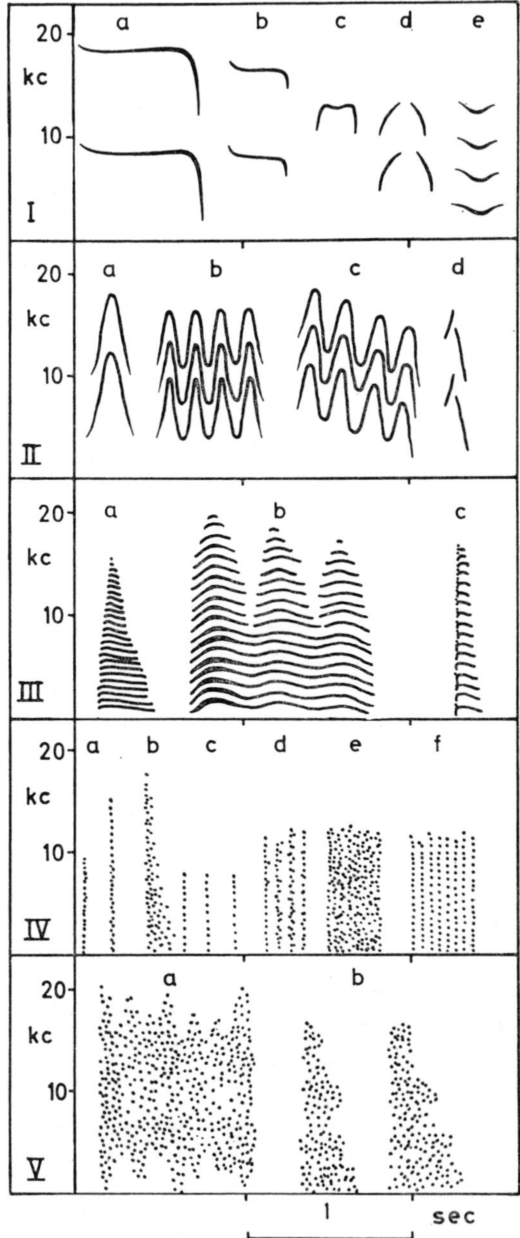

FIG. 4.—Schematic representation of the five fundamental vocalization groups. Only a selection of calls is discussed in the text.

List of vocalizations: I a—isolation squeak; b—peep; c—alarm squeak; d—play squeak; II c—twittering; III b—cackling; c—yapping; IV a—spitting; d—churr; f—purr; V a—shriek.

TABLE 10

CHARACTERISTIC DATA OF THE LISTED SOUND TYPES

Type of Vocalization	*n*	Upper Frequency Limit of the Fundamental (kc)	*m*	Length of the Individual Elements (sec)	*m*	Number of Elements	Total Length of the Sound (sec)	*m*
Isolation squeak	6	9 –11	10.8	...	...	...	0.07–1.3	0.9
Play squeak	12	5 – 9	7	...	...	...	0.03–0.11	0.05
Alarm squeak	6	11 –16	14.5	...	...	...	0.05–0.13	0.09
Twitter	24	4.2– 9.0	...	0.09–0.11	0.13	3–10	...	...
Chuck	16	...	...	0.04–0.09	0.07	1– 2	...	...
Cackling	46	0.2– 0.6	0.45	...	...	...	0.06–0.22	0.14
Yap	15	1	1	...	...	...	0.11–0.25	0.15
Spitting	5	0 –18	...	...	0.02	...	...	...
Chur	3	0 –16	...	...	0.02	...	...	...
Purr	10	0 – 8	...	...	0.02	...	...	...
Shriek	10	0 –18	...	...	...	...	0.3 –0.9	0.6
Chirp	6	...	...	...	...	...	0.28–0.6	0.32
Clicking	7	...	...	0.11–0.15	0.12	3– 6	...	...

n = Number of evaluated spectrograms.
m = Average value.

aggressive elements characteristic of this type of behavior, we believe that some sort of signal is involved here which tells each of the participants that there is no aggressive intent (sounds of reassurance).

All sounds of this group mainly control the relations of members of the group. Therefore, we designate them as sounds governing an inclination to social contact.

Group of Twittering Sounds

Only twittering (see Fig. 4IIc and Plate 10.5IIc) will be mentioned here as an example of this group of sounds. In contrast to the first group, these calls are distinguished by a periodic frequency change encompassing more than an octave. They can be short or long and the pitch level can be constant, increasing, and/or decreasing.

Twittering is associated mainly with food, but external stimuli associated with food can also be answered in this manner.

While eating, the animals remain seated apart from one another, a minimum distance of 10 to 20 cm being maintained at all times. During this period, there is particularly intensive twittering going on.

These observations, as well as others made under experimental conditions, tend to show that the maintenance of a given distance during feeding rather than the food itself gives these and other sounds of the group their functional character. They should, therefore, be designated as sounds meant to maintain distance within the social group. By this we mean that minimum distances must be maintained between individuals of the group while feeding.

Group of Cackling Sounds

The fundamental frequency of the cackling sounds is much lower than those of all other sounds. Above this fundamental is a large number of overtones which can be most readily compared with the vowel elements of human speech. The intensity and structure of individual sounds of this group can, however, vary greatly.

(See Fig. 4IIIb and Plate 10.5IIIb.) Aggressive disposition and behavior of an animal, as well as that of an entire group, are accompanied by cackling. Situations that could bring it about are an unknown stimulus, a stranger to the group and possible rival, food stealing, or laying claim to a desirable sitting place.

(See Fig. 4IIIc and Plate 10.5IIIc.) Yapping is a very characteristic warning sound which is not only recognized by its great intensity but also by its consonant-type initial component. This sound is emitted in the presence of unknown, repulsive stimuli from an external source which enters the visual field of the monkeys. Examples of this would be strange animals and humans and kindred monkeys from a strange group as well. As soon as this sound emanates from one of the animals, excitement is transmitted to the group and results in continued yapping by most or all of the animals. This type of reaction can be compared to the mobbing response of many kinds of song birds at the appearance of carnivorous animals.

In this group, therefore, sounds that express a general mood of aggression are included.

Group of Arr Sounds

The acoustical energy of these short, pulse-type sounds (< 20 msec) is almost uniformly distributed over the entire frequency range. They are emitted individually or in groups and at regular or irregular intervals.

(See Fig. 4IVa and Plate 10.5IVa.) Spectrographic analysis of spitting shows a short, sharply limited pulse ending in an irregularly distributed sound pattern. Spitting is regularly connected with aggressive behavior such as, for example, head grasping, and is always emitted by the aggressor. Spitting is generally limited to an aggressive situation between two animals, a reaction of the group never having been observed as yet.

(See Fig. 4IVd and Plate 10.5IVd.) Churr sounds are heard quite frequently in connection with cackling sounds in situations where there is a mood of aggression.

(See Fig. 4IVf and Plate 10.5IVf.) Purring is characterized by a series of very uniform pulses which can last for more than a second. Most of the time, intensity is very weak. An animal under a year of age purrs directly before and during the time it drinks at its mother's breast. However, grown animals having a close relationship to the young one can be heard purring as well while they, for example, smell and grasp the infant.

All Arr sounds can be included, for the time being, under the principal concept of preparedness for directed aggression. However, in the instance of an infant's purring and purring directed toward the infant, the word aggression has the connotation of strong desire, that is, wanting to have something definite.

Group of Shrieks

Shrieks have the noiselike qualities of the Arr sounds but are distinguished by their much greater length (~ 1 sec).

Only shrieking itself is listed here as an example (see Fig. 4Va and Plate 10.5Va). Shrieking always expresses a high degree of excitement, whether in the midst of a battle, in flight from a superior opponent, or in pain. Shrieking results in a very high degree of excitement in the group.

Group of Combined Sounds

The group of combined sounds includes vocalizations made up of elements of the first to the fifth groups, thereby placing them in the category of formal transitional sounds. To what degree this coincides with content is still to be observed more closely. Two examples of this follow.

(See Plate 10.5VIa.) Chirping consists of a combination of squeaking and twittering, which are constantly heard together. The function of chirping is that of a contact sound, being emitted when the animals move about in surroundings strange to them.

(See Plate 10.5VIb.) Clicking may be heard during a strongly aggressive state of excitement, such as just before a battle with a rival. It is, in the main, twittering interwoven with cackling.

Concerning the Ontogeny of the Sounds

Although no completed observations have ever been made in this connection, several facts may be summarized from observations of behavior made continuously during birth, the second and thirteenth day of life (Bowden, Winter, and Ploog 1966; Ploog, Hopf, and Winter).

Takeshita (1962) describes the vocalizations of a neonate that were emitted even before it had completely left the birth canal. The author describes this sound as "chee."

We observed the first sound 5 minutes after birth (Bowden, Winter, and Ploog 1966). The baby shrieked three times. Before the end of 24 hours, several long squeaks and twittering sounds as well as the typical purring sound during nursing could be heard in a completely developed form. Squeaking was heard the day after birth, initially with the first appearance of genital display (see pp. 158–59 and Fig. 2).

These individual observations lead us to believe that vocalizations of squirrel monkeys are part of innate behavior patterns which mature functionally with each new mode of behavior. Both movements and vocalizations form a greatly varied system of communication having an important bearing on the social life of these monkeys.

DISCUSSION OF VOCALIZATIONS

We may obviously assume that sounds of a formally similar nature have similar meaning. This assumption is supported by the results listed. However, this does not hold true for individual sounds when they have taken on special meanings such as, for example, alarm and warning sounds. In spite of such exceptions, the general characterization of an entire group of sounds seems to present an applicable working hypothesis which permits a clearer comparison of groups with each other. A detailed analysis of vocalizations and behavior in specific situations must show just how far this hypothesis can be applied. As an example of this type of experiment, we have listed the change of vocalization pattern of two animals in three different situations (see Fig. 5). The frequency distribution of the sounds shows a shifting in the direction of increasingly aggressive sound types (cackling and arr sounds). As shown in Figure 5A and B, greater hunger will bring on an increase in the general mood of aggression. There are also sounds which indicate that an animal is

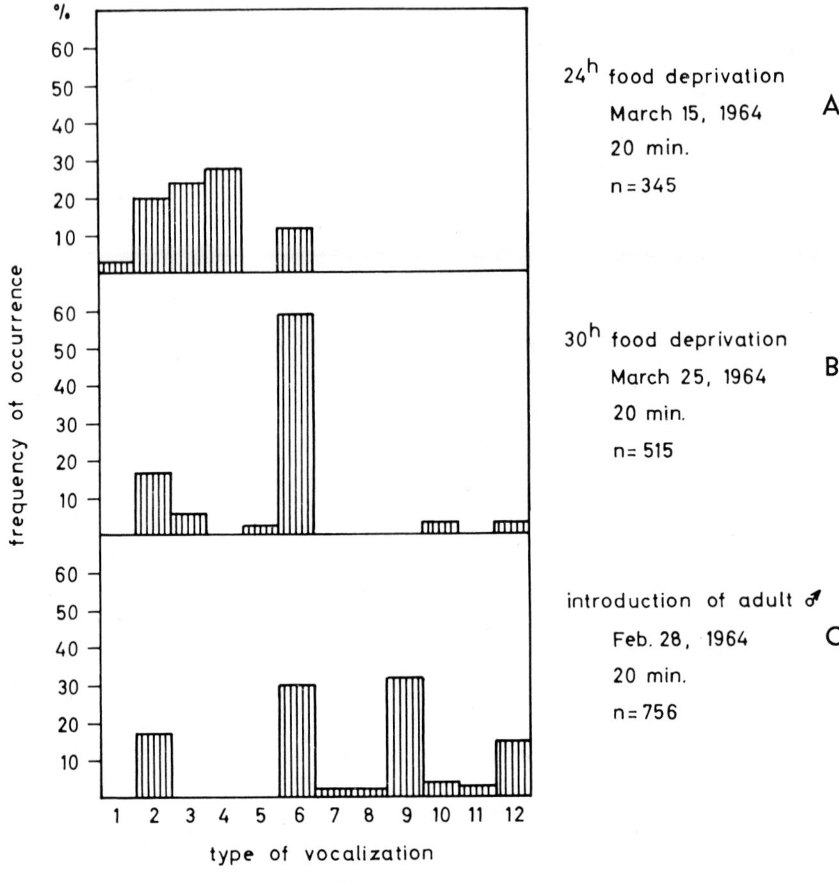

1. isolation squeak
2. peep
3. chirp
4. twitter
5. chuck
6. cackling
7. yap
8. purr
9. spitting
10. shriek
11. champing
12. clicking

FIG. 5.—Distribution pattern of vocalizations (1–12) in three different situations n = number of vocalizations.
A. Contact and food vocalizations (2–4) with few general sounds of aggression (6). B. Decrease of contact sounds (2, 3), disappearance of feeding sounds (4), marked increase in general sounds of aggression (6). C. Displacement of the vocalization pattern toward sounds of directed aggression and excitement (9, 12).

prone to directed aggression upon the introduction of adult males (see Fig. 5C). This analysis can be made even more precise by calculating the transitional probabilities of the sounds following one another.

We can show the meaning of vocalizations in the understanding of behavior by using genital display as an example. The types of display described above (see p. 151 and Fig. 1) are modified by the additional sounds emitted. Three versions of these types are (1) display without vocalization; (2) display with play-squeaking; and (3) display with the churr sound. As already described, the first case occurs between rival males and has an aggressive meaning. Squeaking, we learned, is a sound expressing a mood of play so that, in the second case, it is feasible that the aggressive element of genital display is greatly weakened by play-squeaking and this mode of behavior has assumed a sort of greeting function through "showing oneself." This appears to be confirmed by the behavior of young animals that almost always carry on genital display accompanied by squeaking without subsequent aggressiveness. Squeaking can also be heard during genital display while the animal sees its own image in a mirror (Ploog, Blitz, and Ploog). As a series of observations lead us to suspect, there seems to be a lightly aggressive attitude in the third case. This would also concur with the group characteristic of the churr sound.

A total of twenty-six sounds can be clearly recognized as a consequence of spectrographic analysis. It is most certain that not all sounds, especially all sound combinations, have been established during our observations, but this is expected in view of the very limited conditions offered by our laboratory in comparison with the natural biotope.[2]

LOCALIZATION OF GENITAL FUNCTION

During extensive experimentation with male squirrel monkeys, sites in the brain were located where an electrical stimulus evoked penile erection. These loci of cerebral representation of genital functions are distributed among widely branched brain systems from the medial orbital gyrus to the medulla oblongata and have been shown to be in septal, hypothalamic, and thalamic structures (MacLean and Ploog 1962) as well as caudal in substantia nigra, pons, brachium pontis, superior olive, periaqueductal region, the floor of the fourth ventrical, and anterior medullary velum (MacLean, Denniston, and Dua 1963).

In the section on group behavior (p. 149) it was shown that, aside from the sexual function of the erection in connection with thigh spreading and facing the intended partner, its significance is that of a highly effective social signal stimulus as well (see Fig. 1). The association of this manner of behavior with the genital function in both sexes leads us to expect that not only an erection,

[2] With respect to the natural behavior in the wild see I. T. Sanderson *The Monkey Kingdom* (Garden City: Hanover House, 1957), and S. Eimerl and I. DeVore, *The Primates* (New York: Time Inc., 1965).

but also an enlargement of the clitoris can be elicited by electrically stimulating in the female those brain structures which effected penile erection in the male. To see whether this supposition was correct and to eliminate any possible influence of the female cycle, brain stimulation was carried out on ovariectomized females.

METHODS

The fixation of the stereotaxic platform above the scalp of the animal's skull, the implantation of the electrodes, the exploration of the brain, and the histological techniques involved were executed in the same manner described by MacLean and Ploog (1962). The atlas of Gergen and MacLean (1962) was used for the documentation of the anatomical loci and their designation.

All positive responses (i.e., clitoris enlargement and combined responses) occurred with the following stimulus parameters: stimulus frequency—30/sec, pulse duration—1 msec or stimulus frequency—100/sec, pulse duration—0.1 msec. Currents of 1 ma for a 1-msec pulse and 3 ma for a 0.1-msec pulse were rarely exceeded. The average was between 0.2 and 0.5 ma.

In contrast to observations made of penile erection whereby the degree of erection could be distinguished, no such distinctions could be made in connection with clitoris enlargement due to the size and volume of this organ. Aside from a full enlargement (EC), a slight but definite swelling of the clitoris (ec) could be seen now and then.

RESULTS

The eighty-seven loci found resulted seventy-six times in EC and eleven times in ec. EC could occur separately but also accompanied by various autonomic and/or somatic signs. In twenty-four cases it appeared without other reactions. Of the remaining fifty-two combinations of EC with other modes of behavior, urination was observed thirty-six times, vocalization seventeen times, marked changes in hippocampal activity fourteen times, changes in heart rate seven times, and salivation three times. Responses were frequently paired with motor reactions of varied kind and intensity.

Figure 6 shows schematically all loci of positive stimulus responses with respect to clitoris enlargement in the frontal planes of A 14 to A 6.5. Positive loci are found from the area septalis (A 14; A 12.5; A 11) on the one side to the fasciculus olfactorius hippocampi (Broca's bundle A 14; A 12.5) and on the other side in the medial and lateral nuclei of the hypothalamus (area praeoptica medialis A 12.5; area anterior A 10.5; nucl. paraventricularis A 10.5; area dorsalis A 9.5; A 9; nucl. ventromedialis A 9; area lateralis A 8) and the medial thalamic nuclei (nucl. paraventricularis A 10.5; nucl. antero-medialis A 9.5; nucl. centralis inferior and nucl. centralis superior[3] A 9; A 8). Other

[3] The terminology of Emmers and Akert is used for nucl. centralis inferior and nucl. centralis superior.

positive responses come from the fasciculus mamillothalamicus (A 9.5; A 9) and the frontal thalamic nuclei (nucl. antero-ventralis A 9; nucl. antero-medialis A 9.5; nucl. antero-dorsalis A 9). Finally, the loci of cerebral representation of this female genital function are frequently to be found in the fasciculus telencephalicus medialis (medial forebrain bundle [A 12.5; A 11; A 10.5] and in the nucl. medialis dorsalis thalami [A 8; A 6.5]). Two other loci are found in the pedunculus ventralis thalami (A 10.5).

DISCUSSION

All loci described can be arranged in accordance with the three subdivisions of the limbic system as compiled by MacLean (MacLean 1962; MacLean and Ploog 1962): 1. Positive points coincide with the distribution of known hippocampal projections to parts of the septum, anterior thalamus, and hypothalamus. 2. Positive points have been located in parts of the so-called Papez circuit, comprising the mamillary bodies, the mamillothalamic tract, the anterior thalamic nuclei, and anterior cingulate gyrus. 3. Positive points have been found in parts of the medial orbital gyrus, the medial part of the medial dorsal nucleus of the thalamus, and regions of their known connections.

Those areas not shown in Figure 6 are being studied in experiments still underway.

MacLean linked these systems with functions that guaranteed self-preservation on the one hand and preservation of the species on the other, both of these being basic life principles (MacLean 1958, 1962).

The localizations shown here leave open the question of how natural behavior is coupled with genital functions, since the monkey is restrained in a chair while the stimuli are applied and there is no possibility of free behavioral development.

All stimulus responses observed simultaneously with clitoris enlargement (urination, vocalization, changes in hippocampal activity, changes in heart rate, and salivation) might be elicited without any functional connection by simultaneous stimulation of topographically neighboring systems. On the other hand, there might merely be a functional coupling of several of these stimulus responses. If so, then these coupled symptoms could be attributed to sexual or social behavior.

We know that genital display is coupled with vocalizing and urination (see pp. 152, 178). It is a mode of social behavior connected with self-preservation as well as with the preservation of the species. We suspect that, among the positive loci found, there are also those at which displaying can be elicited under proper circumstances. For this reason we began examining small, freely moving groups of two to four individuals after chronic implantation of electrodes, using a telestimulation method.

The brain map presents structures in the areas shown which are possibly responsible for sociosexual behavior.

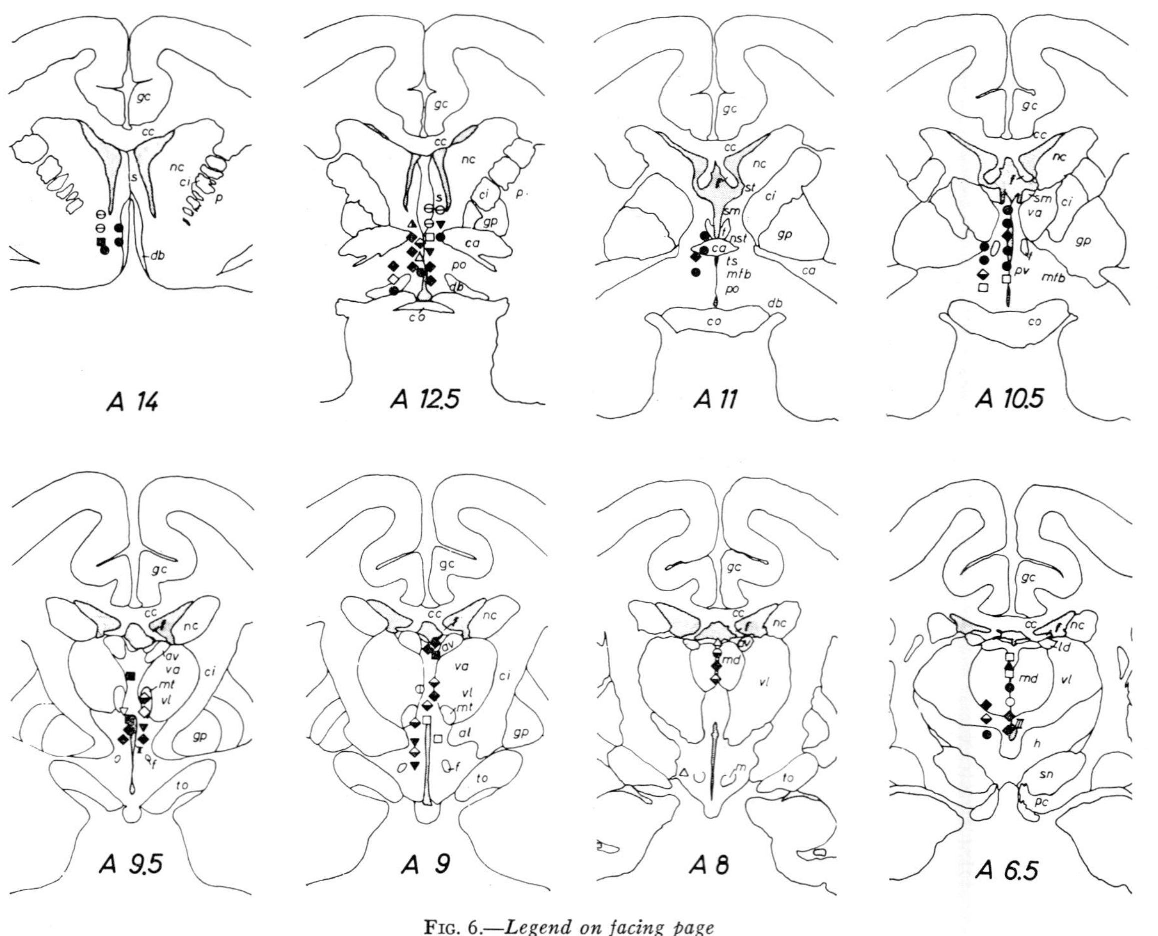

FIG. 6.—*Legend on facing page*

SUMMARY AND CONCLUSIONS

The behavior of squirrel monkeys in captivity was observed in six groups, four of them consisting of five or six members and two of them consisting of eleven and sixteen members. To record quantitatively the actions in these colonies, a list of thirty-seven behavior subpatterns was compiled. As described in 1963, the method of analysis uses these subpatterns to evaluate three measurable phases of an animal's communal existence, that is, self-directed activity, activity directed toward others, and activities directed toward the individual by the others. A specific balance of these phases is characteristic of each animal, and a change of this specific balance points to instability and changing of the group structure.

A survey on the ontogeny of social behavior was given. An especially important type of social behavior is genital display. The types of genital display are described as being (1) display in the open position, (2) display in the closed position, (3) display in the counter-position, (4) display before a mirror, and (5) display toward humans.

Types 1 to 3 appear in communal situations and transmit specific information to definite partners while simultaneously expressing the mood in which

---

FIG. 6.—All loci at which enlargement of the clitoris could be elicited through electrical stimulation are plotted on brain diagrams at representative frontal planes from A 14 to A 6.5 in accordance with the stereotaxic atlas of Gergen and MacLean. Each symbol represents at least one stimulus response with which there is an enlargement of the clitoris.

Key to Abbreviations

al—ansa lenticularis; *av*—nucleus antero-ventralis thalami; *ca*—commissura anterior; *cc*—corpus callosum; *ci*—capsula interna; *co*—chiasma opticum; *db*—fasciculus olfactorius hippocampi (diagonal band of Broca); *f*—fornix; *gc*—gyrus cinguli; *gp*—globus pallidus; *h*—area tegmentalis; *ld*—nucleus lateralis dorsalis thalami; *m*—corpus mamillare; *md*—nucleus medialis dorsalis thalami; *mfb*—fasciculus medialis telencephali (medial forebrain bundle); *mt*—fasciculus mamillothalamicus; *nc*—nucleus caudatus; *nst*—nucleus stria terminalis; *p*—putamen; *pc*—pedunculus cerebri; *po*—area preoptica; *pv*—nucleus paraventricularis hypothalami; *s*—septum pellucidum; *sm*—stria medullaris; *sn*—substantia nigra; *st*—stria terminalis; *to*—tractus opticus; *ts*—nucleus triangularis septi; *va*—nucleus ventralis anterior thalami; *vl*—nucleus ventralis lateralis thalami; *III*—ventriculus tertius. *EC*—full enlargement of clitoris; *ec*—slight but definite swelling of clitoris with or without other responses; *HA*—marked changes in hippocampal activity; *U*—urination; *V*—vocalization; *HR*—changes in heart rate; *S*—salivation.

Symbols

♦	*EC*	○	*EC* + *U* + *S*	⊖	*EC* + *U* + *V* + *HA*
⬙	*ec*	□	*EC* + *U* + *V*	⦶	*EC* + *U* + *V* + *S*
■	*EC* + *HA*	△	*EC* + *U* + *HR*	◒	*EC* + *U* + *HR* + *HA* + *S*
●	*EC* + *U*	◊	*EC* + *U* + *HA*	◭	*EC* + *U* + *HR* + *HA* + *V*
▼	*EC* + *V*	▽	*EC* + *V* + *HA*		
▲	*EC* + *HR*				

the displaying animal finds itself. According to the standing and role of the animal, its age and sex, displaying has different meanings and effectiveness.

In stable groups, adult males display toward the alpha animal only rarely. Infants and adolescents of both sexes, on the other hand, display most often toward the top animal and toward their mothers as well. The transition from an infantile distribution pattern of display to that of the adult monkey that has found its role in the group continues for at least 2 years and probably 3. Transition takes place gradually as certain members of the group increase the frequency of their threatening behavior toward the young animal. The development of displaying in the young animal lends itself particularly well to the demonstration of how this social signal stimulus has an effect on the group on the one hand and how the group modifies this behavior on the other.

Observations of a female infant which remained with the group and which displayed on the day after birth, and observations of other infants that grew up alone with their mothers lead us to assume that displaying is an innate behavior pattern.

The general meaning of this social signal stimulus can be described with the words *demanding, self-assertion, courting,* and *desiring closer contact.* Dominance is accentuated in the closed position with the aggressive component appearing more markedly. In young animals and females, frustration and defense as well are expressed in the open position. In groups where there are more females per male there is, on the average, more frequent displaying by all members of the group toward all other members of the group than there is in groups with a balanced ratio of the sexes. This supports the supposition that displaying expresses both competitive and courtship behavior.

The large repertoire of motor patterns serving as a means of social communication is accompanied by an extensive vocal repertoire. The list compiled to date shows twenty-six sounds that can be clearly distinguished by spectrographic means. Most sounds transmit information to others in the group, which then exhibit either motor or vocal responses. The sounds simultaneously express the moods in which the animals find themselves at the time they are uttered.

Six classes of sounds could be distinguished: (1) squeaking as an expression of contact mood, (2) twittering as an expression of the desire to maintain a minimum distance, (3) cackling as an expression of a general mood of aggression, (4) arr-sounds as an expression of the disposition toward directed aggression, (5) shrieking as an expression of a high degree of excitement, and (6) combination sounds with elements from classes 1 to 5.

Special consideration was given to the ontogeny of vocal behavior. The initial sound was heard in the form of three shrieks, 5 minutes after birth. Within the first 24 hours after birth, squeaking and twittering as well as characteristic purring during nursing were heard in a fully developed form. Thus, these individual observations lead to the assumption that vocal behavior is also innate and is functionally connected with maturing modes of behavior.

The meaning of vocalization in connection with the understanding of behavior modes can again be demonstrated very well in genital display. Squeaking accompanies genital display during play, so the aggressive tendency of displaying is weakened by these contact sounds. However, if displaying is accompanied by purring, the aggressive element is accentuated.

Motor patterns and vocalizations together form a very extensive system of communications through which the highly organized social behavior of the squirrel monkey is revealed.

The attempt is made to connect elements of social behavior with specific brain structures and systems. In earlier brain stimulation experiments, the cerebral representation of penile erection could be proved. Various types of vocalization in association with erection were also observed during these experiments. Ovariectomized females were used to prove the cerebral representation of clitoris function. The cerebral loci found to date at which an enlargement of the clitoris could be elicited by means of electrical stimulation are the same as those for penile erection. At a fifth of these loci, the genital stimulus response appeared in combination with various vocalizations.

For now, the loci studies leave the question of natural behavior coupled with genital function unanswered until sufficient experience can be obtained with stimulation experiments on animals moving about freely in the community. It is suspected, however, that the loci found include those at which displaying can be elicited.

It is probable, from the phylogenetic point of view, that the development of sociosexual behavior corresponds to a brain development which makes this behavior possible. The designated brain structures in the septum, hypothalamus, and thalamus are probably part of a widely branched cortico-subcortical brain system which determines the social behavior specifically associated with the species of squirrel monkeys.

The social behavior of these primates might indicate a landmark where sexual behavior patterns begin to pertain to other social patterns so that social life is modified and developed in a phylogenetically new direction.

REFERENCES

Altmann, S. A. 1962. A field study of the sociobiology of rhesus monkeys, *Macaca mulatta*. *Ann. N.Y. Acad. Sci.* 102:338–435.

Andrew, R. J. 1963. Trends apparent in the evolution of vocalization in the Old World monkeys and apes. *Symp. Zool. Soc. Lond.* 10:89–101.

———. 1963. The origin and evolution of the calls and facial expressions of the primates. *Behaviour* 20:1–109.

Bowden, D., Winter, P., and Ploog, D. 1966. Pregnancy and delivery behavior in the squirrel monkey and other species. *Folia Primatol.* (In press.)

Carpenter, C. R. 1934. A field study of the behavior and social relations of howling monkeys (*Alouatta palliata*). *Comp. Psychol. Monogr.* 10:1–168.

———. 1958. Soziologie und Verhalten freilebender nichtmenschlicher Primaten. In *Handbuch, Zoologie,* ed. Kükenthal 8 10(11):1–32. Berlin: W. de Gruyter.

Castell, R., and Ploog, D. Zum Sozialverhalten der Totenkopfaffen (*Saimiri sciureus*): Auseinandersetzung zwischen zwei Kolonien. (In preparation.)

Chance, M. R. A. 1956. Social structure of a colony of *Macaca mulatta. Brit. J. Animal Behaviour* 4: 1–22.

Emmers, R., and Akert, K. 1963. A stereotaxic atlas of the brain of the squirrel monkey (*Saimiri sciureus*). Madison, Wisconsin: The University of Wisconsin Press.

Gergen, J. A., and MacLean, P. D. 1962. A stereotaxic atlas of the squirrel monkey's brain (*Saimiri sciureus*). Washington, D.C.: Public Health Service Publication No. 933.

Hinde, R. A., and Rowell, T. E. 1962. Communication by postures and facial expressions in the rhesus monkey (*Macaca mulatta*). *Proc. Zool. Soc. Lond.* 138:1–21.

Hopf, Sigrid, and Ploog, D. Soziometrische Analyse der Ontogenese des Verhaltens von Totenkopfaffen (*Saimiri sciureus*). (In preparation.)

Itani, I. 1963. Vocal communication of the wild Japanese monkey. *Primates* 4:11–66.

MacLean, P. D. 1958. The limbic system with respect to self-preservation and the preservation of the species. *J. Nerv. Ment. Dis.* 127:1–11.

———. 1962. New findings relevant to the evolution of psychosexual functions of the brain. *Ibid.* 135:289–301.

MacLean, P. D., Denniston, R. H., and Dua, S. 1963. Further studies on cerebral representation of penile erection: caudal thalamus, midbrain and pons. *J. Neurophysiol.* 26:273–93.

MacLean, P. D., and Ploog, D. W. 1962. Cerebral representation of penile erection. *J. Neurophysiol.* 25:29–55.

Maurus, M., Mitra, J., and Ploog, D. 1965. Cerebral representation of the clitoris in ovariectomized squirrel monkeys (*Saimiri sciureus*). *Exper. Neurol.* 13:283–88.

Ploog, D. 1963. Vergleichend quantitative Verhaltensstudien an zwei Totenkopfaffen-Kolonien. *Z. Morphol. Anthropol.* 53:92–108.

Ploog, D., Hopf, S., Winter, P. Zur Ontogenese des Verhaltens von Totenkopfaffen (*Saimiri sciureus*). (In preparation.)

Ploog, D. W., Blitz, J., and Ploog, F. 1963. Studies on social and sexual behavior of the squirrel monkey (*Saimiri sciureus*). *Folia Primatol.* 1:29–66.

Ploog, D. W., and MacLean, P. D. 1963. Display of penile erection in squirrel monkey (*Saimiri sciureus*). *Animal Behaviour* 11:32–39.

Schleidt, W. M. 1964. Eine Apparatur zur Tonfrequenzspektrographie aus Bausteinen. Tierstimmen optisch dargestellt. *Rohde & Schwarz Mitt.* 18:155–58.

Takeshita, H. 1962. On the delivery behavior of squirrel monkeys (*Saimiri sciurea*) and a mona monkey (*Cercopithecus mona*). *Primates* 3:59–72.

Winter, P., Ploog, D. W., and Latta, J. Vocal repertoire of the squirrel monkey (*Saimiri sciureus*), its analysis and significance. *Exper. Brain Res.* (In press.)

DISCUSSION OF CAUSAL MECHANISMS

DAVID MC K. RIOCH

The chapters in this section of the volume introduce a number of general problems of research on behavior in addition to the specific problems of the experiments reported.

An analysis of the functional structure of the Division of Neuropsychiatry at the Walter Reed Army Institute of Research was recently conducted by Mr. A. K. Rice of the Tavistock Institute of Human Relations. He found it desirable to classify the various projects and laboratories systematically, and a variety of criteria were tested for this purpose, such as the major technical procedures used in the projects, work on animals vs. work on humans, location of the work, and so forth. The most satisfactory criterion found was the *degree of control* which the investigator had over the contingencies influencing the course of the event or events under investigation. The degree of control is a unified dimension over the whole range of work in progress in the Division, and it is correlated with certain general characteristics of research strategy. It is obvious that one can more readily exercise control over the situation in a laboratory than in a restricted territory (such as Santiago Island), and even the latter is more easily controlled than most of the naturalistic conditions. This is not to say that control and the laboratory can be equated. On the one hand, every new technique and new phenomenon represent situations for which controls have to be developed in the laboratory. On the other hand, many problems cannot be brought into the laboratory at all. These include such problems as the function of the particular animal group in the economy of the game reserve or jungle—as a competitor for food and water, as a source of food for predators, as a host for infectious agents, and so forth. The internal organization of the group will also vary, depending on the surroundings whether natural or artifactually imposed. Controls for certain fac-

David McK. Rioch, Division of Neuropsychiatry, Walter Reed Army Institute of Research, Washington, D.C.

tors in these problems must be developed in the field. With this viewpoint, we are learning to look at behavior not as "normal" or "abnormal," but as the interaction with the environment of an adaptive mechanism, controlled by feedback of information on the course of the interaction. Instead of thinking in terms of normal, natural, and abnormal, we need to note differences in behavior which can be correlated with definable historical, environmental, or intrasomatic contingencies, whether experimentally or naturalistically determined.

The objective of laboratory studies, thus, is not merely to demonstrate that certain phenomena, well-known in the field, can be evoked in the same or in modified form in the laboratory—though this is a necessary condition. The objective of laboratory experimentation is to facilitate analysis of the course of behavior and of the mechanisms mediating that behavior. Laboratory facilities now permit a great deal of control of the life history of the organism and some control of its genetic history. They also permit quantitative (that is, precise) manipulation of the environmental contingencies which direct behavior. Consequently, they permit determination of the range over which changes in these contingencies are correlated with quantitative or qualitative changes in the response patterns. Of greater importance to the study of behavior is the possibility the laboratory provides for control of the consequences of behavior under a set of reproducible contingencies. The resulting adaptations of behavior—which are generally classified as learning, remembering, and forgetting—are among the most important phenomena which determine survival.

With regard to memory, however, it is important to bear in mind that the brain does not store data, as it were, in a card-index file or on a memory drum in the manner of storage mechanisms of computers. Rather, the units (whatever they may be) of the communicating networks in the brain are themselves changed in the course of behavioral events. As a result, all data are to a greater or lesser extent differently processed as experience continues, although such changes in the covert cerebral processes may be more overtly expressed in certain situations than in others.

The experiments reported by Dr. Miller are a nice example of the carefully considered use of the facilities of the laboratory. This report concerns the most recent in a long series of studies directed toward elucidating the modes and the consequences of communication between monkeys. It is clear that facial expression is a sufficient signal to evoke behavior appropriate to the situation, and the groundwork is thus laid for comparing this signal with other signals (such as auditory), as well as for looking at the behavior which may be evoked by introducing discrepant signals through different input channels. Dr. Miller's studies also illustrate the value of using intrasomatic responses as indicators. In spite of the potentialities that miniaturized electronic amplifiers and transmitters provide for such studies in the free-ranging situation described by Dr. Robinson, the monkey in the restraining chair is still preferred for such research. The major problem is maintaining a sufficient number of transducers in proper relation to blood vessels and other organs to provide data on patterns of response, since the autonomic output seldom shows changes which are clearly associated with one or another attitude or emotional disturbance. Nevertheless, autonomic responses may provide very sensitive indicators of changes in the interaction between the organism and the environment in the absence of overt manifestations.

It seems to me that the study of the intrasomatic events in relation to ongoing behavior is now one of the more fruitful areas for collaboration between the laboratory and the field. To survive, the organism must not only interact with the environment adequately, but must also maintain the mechanisms of interaction within an appropriate range of efficiency. To a considerable extent, laboratory studies have been limited to short time intervals, and naturalistic studies have taken fatigue, sleep, tension, and so forth more or less for granted. In human studies, it is becoming clear that the interrelationships between the activity we call thinking (including perception, storage, retrieval, symbolic operations) and states of fatigue, loss of sleep, inadequate diet, and other conditions of reduced capacity are of very considerable importance. Certain functions, including vigilance, which are commonly included in the concept of "thinking" may be markedly impaired. It would thus be highly desirable to learn more of the social behavior and group structure of nonhuman primates during and following naturally occurring stressful events in parallel with laboratory studies of isolated loading of the organism, including the biochemical and physiological adjustments, and its ability to discriminate, learn, and so forth.

It may well be that in some instances the laboratory may play another important role, such as in differentiating patterns of behavior which superficially appear similar. For example, in a series of experimental studies on "emotional" behavior, Mason and his associates (1961) found a marked difference in the responses of the 17-hydroxycorticosteroids and of norepinephrine as compared with the responses of epinephrine in situations in which general observations of the animals showed no differences. Thus, the blood levels of the former two hormones increased significantly in all situations evoking emotional disturbance. In contrast, the blood levels of epinephrine only rose in completely novel, in ambiguous, and in unexpected situations. It appears that even if the overt response is equivalent or similar, the covert response of the animal to the situation in which it has had previous experience (and so "knows" the probable course) is different from that in the situation which may be said to be "open ended." These observations lead to some very interesting questions concerning so-called clinical anxiety in the field of psychiatry. It would be interesting to determine in naturalistic studies whether equivalent classes of patterns of disturbed behavior could be differentiated, either on the basis of differences in immediate responses in the free-ranging situations or on the basis of later events, such as the rate of return to normal activities. Experimental methods for differentiating patterns of covert behavior which are not being overtly expressed will no doubt be found and, when introduced into the field or into the limited-range situation, will provide an additional analytic tool.

Miniaturization of electronic circuits and the development of insulating and protective materials which are chemically inert when implanted in tissue over long periods of time have focused attention on new possibilities of correlating events in the brain and body with events observed in behavior. Dr. Robinson has given an excellent description and discussion of the ways in which newer techniques may be applied in field studies. He has also presented data on the association of vocalization with other patterns of behavior in response to intracranial stimulation. This is a very difficult field and will require long study with very rigidly controlled methods before we can clearly formulate what goes on in the "black box." In the meantime, it may well to emphasize that electrical stimulation of the brain, except at the final common path, does not evoke the functions of the structures which are stimulated. Rather, it introduces a pattern of activity in certain fibers and cells which, in its frequency and spatial distribution, is artifactual. This may occlude ongoing normal functions in the manner in which Dr. Wilder Penfield showed that electrical stimulation in the speech areas of the cortex produced aphasia during the period of stimulation in patients under local anesthesia. When overt patterns of behavior are evoked by electrical stimulation, it is clear that the coordinated activity of the overt behavior is not a direct function of the stimulus introduced. Rather, it would seem that the structures stimulated under the conditions of the experiment are able to evoke organized patterns of activity in the mechanisms to which they project. It is thus of interest that the forebrain structures from which Dr. Robinson evoked vocalization project in major part to the subthalamus and to the mesencephalon, but not to the more caudal areas where the final motor mechanisms are located. We must assume, then, that the electrical stimulation in these experiments set up activity in highly complex, reciprocal, forebrain-midbrain systems which mediate the complex patterns shown in the overt behavior. It would further appear that the pattern evoked by stimulation, or any modifications of the pattern, would be strongly influenced by other excitatory or inhibitory impulses concurrently coming into the complex coordinated system which mediates the overt pattern. This may be illustrated by some experiments conducted by Wasman and Flynn (1962). They stimulated cats through electrodes implanted in the lateral hypothalamus and evoked either a stalking or an affective pattern of aggressive activity. The latter included vocalization—growling, spitting, yowling, and so forth—as well as piloerection and clawing and biting. Which pattern was evoked seemed to depend on the placement of the electrodes. In both patterns the behavior of the cat during stimulation was markedly controlled by the environmental stimuli present. The presence of another animal—rat, guinea pig or cat—resulted in a sudden, well-directed, vigorous attack. A dead or a stuffed animal had less effect. During stimulation in a bare cage, the cats moved about, occasionally pushing at the door of the cage. Since the stimulus was the same in these different circumstances, one must conclude that the difference in the patterns evoked was related to the other activities of the nervous system determined by the environmental contingencies. It would be of great interest, under free-ranging conditions, to examine the interaction of the experimental and environmental factors and, further, to correlate such data with appropriate ablation studies in order to ascertain the necessary and sufficient, intrasomatic and extrasomatic conditions determining different patterns of interaction of the organism with the environment.

Dr. Ploog's paper presents data from three segments of a large, very comprehensive study of the social behavior of squirrel monkeys. His work illustrates the value of utilizing and coordinating a variety of different experimental situations and techniques. The suggestion that a single pattern of interaction—in this instance display behavior—may be sufficient for defining the social structure of the group is of interest. Observations on other species, for example, baboons, under naturalistic conditions, indicate that there is greater flexibility to the social structure than would be demonstrated by observations on a single pattern of interaction. It would be of interest to know whether the greater rigidity of social structure described for rhesus monkeys on Santiago Island and in Dr. Ploog's experiments represents a difference in species or whether it is a function of limited territory. The accuracy with which vocalization now can be recorded and either visually displayed for analysis or played back as a stimulus to the animal is well illustrated by the material Dr. Ploog presents. These methods provide an extremely powerful tool for a very wide variety of studies, and one hopes that atlases of vocalization of various species will become available for general use in the near future.

Dr. Ploog's findings that there are similar mechanisms in male and female brains, and that very limited aspects of more complex behavior patterns can be evoked by stimulation of the hindbrain are in keeping with a good deal of experimental, physiological, and clinical observations. It appears that sub-routines, which are parts of several more complex behavior patterns, may be mediated by nerve nets in close association with the final motor neurons. Such sub-routines are likely to appear in more stereotyped form in those reactions involving the viscera, the blood vessels, and so forth, than in those involving the skeletal muscles. In the latter, interaction with the environment requires greater modifiability and flexibility than in the former.

Although there seems to be no question that the judicious coordination of laboratory and naturalistic studies will provide a great deal of precise data on behavior and on the mechanisms mediating behavior, it would probably be an error to overemphasize this strategy of research. Laboratory studies necessarily focus attention on limited aspects of behavior and are directed toward differentiating such aspects into their component parts. Inevitably, one runs the danger, according to the old adage, of "learning more and more about less and less."

It has occurred to me that sufficient data have already been collected on different groups of organisms—with a wide variety of behavioral capabilities, living in a variety of habitats—to justify considering the approach to a number of more general problems by analyzing the group as a system in interaction with its environment. Broadly speaking, this approach places the natural environment in the same category as laboratory equipment *vis-à-vis* the organism. Since the various groups of primates which have been studied are extant, it seems to me we would be justified in asking what functions have to be performed to permit the survival of the group, and what individual and social behaviors in a group are involved in performing these functions. This type of system analysis has the advantage of drawing attention to long-term patterns which are difficult to keep in mind when one is engaged in making day-to-day observations or conducting the immediate experiment. Attention is also directed toward examining the interaction of different functions and factors—such as food supply, reproduction rate and size, and organization of groups. If one takes mere survival as the baseline level for considering groups, one can then (particularly in human groups) proceed to consideration of optimal levels of certain functions for defined objectives under a particular set of circumstances. Further, consideration of the functions to be performed provides a basis for comparing different groups or species on the basis of the mechanism used. Thus, some species use counterattack and others flight or hiding in response to predators, raising the interesting question whether other parts of the behavioral repertoire show correlated differences. In view of such considerations, it may be of interest to list some of the functions which, if only on a speculative basis, would appear to be necessary for the survival of a group.

A number of the necessary functions are concerned with interactions of the individual members of the group and the environment. These include (1) acquisition of food and water; (2) provision of shelter for sleep, rest, recreation, and recuperation (from non-fatal wounds, infections, and so forth); (3) protection against predators by one or other forms of escape using the environmental potentials or by counterattack; (4) control of potential diseases by immune reactions or other defenses; (5) maintenance of adequate rates of reproduction; and (6) control of the environment, including exploratory and manipulative functions. The first five of these general classes of function appear to be self-evident. One might well consider the control of the environment as also a necessity in view of the universality of the patterns of behavior referred to as territoriality. However, knowledge of the environment and a store of appropriate programs of action already adjusted to the home range are of such significance in the first three functions listed that territoriality might be regarded as a derivative of them rather than an additional requirement. Certainly territoriality depends on ability to store and retrieve information, and some of its characteristics may be correlated with those activities generally classified under the heading of "intelligence." In human groups (see Parker 1920) the functions of manipulation (that is, something which can be repeated and improved) and of exploration may be regarded as including the technologies and the arts and appear to be of fundamental importance. This development is clearly dependent on symbolic behavior. It is likely that these functions are also necessary for other mammalian species in that the environment is continually undergoing change. It may also be that in a system of adaptive feedback mechanisms it is necessary to have an output which is continually feeding back a response that is variable—that is, unexpected—in order to prevent the establishment of a lethal stereotype. Speculating on some of these problems seems to me to be worthwhile, since it directs attention to a number of the less dramatic and obvious forms of behavior. Thus it leads us to ask what nonhuman primate groups do and how they act when their behavior is not under the control of specific interoceptive or exteroceptive signals. Further, could such activity tend to improve living conditions?

Another set of functions which seems necessary for group survival concerns social behavior rather than individual interactions with the physical environment. These include (1) a system of communication; (2) a set of signs or signals identifying the members of the group; (3) a mechanism for planning and decision making; (4) a policing system for maintaining group structure; and (5) an educational system, at least at a rudimentary level. It will be noted that the terms used in this list are taken from human culture. This seems justified insofar as one is dealing with the formal characteristics of group structure and not with the particular mechanisms used for maintaining the structure. Thus, it is quite apparent that there is no separated class of police monkeys, though a considerable part of the social behavior seems to concern policing (much as in technologically unsophisticated human cultures).

Whether or not behavior is hierarchically organized in the functional structure of the nervous system, it is obvious that the formulation of the data of behavior requires hierarchical organization. For such a purpose one needs a unified theory and formal criteria, independent of the mechanisms (and, therefore, of the energy) utilized in the expression of the behavior. Communication theory, including cybernetics, provides a useful conceptual tool for this purpose. It is of particular value in that it allows analysis of the organization of the group (or of a single organism) in the temporal dimension—that is, in the course of particular behavioral patterns. Since study of the temporal course of behavior requires the definition of an end state, it seems to me that survival of the group provides a reasonable baseline for consideration of other end states and of the range of variability of the functional structure which is compatible with survival.

This type of approach to the study of behavior requires much closer relationships between the laboratory and the field than has commonly been practiced. We need more experience in taking data and problems from one to the other in order to exploit fully the potentialities of both. The inclusion of several chapters on the experimental analysis of behavior in this volume is a significant step in this direction. I hope that the precedent will be followed in future volumes and will become a tradition.

REFERENCES

Mason, J. W., Mangan, G. F., Jr., Brady, J. V., Conrad, D., and Rioch, D. McK. 1961. Concurrent plasma epinephrine, norepinephrine and 17-hydroxycorticosteroid levels during conditioned emotional disturbances in monkeys. *Psychomat. Med.* 23:344–53.

Parker, Carlton H. 1920. *The casual laborer and other essays.* New York: Harcourt, Brace, and Howe, pp. 125–65.

Wasman, Marvin, and Flynn, John P. 1962. Directed attack elicited from hypothalamus. *Arch. Neurol.* 6:220–27.

Part IV

SOCIAL DYNAMICS

11

THE AYE-AYE OF MADAGASCAR

JEAN-JACQUES PETTER AND ARLETTE PETTER

Daubentonia madagascariensis, or the Aye-aye, is undoubtedly the strangest living mammal of Madagascar and the most valuable from a scientific point of view.

Since its discovery, it has continued to interest zoologists, and its anatomical characteristics are now well known. It belongs to the group of lemurs, where it occupies a special place because of some exceptional characteristics, particularly its rodent-like dentition, the filiform appearance of the third digit, and claws instead of nails.

It has never been common in Madagascar, and for several years, no trace of it had been found, although no special effort had been made to do so. The cloud of mystery and superstition in which it was enveloped caused it to be considered one of those fabulous extinct animals of the old Malagasy forests.

It was then with great joy, during an expedition to Madagascar in 1957 (Petter and Petter-Rousseaux 1959), that we were able to ascertain the existence of this animal in the neighborhood of Mahambo, 13 km south of Fénérive, on the east coast of Madagascar. Two specimens were seen 82 km from Tamatave along the old road between Tamatave and Fénérive, and distant cries revealed the existence of others.

Unfortunately, this happened at the end of our stay, and we were unable to stay any longer to observe the animals we had just discovered. Wishing to return for a more complete study, we asked for this region to be protected, and thanks to Dr. Paulian, director of the Institut de Recherches Scientifiques and of Mr. Saboureau, curator of the Service de Protection de la Nature des Eaux et Forets, we were able to obtain its protection.

Jean-Jacques Petter and Arlette Petter, Museum National d'Histoire Naturelle, Brunoy (S.-et-O.), France.

We were not able to return to this region until 1963.[1] In the meantime, it had unfortunately undergone considerable changes: the coastal forest had largely been destroyed in order to build the new road. There were still traces of the Aye-aye's nests, but we could not find any sign of recent presence of these animals, and it is reasonable to believe that they have deserted the spot.

The few Aye-ayes that were living there in 1957 were probably scared off or killed when the road was being built. Furthermore, besides the systematic destruction of part of the forest for the building of the road and new villages along it, some permits for felling trees were issued. The rest of the forest that was inhabitable for the Aye-ayes probably became too "poor" to allow them to subsist, the exploitation having brought about the suppression of the major part of the trees likely to bear fruit. The same phenomenon has occurred during the last few years in the remaining parts of the forest along the east coast.

There was, until recently, a habitat which the Aye-aye could still frequent: the land surrounding the villages. They still found a few large trees in which to build their nests, and coconut, mango, litchi, and other fruit tree plantations provided enough food all the year round. So far, the popular superstitions forbid killing or even disturbing the Aye-aye; so they could live in peace in the vicinity of human beings.

As might be expected, the superstitions are decreasing in effect from day to day. There are two reasons for this: the cultural evolution of the primitive natives and the end of colonial rule. The people are more and more in contact with the pleasures of town life and technical civilization and lose their old superstitions. Furthermore, with the end of European settlement, a number of plantations used by the settlers are being abandoned. At first there was more fruit than required, whereas now there are not enough mangos or litchis for the ever increasing number of natives of the local villages who have taken the place of the European settlers.

With competition becoming greater and greater, it is natural that the Aye-ayes have had to give way. In many villages, these lemurs have been killed with sticks; the few specimens that are left will completely disappear in a few years.

At the beginning of 1964, there were still two Aye-ayes in the village of Mahambo. According to the information we were able to collect, at least three animals were killed during the preceding year. During two visits of 20 days each in the region of Mahambo, in October and November, 1963, and January, 1964, we tried to observe these animals as often as possible, and we collected as much information as we could on their behavior. Unlike all the other large lemurs we studied before (true lemurs which form a complex social group and

[1] Expedition carried out with the financial help of the CNRS (France) and UNESCO.

Indriidae which live in family groups), the Aye-aye, which is able to build a complicated nest and is very active, seems to be strictly solitary and the intercommunications between individuals seem to be very limited.

HOME RANGE

The home range of the Aye-aye was, when we observed it, about 12 acres. This area was almost entirely covered with fruit trees and included the village of Mahambo, which is exceptionally rich, and apparently was not connected to the neighboring forest. Most of the trees were tall and well developed. There were coconut, mango, litchi, *Terminalia, Eugenia,* and breadfruit trees. The places the farthest away from the village included tracks which were often frequented; furthermore, the home range also included clove and coffee plantations and small enclosures planted with maize.

One adult female and one male not yet fully developed were found on this range. Neither appeared to keep to any specific part of this zone, and they were never observed together.

This range appeared to us to be very large; indeed, owing to the absence of competition, the animals moved freely over the whole area in quest of the food they preferred, only exploiting a tiny part of the resources of each tree. The natural obstacles which limited them were often small: patches of trees which did not interest them and large clearings. In more natural conditions, it is probable that this home range is limited by the neighbors, against the incursions of which they probably defend themselves. In 1957, we observed the encounter and the fight of two animals; our presence probably drove one of them out of the usual limits (Petter and Petter-Rousseau 1959, Petter 1962*a*).

NESTS

Nests were very numerous on this range: we counted about a score. They were grouped in certain preferred places with particularly tall trees or in the vicinity of an important source of food.

The first zone of nests was located in the garden of a farmer near the village. The trees were not very tall but there were several coconut trees frequented by the Aye-ayes. This garden had five nests, four of which were in litchis and one in a coconut tree.

These nests were built about 40 feet from the ground. They were never less than 30 feet from each other. Three of them were occupied in November by two Aye-ayes; none was occupied the following month. In the meantime, the gathering of the litchis caused a great deal of noise (since sticks are thrown up in the trees) and damaged the branches and the nests considerably. One of the nests completely disappeared.

A second zone of nests, also near the village, was located in the extreme south of the range and consisted of land with tall trees, and litchis and mangos with very dense foliage. In this area of about 80 by 50 m, there were six nests about 70 feet from the ground in four mango trees. According to the dryness of the leaves, we were able to calculate that two of the nests were old and that four had been built in November or December.

Finally a third zone, frequented in January, was found in the northeast of the range, the farthest away from the village. There were four recently built nests at least 60 feet from the ground in mango trees and *Albizzia*. This zone, which was relatively less rich in fruit, was undoubtedly an area of retreat in the face of ever increasing disturbances due to the gathering of litchis and mangos in the neighborhood of the village.

Outside these three zones, we found two old, abandoned nests in large litchis close to each other, one nest in a coconut tree near the village, and an isolated nest in a large *Eugenia* about 50 yards from one of the groups of nests, which was built and inhabited in December.

It is difficult to estimate when the oldest of these nests was built. In forest land, they can remain intact for a long time. We found nests that were intact 7 years later (Petter-Rousseaux 1962). Close to the village they are built in fruit trees that are exploited by the natives; there the nests are fairly quickly destroyed. A newly built nest (Fig. 1) is a large green mass for several days, but after a week there is nothing to distinguish it from below from an old nest. A very old nest which has not been inhabited for a long time is smaller in size, having gradually lost some of the original twigs.

The nest is built in a strong fork of a large tree (Fig. 2). The fork must have several branches, the general aspect of which suggests a bowl, and which can support a nest with a diameter of about 50 cm. The frame of the nest consists of twigs about 30 cm long taken from the tree itself or from neighboring trees, and these are entwined in the branches of the selected fork. The completed nest is closed at the top by a roof made from twigs laid flat, and has a lateral opening. The interior cavity is small, allowing for only one Aye-aye. The bottom is covered with a thick layer of shredded leaves.

The nest, then, is a complicated dwelling place; it probably takes at least 24 hours to build. Why the Aye-aye builds a new nest is unknown.

We have followed an Aye-aye's activities at different hours of the night, and we have seen it entering the nest, which it recognized, by a known route. Several times, we saw it leave the nest at night on awaking, choose some twigs from the immediate surroundings, and return with them to complete the nest; thus we often saw a nest with dry leaves and a few green twigs.

In the nests we observed, the foliage used was principally that of litchis. A nest in a coconut tree was partially built with twigs from a neighboring litchi. In the wall of a nest built in a *Eugenia*, we found a piece of palm leaf which had been carried from a tree more than 10 yards away. It is possible

Fig. 1.—An Aye-aye's nest

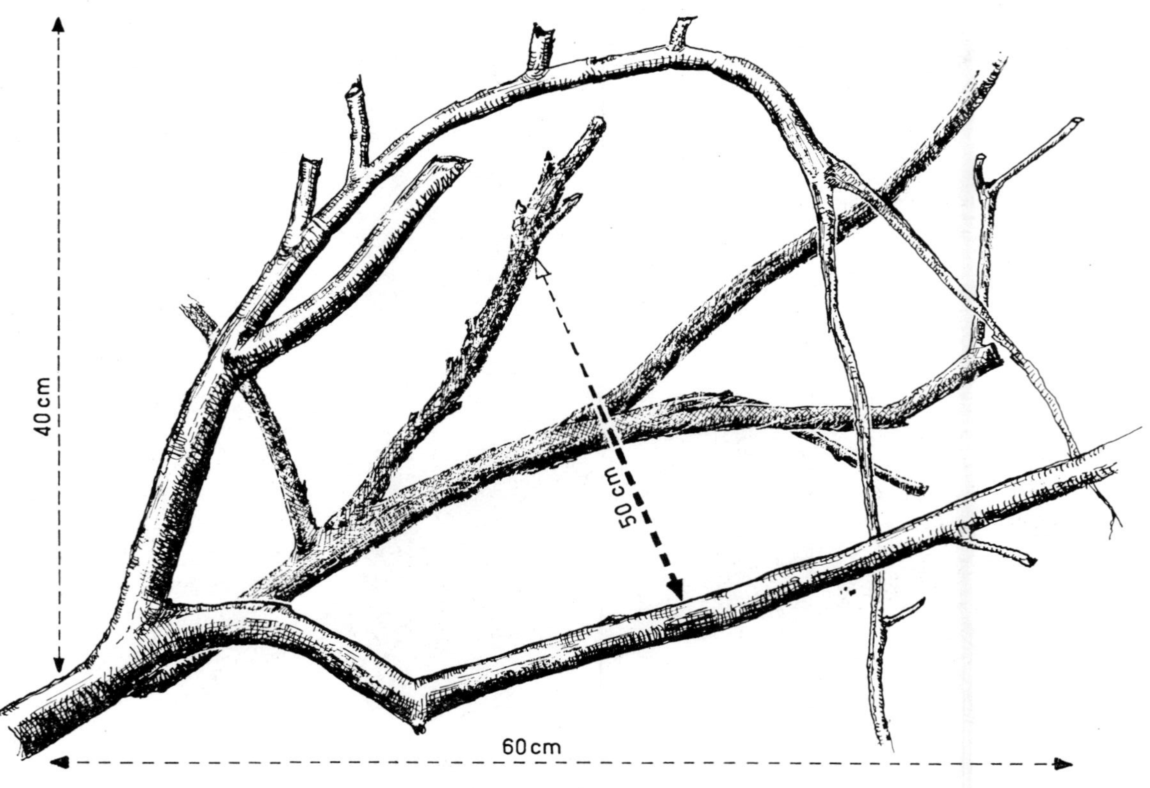

FIG. 2.—Fork of tree used to lodge nest

that the animal, upon seeing a bowl-like formation of branches (more frequent in litchis), is stimulated to build a nest.

Contrary to what is sometimes said, the nest is inhabited for several consecutive days; it may even be abandoned temporarily and used again. An accident may cause the Aye-aye to abandon the nest. We observed an Aye-aye in a nest in a coconut tree during two consecutive days. In the evening of the second day, a big, dry palm frond fell from the tree, making a lot of noise just when the animal was leaving its nest. It did not return the following morning. On another occasion, however, a nest that we had shaken and slightly damaged to get the animal to come out during the day, was reoccupied the following morning. That morning the Aye-aye that we had disturbed came out of its nest and went back in again almost immediately. It woke at about 6 P.M. as usual and then partially repaired the nest by adding a few new twigs.

It seems to us that the same nest might have been used by two Aye-ayes; this, moreover, was confirmed by an experiment we made in the daytime: one of the Aye-ayes, forced out of its nest, went to the nearest nest, which was occupied by the other Aye-aye, and only returned to its own nest after having been driven away by the owner of the second nest.

A nest, once abandoned, is frequently invaded by ants which are abundant everywhere. It may, moreover, be inhabited by rats, which are very numerous near the villages, or even by another species of nocturnal lemur: *Cheirogaleus major*.

FOOD

The Daubentonia feeds on various fruit and larvae (Petter 1962*a*, 1962*b*; Petter and Petter-Rousseaux 1959). All its nocturnal activity is devoted to finding food; it uses up a surprising amount of energy collecting tiny larvae.

In the vicinity of the villages, fruits provide a very important part of the diet. Coconut is always available and is a fruit which the Aye-aye falls back on as soon as the seasonal fruit have disappeared. The nut is always attacked when it is still on the tree; the Aye-aye always chooses unripe fruit. It is then full of juice, and its soft pulp is only about ½ cm thick. With violent movements of the head, the Aye-aye gnaws and quickly tears off the fiber from a large area; the nut itself is then gnawed. The Aye-aye makes a round hole 3 to 4 cm in diameter with its incisors. This very noisy process takes about 2 minutes. It extracts first the juice, then the pulp—as much as is possible considering the smallness of the hole. To do that, it inserts its third finger, then brings the digit to its mouth, which is always near the hole, with extremely fast movements. In this way it sometimes manages to claw out almost all the pulp, only leaving that opposite the hole. The nut emptied in this way is left attached to the tree, where it may remain 3 or 4 days before it falls. A considerable number of emptied nuts, with the characteristic round hole, are found at the base of coconut trees which have been attacked by Aye-ayes.

During the fruit season, the Aye-aye eats the fruit of litchi and mango trees too. In the latter, it only consumes the ripest part. It often carries the mango between its teeth for a few yards and then, hanging by its feet, it can easily eat the fruit, turning it round and round in its hands.

The Aye-aye spends a considerable part of the night hunting for larvae in old tree trunks. The animal walks slowly along the branches, sometimes hanging from a horizontal branch, scrutinizing every square inch. It is difficult to say if the animal detects the grubs by hearing or by smell: it keeps its nose constantly up against the branch, but when it slows down, it seems to turn its head to one side and put one of its large ears toward the bark (Plate 11.1).

This process of detection along dead branches may bring it almost down to the ground. One night we saw an Aye-aye carefully go around an enclosure, walking slowly along a small wooden fence, about 3 feet high.

When an Aye-aye discovers a larva, it gnaws the bark ferociously with its incisors (Plate 11.2). One it has made a hole, it inserts its third finger (Plate 11.3), which it moves exactly like a pipe cleaner, sometimes turning it, pulling it out and pushing it in many times; then it puts its finger quickly in its mouth before beginning all over again. It seems to squash the larva, bringing out bits to put in its mouth rather than extract the larva whole.

Another tidbit is the parasitic larvae in the stone of *Terminalia* fruit. In January, every evening when the rainfall was not too heavy, at least one Aye-aye was seen among the *Terminalia* fruit, which was still green and about the size of a small apricot.

For several nights, we watched an animal which appeared to eat much of this fruit. In fact, the heap of debris on the ground and closer observation showed us that the Aye-aye opened the fruit and the stone only to eat a larva inside.

The animal walked along the branches of *Terminalia,* picked a fruit with its teeth, and then settled down to gnaw it. Sometimes it hung upside down, holding onto the branches by its hind feet. Holding the fruit in its hands, it first tore off the green pulp from one end and then vigorously gnawed the end of the hard stone inside. This gnawing lasted from 20 to 30 seconds and made a noise which could be heard at quite a distance. This phase was followed by a period of silence, during which we saw the Aye-aye holding the fruit in one hand and thrusting the third finger of the opposite hand into the hole and turning it at the same time. It then put the finger quickly and laterally into its mouth, probably in the space behind the incisors. (The other fingers, especially the fourth and fifth, were bent back and separated from the third.) It repeated the same process two or three times. The period of silence was ended by the noise of fruit falling on the ground, and the animal then went to find another fruit. From the movements of its finger, it would seem that the Aye-aye squashes the larva in the stone and puts it bit by bit in its mouth.

PLATE 11.1.—Aye-aye looking for larvae in a hole under the bark. It focuses its attention, bends its large ears toward the hole, and is about to introduce its third finger.

PLATE 11.2.—An Aye-aye tearing off bark with its powerful incisors

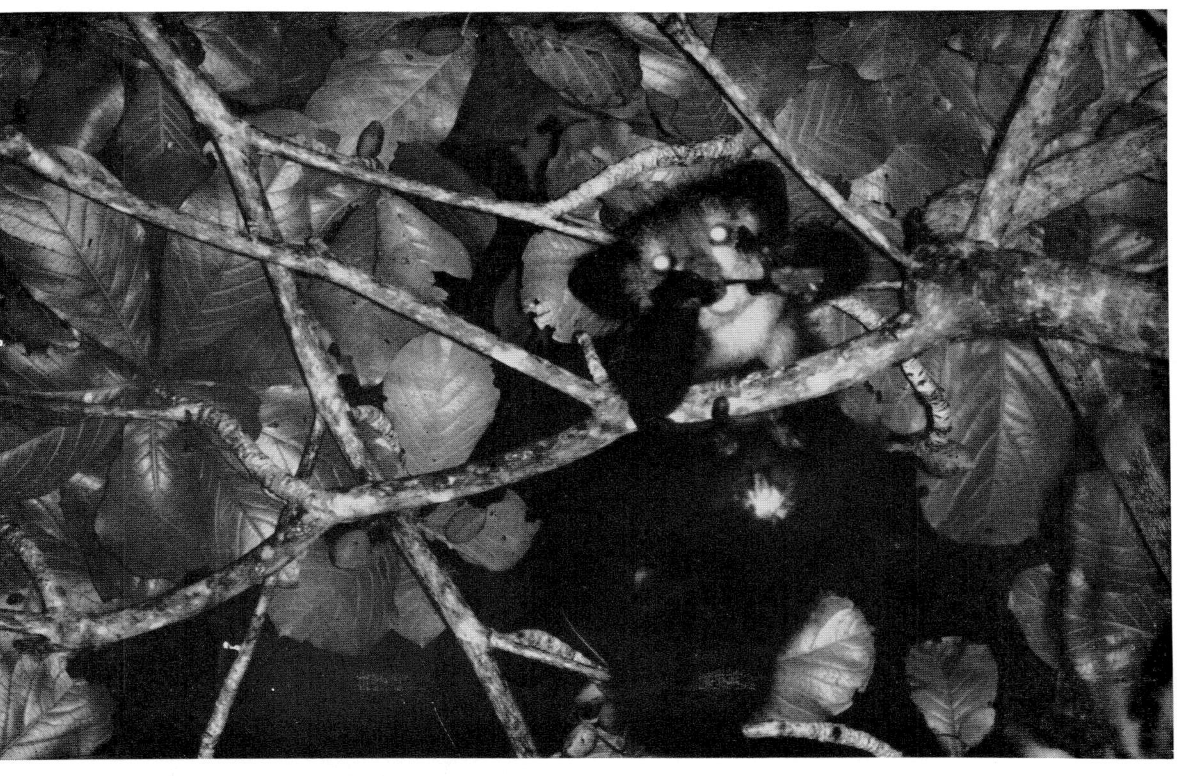

PLATE 11.3.—Aye-aye introducing its third finger into a *Terminalia* fruit

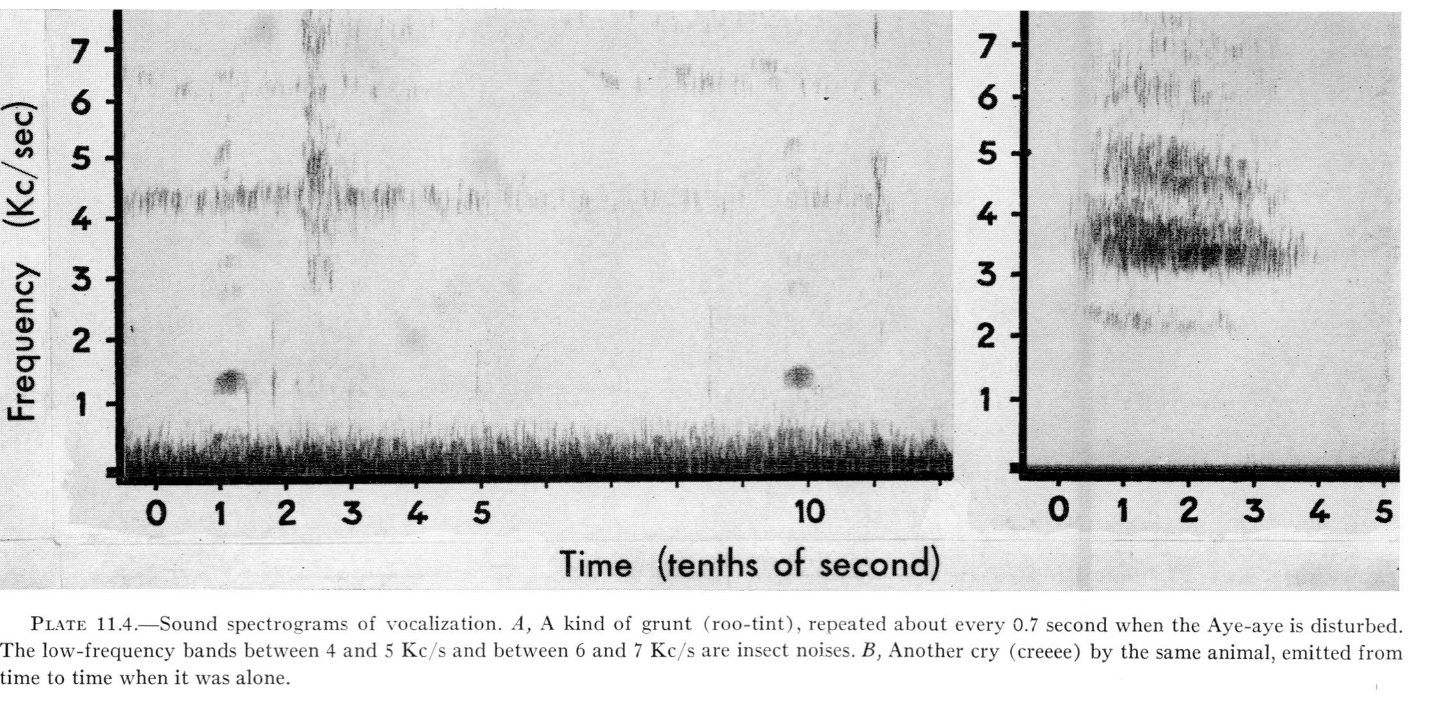

PLATE 11.4.—Sound spectrograms of vocalization. *A*, A kind of grunt (roo-tint), repeated about every 0.7 second when the Aye-aye is disturbed. The low-frequency bands between 4 and 5 Kc/s and between 6 and 7 Kc/s are insect noises. *B*, Another cry (creeee) by the same animal, emitted from time to time when it was alone.

At the foot of the tree there was a heap of fruit gnawed at one end and emptied of their kernels and some untouched fruit which had probably been picked and dropped by the Aye-aye. Although apparently healthy, each fruit contained a small grub, which occupied the space of the kernel, of which only a thin brown skin remained. This larva was about 7 × 3 mm.

We observed larvae torn from at least forty fruit, which seems to represent an amount of work disproportionate to the amount of food collected. The Aye-aye may stay longer than 1 hour in one *Terminalia* tree.

On a number of occasions, we observed an Aye-aye staying a long time in a breadfruit tree. It seemed to suck the leaf buds on the end of twigs. We were unable to find out if it was really interested in the plant itself or in a parasite, as in the *Terminalia* fruit. None of the buds that we collected contained insect larvae, but the time spent in the tree by the animal and the care it took gives the impression that it was looking for grubs.

ACTIVITIES

The Aye-aye leaves its nest regularly at nightfall, if it is not raining too hard. (They came out shortly after 6 P.M. in November, about 6:30 P.M. at the end of December.) It is then still possible to see the animal and follow it for about a quarter of an hour. If the moon is not up, further observation becomes practically impossible, for the animal moves quickly and silently in the treetops. An electric torch is very useful, but its continual use causes the animal to run away. By moonlight, the observations are relatively easy because artificial light is only necessary from time to time (Petter 1959).

The Aye-aye returns to its nest with the same regularity as it leaves, just before daybreak; at the beginning of November, this was about 4 A.M. and at the end of December, 3:30 A.M.

Except during periods of heavy rainfall, we have never observed the Aye-aye motionless for a long period at night, as other nocturnal lemurs are (Petter 1962*b*). When leaving the nest, it often stops for a few minutes on the neighboring branches and often, hanging from a branch with three limbs, scratches its side, armpits, and under its chin for a long time with its remaining limb. Sometimes it busies itself fetching back to the nest some twigs that it carries in its mouth from the same tree or a neighboring tree. Then it goes farther afield to reach some fruit trees, mango or coconut, where it generally has its first meal.

An hour or two later it starts looking for larvae, quickly heading for a *Terminalia* infested with larvae; then it goes on its way to find some more fruit. It explores some old, dead branches, moving slowly forward with its body very close to the tree and throwing its arms outward and forward rather like a person swimming. Finally, before anyone could possibly detect the coming dawn, the Aye-aye goes back to its nest. The nocturnal wandering of the Aye-

aye is rather like a stroll with constant inspection; the animal will take an hour to travel a hundred yards. However, it can move much faster. It is impossible to follow an Aye-aye in the dark, even if there is not much undergrowth.

The animal appears to recognize each branch on its itinerary. Furthermore, it seems to know all the details of the trees it travels on, having several possible routes to go from one point to another. It is probably able to find its way by leaving its scent as the other lemurs do (Petter 1962*b*), but unlike many other lemurs it has no special gland for this purpose. We once saw an Aye-aye moving along a branch and rubbing its anogenital region, rather like *Cheirogaleus*, but more discretely. Unlike the latter, it never leaves any trace of excrement on the branches where it has been.

Contrary to our observations of 1957 (Petter 1962*a*, Petter and Petter-Rousseaux 1959), we rarely heard the cries of the Aye-ayes during this visit. This is probably because the two animals were alone and probably of the same family. The animals were sometimes heard in the evening when they were fairly close together; the call of one was always followed shortly by the call of the other. The call, of short duration and repeated, could be heard at a distance of 100 yards at least and resembles metal sheets being rubbed together (Plate 11.4). In 1957 (Petter 1962*a*, Petter and Peter-Rousseaux 1959), the animal observed in more natural conditions emitted this type of call more frequently, sometimes every half minute, and the call was followed by the call of another in the forest. The only other kind of call ever heard is a type of grunt it emits when disturbed (Plate 11.4).

The only other kind of lemurs seen in this area were *C. major*, which were very numerous. Encounters were quite frequent in the fruit trees. The *Cheirogaleus* seemed somewhat more aggressive than the Aye-ayes and apparently tried to follow them. *Cheirogaleus* did not try to get away, whereas the Aye-ayes sometimes did. However, our presence may explain that.

SUMMARY

To conclude, these observations were carried out in a rather special habitat, plantations and villages, which is not the natural habitat of the Aye-aye. It seems unlikely, however, that observations could be made in better natural conditions at present, since there are very few Aye-ayes.

In the forest, the Aye-aye eats neither mangos nor coconuts, but other fruit, and being less disturbed it no doubt builds fewer nests. The trees in the forest are of different species from those near the villages and have fewer leaves and more branches. Its home range in the forest is limited by its neighbors against the incursions of which it probably defends itself, judging by what we saw in 1957.

The *Daubentonia* seems to be a nocturnal and strictly solitary animal. The simplest explanation of its very simple call is perhaps to avoid contacts with other *Daubentonia* during its wanderings in the forest.

REFERENCES

Petter, J. J. 1959. L'observation des Lémuriens nocturnes dans les forêts de Madagascar. Utilisation des rayons I.R. *Naturaliste malgache* 11 (1–2): 165–73.

———. 1962*a*. Recherches sur l'écologie et l'éthologie des Lémuriens malgaches. *Mem. Mus. Nat. Hist. Nat.* (A)27: 1–146.

———. 1962*b*. Ecological and behavioral studies of Madagascar lemurs in the field. *Ann. N.Y. Acad. Sci.* 102 (Art 2): 267–81.

Petter, J. J., and Petter-Rousseaux, A. 1959. Contribution à l'étude du Aye-aye. *Naturaliste malgache* 11 (1–2): 153–64.

Petter-Rousseaux, A. 1962. Recherches sur la biologie de la reproduction des primates inférieurs. *Mammalia* 26 (Suppl. 1): 1–88.

12

NEWLY ACQUIRED BEHAVIOR AND SOCIAL INTERACTIONS OF JAPANESE MONKEYS

ATSUO TSUMORI

INTRODUCTION

Itani (1958) analyzed the process whereby caramels were gradually introduced to the Takasakiyama troop in 1954 and 1955 and found that the infants and juveniles under 3 years readily accepted the caramels. This candy test indicates that the younger the monkeys are, the more willingly they get used to new situations. His data suggested that Japanese monkeys stagnate in their adaptability to new situations when they reach 3 years old (Fig. 1).

Kawai (1964, 1965) observed the Koshima troop for 10 years. He recorded the adaptive behaviors that the troop members acquired in a new feeding ground. The newly acquired behavior patterns of the Koshima troop were as follows: (1) sweet potato washing, (2) dabbling in water and swimming, (3) "placer-mining" selection of wheat, and (4) give-me-some gesture. The sweet potato washing and "placer-mining" selection of wheat are good examples of the level of practical intelligence of wild Japanese monkeys. Such behavior patterns seem to be smoothly transmitted among individuals in the troop and handed down to the next generation. According to Kawai's remarks, these behavior patterns may be characterized as "precultural." For instance, in 1953 a female first tried potato washing, and by 1962, this behavior had been picked up by all the troop members except newly born infants, 1-year-old infants, and adults more than 12 years old (Fig. 2).

Wheat was also given to the Koshima troop. When the wheat is scattered over a beach, it becomes buried in the sand, and it is difficult for the monkeys to select it. They had been picking up the grains one by one, but recently they have begun to scoop them up with the sand and carry the mixture to the sea or to a stream of water. This is called the "placer-mining" selection method. The

Atsuo Tsumori, Japan Monkey Centre, Institute of Primatology, Aichi-Pref., Japan.

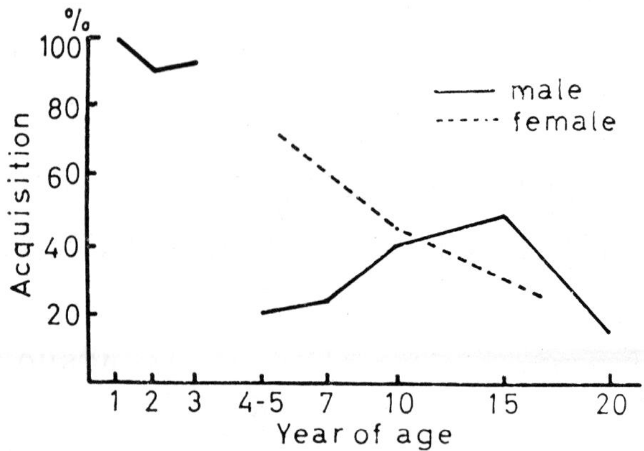

FIG. 1.—Sex difference of acquisition of candy feeding in the Takasakiyama troop in September, 1955 (Itani 1958).

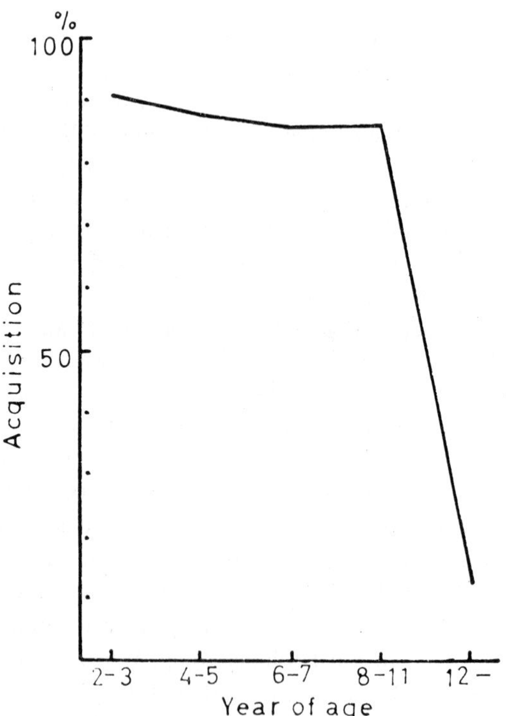

FIG. 2.—Percentage of monkeys that performed sweet potato washing as a function of age in the Koshima troop in August, 1962 (Kawai 1965).

behavior pattern is now spreading among the troop members, and in 1962 it had been acquired by 38.7 per cent of the troop (Fig. 3). In short, 6- to 7-year-old monkeys willingly acquire the "placer-mining" selection of wheat.

THE PURPOSE OF THE RESEARCH

Lively and playful, 2- and 3-year-old Japanese monkeys are highly curious and flexible in their behavior, possessing high imitative ability and adaptability compared with the older ones. But when more complicated tests are given to the monkeys, an accumulation of experiences and understanding abilities are required—just as in the "placer-mining" selection in the Koshima troop. Could it be said that the adult is superior to the infant and juvenile in this respect?

Specifically, the questions are (1) what are the factors which determine the behavior change observed in the problem solving among the Japanese monkeys in a natural troop and (2) what are the ecological or sociological factors in their natural troop that seem to determine the progress of their behavior changes?

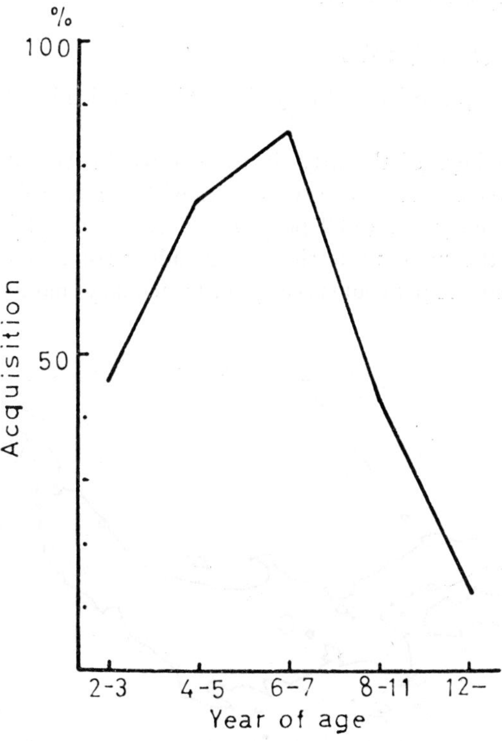

FIG. 3.—Percentage of monkeys that performed "placer-mining" selection of wheat as a function of age in the Koshima troop in August, 1962 (Kawai 1965).

Sand-digging Test

The sand-digging test is a simple test that determines whether a monkey can reach the buried feed. We first dug a hole 6 to 7 cm deep in the sand in front of a monkey and put two or three peanuts in the hole. We then covered the hole and smoothed the sand, keeping the monkey more than 2 m away from the hole during all this time. We then allowed the monkey to approach the hole. The sand-digging test is a kind of delayed-response experiment (Köhler 1921, Yerkes and Yerkes 1929). To reach the peanuts, the monkey has to find an implicit cue instead of an external one; so from this we expected to understand the Japanese money's psychological structure—namely, the symbolic process which plays an important role in problem solving (Plate 12.1).

RESULTS OF THE SAND-DIGGING TEST

The sand-digging test was given to two hundred fifty-seven monkeys in three troops: Koshima troop (Miyazaki Prefecture; Tsumori, Kawai, and Motoyoshi 1965), Takasakiyama troop (Oita Prefecture), and Ohirayama troop (Aichi Prefecture; Fig. 4).

Koshima Troop (I) Original Test

In 1962 the troop consisted of fifty-six monkeys which had been accustomed to peanuts.

Forty-one monkeys of the fifty-six were tested in August, 1962, and the criterion was three consecutive performances with success (Table 1). Eighteen monkeys of the forty-one (43.9 per cent) were successful on the first trial (Table 2), and the success rate rises as age advances. The 6- and 7-year-old monkeys achieved 100 per cent success, but for monkeys more than 8 years old,

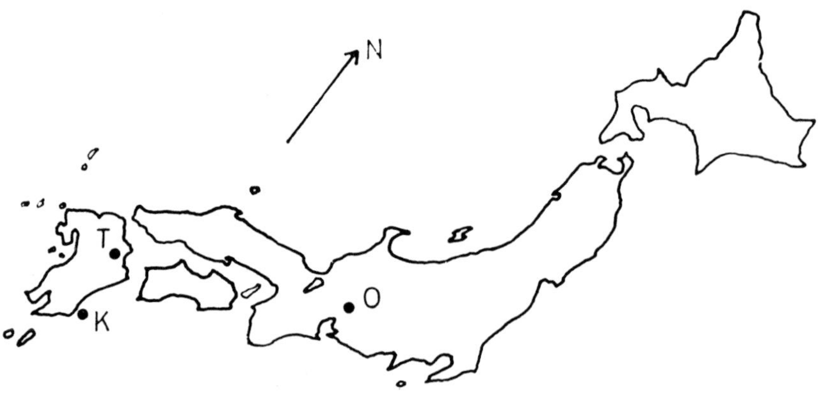

FIG. 4.—Distributions of test troops (K = Koshima troop, O = Ohirayama troop, and T = Takasakiyama troop).

PLATE 12.1.—A female digging for peanuts in sand at Ohirayama. (Aichi Prefecture, 1963)

the success rate dropped, and every one more than 12 years old failed on the first trial (Fig. 5).

Twenty-three monkeys that failed to show a correct response on the first two trials were as a rule given the special training trials, which involved giving some helpful external cues to them. After additional one hundred ninety-two tests, which involved some special training trials, all the monkeys of this troop reached a success rate of 78.0 per cent on the last trial (Fig. 6).

TABLE 1

NUMBER OF TEST SUBJECTS IN THREE TROOPS

Troop	Test Period	No. of Tests	No. of Subjects			Per Cent of Sampling	Training Tests per Subject
			M	F	Total		
Koshima							
I...........	Aug. 13–29, 1962	274	19	22	41	73.2	Several
II...........	Feb. 23–Mar. 27, 1963	132	23	21	44	80.0	3
Takasakiyama..	Dec. 14–19, 1962	127	78	49	127	16.4	1
Ohirayama.....	Feb. 3–20, 1963	258	34	52	86	81.1	3

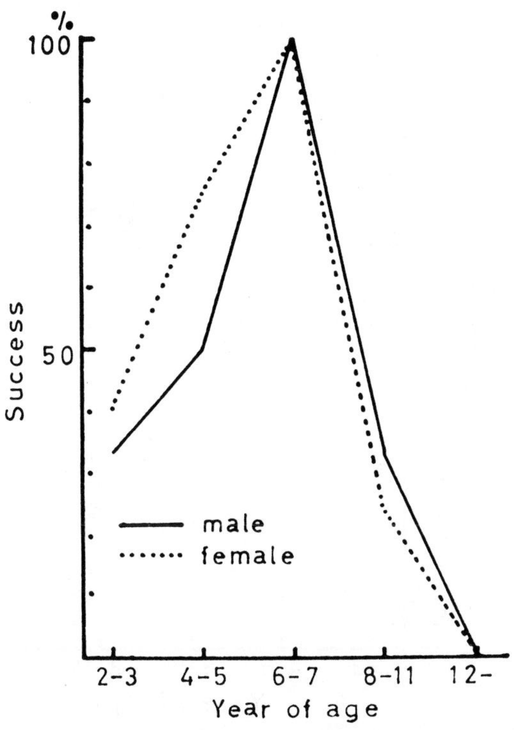

FIG. 5—Percentage of males and females that succeeded on the first sand-digging test in Koshima troop in August, 1962 (Tsumori, Kawai, and Motoyoshi 1965).

The tested monkeys were placed in five categories according to the rate of acquisition (A>B>C>D>E).[1] The results are shown in Figure 7. It must be understood that the progress of the performance level is not parallel to the age advances as shown in the success-rate curve (Figs. 5 and 6).

[1] A = Individuals that succeeded in the first trial; B = Individuals that succeeded in the second trial; C = Individuals that reached the criterion only after having had one special training; D = Individuals that reached the criterion after having had two special trainings; E = Individuals that reached the criterion after having had three or more special trainings and also those that failed in all the trials.

TABLE 2

THE PERCENTAGE OF SUBJECTS OF THE KOSHIMA TROOP THAT PERFORMED THE CORRECT RESPONSE IN EACH TRIAL

Trial	Male	Female	Total
First trial..................	36.8	50.0	43.9
After training.............	68.4	86.4	78.0

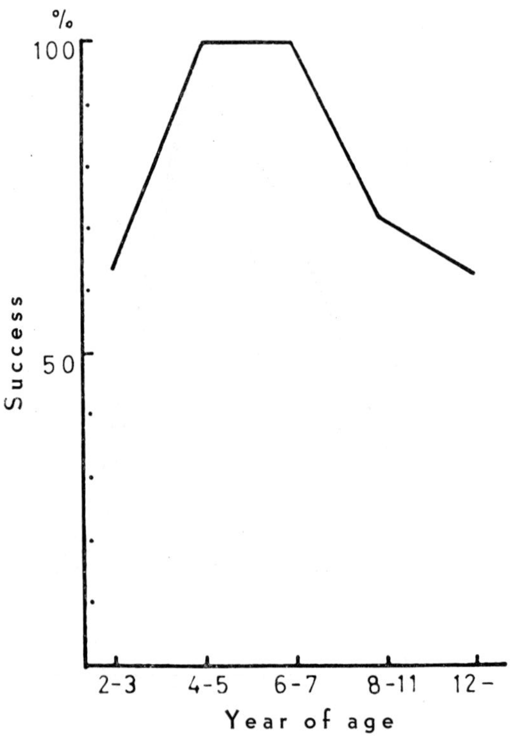

FIG. 6.—Percentage of monkeys that succeeded after the several trainings in the Koshima troop in August, 1962 (Tsumori, Kawai, and Motoyoshi 1965).

Koshima Troop (II) Retention Test

Six months after the original sand-digging test, the test for the retention of digging behavior was given to forty-four monkeys of the fifty-five. Each subject was given three tests (Table 1), and the success rates were as follows: the first trial, 79.5 per cent; the second trial, 75.0 per cent; and the third trial, 81.8 per cent (Table 3). In these numbers of success rates were included that of four solitary males which had not been tested previously. They did not succeed in the test at all. Eighty-seven point five per cent of the monkeys that had participated in the original test succeeded in the first trial, excluding the results of solitary monkeys, and it can be said that the sand-digging behavior acquired 6 months earlier had been latently reinforced rather than forgotten during the rest period (Fig. 8).

Takasakiyama Troop

In 1952, when the troop was first given peanuts, the population numbered two hundred twenty. In December, 1962, when the test was given, the troop was divided into three troops sharing a feeding place; then the total population amounted to seven hundred seventy-one (Itani *et al.* 1963).

TABLE 3

THE PERCENTAGE OF SUBJECTS IN THE KOSHIMA TROOP THAT PERFORMED THE CORRECT RESPONSE IN EACH TRIAL

Trial	Male	Female	Total
First........	60.9 (73.7)	100.0	79.5 (87.5)
Second......	60.9 (73.7)	90.5	75.0 (82.5)
Third.......	65.2 (78.9)	100.0	81.8 (90.0)

( ) shows the value disregarding results from four solitaries who appeared during the survey.

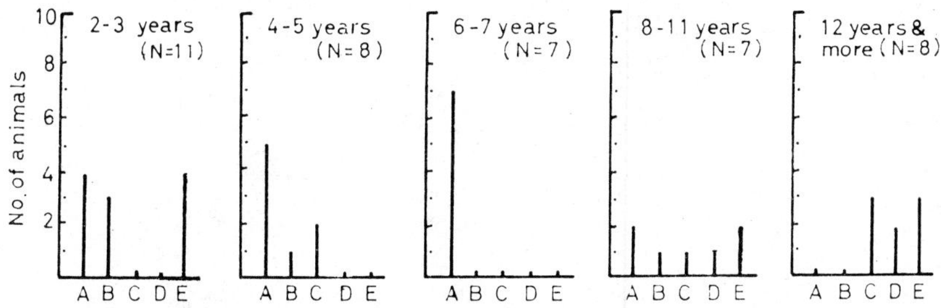

FIG. 7.—Categories representing learning rates. Distribution of monkeys in the "acquisition" categories at each age stage in the Koshima troop in August, 1962. (A is better than B at the sand-digging test learning and so on: B > C > D > E) (Tsumori, Kawai, and Motoyoshi 1965).

The sand-digging test was given once to one hundred twenty-seven individuals, 16.4 per cent of the total. The success rate was 31.5 per cent (Tables 1 and 4). No difference among the three troops was found.

As with the Koshima troop, the percentages dropped suddenly over the age of 8. This tendency was especially noticeable in males (Fig. 9). In contrast with the Koshima troop, the percentages did not improve as age advanced. In particular, the remarkable progress of the monkeys from 6 to 7 years old was not found at all.

Ohirayama Troop

Eighty *Macaca fuscata yakui,* captured separately at Yakushima, were released at Ohirayama in 1957, after their group formation was completed. By 1963, when the first test was given, the troop consisted of one hundred seven monkeys.

TABLE 4

PERCENTAGE OF SUBJECTS OF THE TAKASAKIYAMA TROOP THAT PERFORMED THE CORRECT RESPONSE

Trial	Male	Female	Total
First trial.....	26.9	38.8	31.5

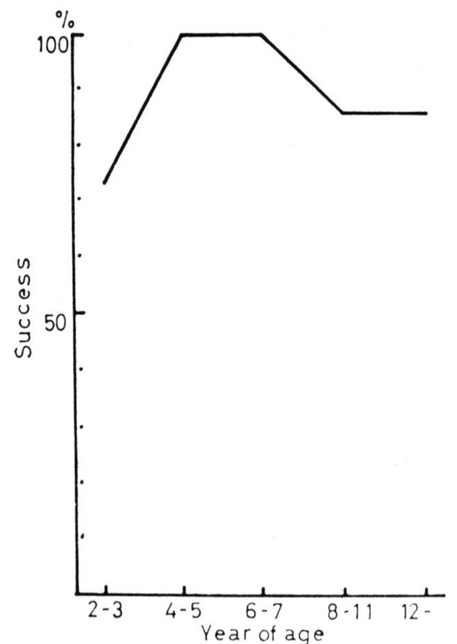

FIG. 8.—Percentage of monkeys that succeeded a half year after the first survey in the Koshima troop (February to March, 1963).

The sand-digging test was given three times to each of eighty-six of the one hundred seven monkeys. Twenty-three point three per cent was the average success rate at the first trial. This was the lowest among the three troops, but the results of the successive trials, 36.0 per cent and 47.7 per cent, show gradual progress (Tables 1 and 5). The percentage of successful monkeys rises as age advances, and 6- to 7-year-old monkeys score 50 per cent; but the rate falls suddenly for individuals more than 8 years old (Fig. 10). The percentage of success according to sex was male, 17.6 per cent, and female, 26.9 per cent. The patterns of their successful curves apparently differ. Males show a Takasakiyama-male type of success pattern, which has no peak at the age of 6 or 7, and females show a Koshima type, which has a peak.

TABLE 5

PERCENTAGE OF SUBJECTS OF THE OHIRAYAMA TROOP THAT PERFORMED THE CORRECT RESPONSE IN EACH TEST TRIAL

Trial	Male	Female	Total
First.	17.6	26.9	23.3
Second.	26.5	42.3	36.0
Third.	38.2	53.8	47.7

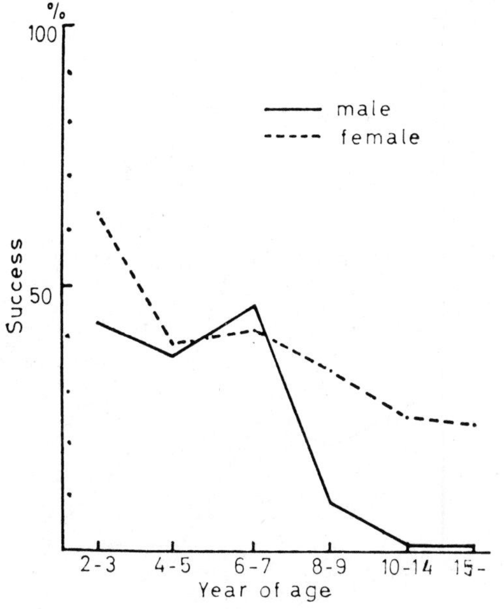

FIG. 9.—Percentage of males and females that succeeded on the first sand-digging test in the Takasakiyama troop in December, 1962.

DISCUSSION

The problem-solving behavior in the sand-digging test consists of four components: the motivational factor directly caused by an incentive (that is, the peanuts); the factor of perceptual differentiation formed when a stimulus object is given; and the factor of ability to retain this cue during the delay period; and the factor of ability to use its hand as a tool to dig the peanuts out of the sand.

From observation of behavior, it seems that there is no difference in the strength of monkeys' motivation by peanuts; and the ability to differentiate the perceptual cue has also been observed, except for the infants; and no difference in the technique of sand digging is observed among the monkeys over 2 years old that were tested. Therefore, the result of the sand-digging test is relevant primarily to the factor of ability of wild Japanese monkeys to retain a cue during the delay period, and it seems to me that this factor implies a symbolic process and accounts to some extent for the apparent differences varying with age.

Candy Test and Sand-digging Test

The greatest difference in the performance level for the candy test was shown between the ages of 3 and 4 (Fig. 1). But generally the difference

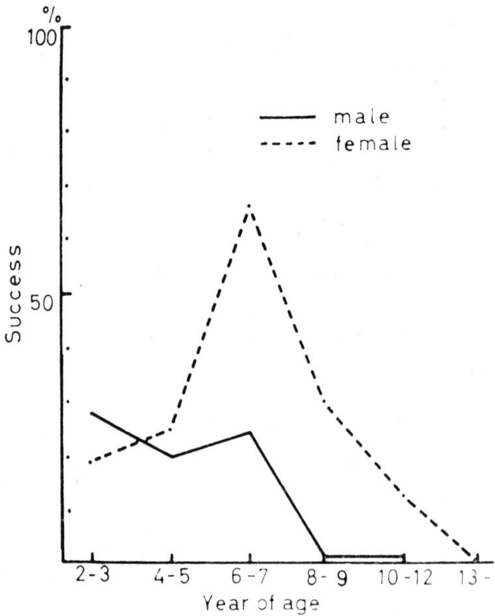

FIG. 10.—Percentage of males and females that succeeded on the first sand-digging test in the Ohirayama troop in February, 1963.

shown in the sand-digging test appears some time after 6 or 7 years of age in the members of every troop. Between the candy test and the sand-digging test there is 3 to 4 years' difference between the stabilization of the results. The disparity seems to be caused by differences in the mechanisms of learning in the two tests.

The habit that is newly acquired through the candy test is the process of conditioning so that a simulus, being unfamiliar and of no value to the monkey, gradually acquires value through curiosity and exploratory reactions. This is the formation of a new attitude. The sand-digging test is learning of a problem-solving type—how to gain the peanuts that already have value to the monkey. Therefore, the result of the candy test can be considered to reflect directly a tendency of wild Japanese monkeys to accept novelties, while the sand-digging test requires an indirect process, or an intellectual process, of forming and retaining the implicit cue mentioned above.

However, the candy test requires an adaptation to new food, and the sand-digging test, an adaptation to a problem situation. Strangely enough, the adaptation to the more complex and difficult situation in the sand-digging test produces better marks from young to advanced ages than does the candy test. Suppose the motivation to seek a hidden object is stronger than the motivation caused directly by an incentive. The former is a process based mainly upon a symbolic process. This problem has not yet been resolved to our satisfaction.

Age Difference

The results of the sand-digging test indicate through the analysis of seeking behavior that as age advances, monkeys make rapid progress up to 6 or 7 years of age in locating the hole, choosing and repeating a helpful behavior to solve the problem, keeping their interest in the problem, and shortening the time required to solve it. Monkeys of 6 or 7 years of age that tried digging but failed made precise, quick, and sharp responses. In many instances, however, they gave up digging within 20 seconds and simply left the test site—they began to lose interest in and motivation for the problem situation (not for the peanuts) by the age of 6 and 7. The decline in the results seen from adolescent to adult is considered to be caused by their losing the motivation which is a main factor in solving the problem. The sand-digging test does not make clear whether the motivation regresses independently or in accord with inferiority of retention of the cue.

Sex Difference

The female is less active than the male. As a result, during the delay period, the male's frequency of intereference to cue retention occurs more often than the female's. During the sand-digging test, all adult females showed more enduring attention to the test situation than the males, even when they failed, except for extremely emotional and unstable individuals. Sex seems not to exist in the cue-producing and cue-retaining processes themselves, but it appears during the motivational processes that support them. Whether a definite answer can be given or not is a future task.

Troop Difference

There are no data proving that the difference in intellectual ability depends upon a troop and a habitat among the species *M. fuscata*. As pointed out before, however, the percentage of success according to troops is the Koshima troop, 43.9 per cent; the Takasakiyama troop, 31.5 per cent; and the Ohirayama troop, 23.3 per cent on the first trial. The pattern of success as a function of age differs according to troops, and may show either the Koshima or the Takasakiyama type of curve. We can assume that some ecological and sociological factors are working differently upon troops, influencing decisively the sand-digging behavior.

Effects of Population and Social Organization

If the population is unusually large, as in the Takasakiyama troop, the tension among individuals also increases, making social organization naturally firm and the inhibition of a troop upon individuals powerful (Itani 1958). The results of the sand-digging test in this troop improve as age advances, but the adolescent does not reach the peak in the curve. One reason for this may be not only the low test density, but also the group pressure that seems to work adversely upon the individual's attitude and behavior in the sand-digging test.

Effects of Kinship and Communication

Members of the Koshima troop are strongly tied by kinship. At the time of feeding, grooming, and sleeping, individuals linked by kinship often flock together. Sons who have left the central part and now belong to the ordinary male class are often seen taking part in kinship group activities. Besides, communication among individuals is especially frequent in a kinship group, and it becomes a main factor in determining the type of social structure of this troop. A similar relationship is observed in the Ohirayama females. Kawai (1964) reports that the condition of acquisition of "precultural" behavior patterns in the Koshima troop differs according to kinship. This tendency can be found in the sand-digging test too. Many cases are recorded at Koshima and Ohirayama in which, through kinship, the seeking and digging behaviors have been imitated and facilitated. Kinship as seen in the Koshima and Ohirayama troops is likely to make communication among individuals of the same kinship group frequent and to provide a social field that promotes seeking and digging behavior.

Effects of Environmental Conditions of Feeding Ground

It is interesting to observe that a successful curve on the sand-digging test can be identified with that of acquisition rate in "placer-mining" selection of wheat in the Koshima troop (Figs. 3 and 5). And, within individuals, a positive correlation can be found between the results of the sand-digging test and such "inventive" behavior in this troop as "placer-mining" selection and potato washing; that is, an individual that has more newly acquired behavior patterns makes a better record in the sand-digging test.

These newly acquired or "inventive" behavior patterns have been observed among a few individuals in the Takasakiyama and Ohirayama troops and other troops, but they have not spread. The sea, the sand, and the deep bushes of Koshima give more variety to its feeding ground than that of other troops. The formation of newly acquired behavior patterns is so closely connected with water and the sand beach that the environmental conditions of Koshima have promoted the acquisition and propagation of these newly acquired behavior patterns among individuals in this troop (Kawai 1964, 1965). The sand-digging test provides the monkeys with advantageous conditions for problem solving because the test is given on the feeding ground in which they have lived. Especially, it should be noted that the digging behavior, which is one of the important components of the problem solving in the sand-digging test, has been acquired by the monkeys that displayed the wheat-washing behavior at the test time.

REFERENCES

Itani, J. 1958. On the acquisition and propagation of new habit in the natural group of the Japanese monkeys at Takasakiyama. *Primates* 1:84–98.

Itani, J., Tokuda, K., Furuya, Y., Kano, K., and Shin, Y. 1963. The social construction of natural troops of Japanese monkeys in Takasakiyama. *Primates* 4:1–42.

Kawai, M. 1964. *Ecology of Japanese monkeys.* Tokyo: Kawade-shobo-shinsha.

———. 1965. Newly-acquired pre-cultural behavior of the natural troop of Japanese monkeys on Koshima islet. *Primates* 6:1–30.

Köhler, W. 1921. *Intelligenzprüfungen an Menschenaffen.* Transl. K. Miya. Tokyo: Iwanami-shoten.

Tsumori, A., Kawai, M., and Motoyoshi, R. 1965. Delayed response of wild Japanese monkeys by the sand-digging test (I)—Case of the Koshima troop. *Primates* 6: 195–212.

Yerkes, R. M., and Yerkes, A. W. 1929. *The great apes.* New Haven: Yale Univ. Press.

13

SOCIAL ORGANIZATION OF HANUMAN LANGURS

YUKIMARU SUGIYAMA

INTRODUCTION

The present study is in some measure the outcome of a long-felt desire to study the social life of monkeys in other parts of the world under natural conditions for comparison with the life of the Japanese monkey *Macaca fuscata* (Itani 1954, 1961; Frisch 1959, 1963; Imanishi 1960).

India is an excellent locale for intensive studies of monkeys living in natural and seminatural conditions (Plate 13.1). In spite of this, few studies of the socioecology of Indian nonhuman primates have been made. These few include studies of the life of the bonnet monkeys (*Macaca radiata*) by Nolte (1955) and the population survey of rhesus monkeys (*Macaca mulatta*) by Southwick, Beg, and Siddiqi (1961*a*, *b*). Jay (1962, 1963*a*, *b*) recently studied the ecology of hanuman langurs (*Presbytis entellus*) for 18 months. This was also the first intensive study of Colobinae in their natural condition, except for the short period of observation of the leaf monkey (*Trachypithecus cristatus*) in Malaya by Furuya (1963). A recent study on langurs by Ripley is included in this volume.

Although my project included an intensive study of both the hanuman langur and the bonnet monkey, this chapter is based on the social life of langurs only. The hanuman langur is common throughout India except in a few places where corresponding species of the same genus occur. The hanuman langur has dark, silver-gray hair on the body, a completely black skin, and a long tail. An adult male may weigh about 11 kg (Plate 13.2). Their habitat extends from high evergreen forests to temple buildings of the arid region.

Our ecological and sociological studies of the hanuman langur in its natural condition were carried out between June of 1961 and April of 1963, mainly at

Yukimaru Sugiyama, Laboratory of Physical Anthropology, Kyoto University, Sakyo-ku, Kyoto, Japan.

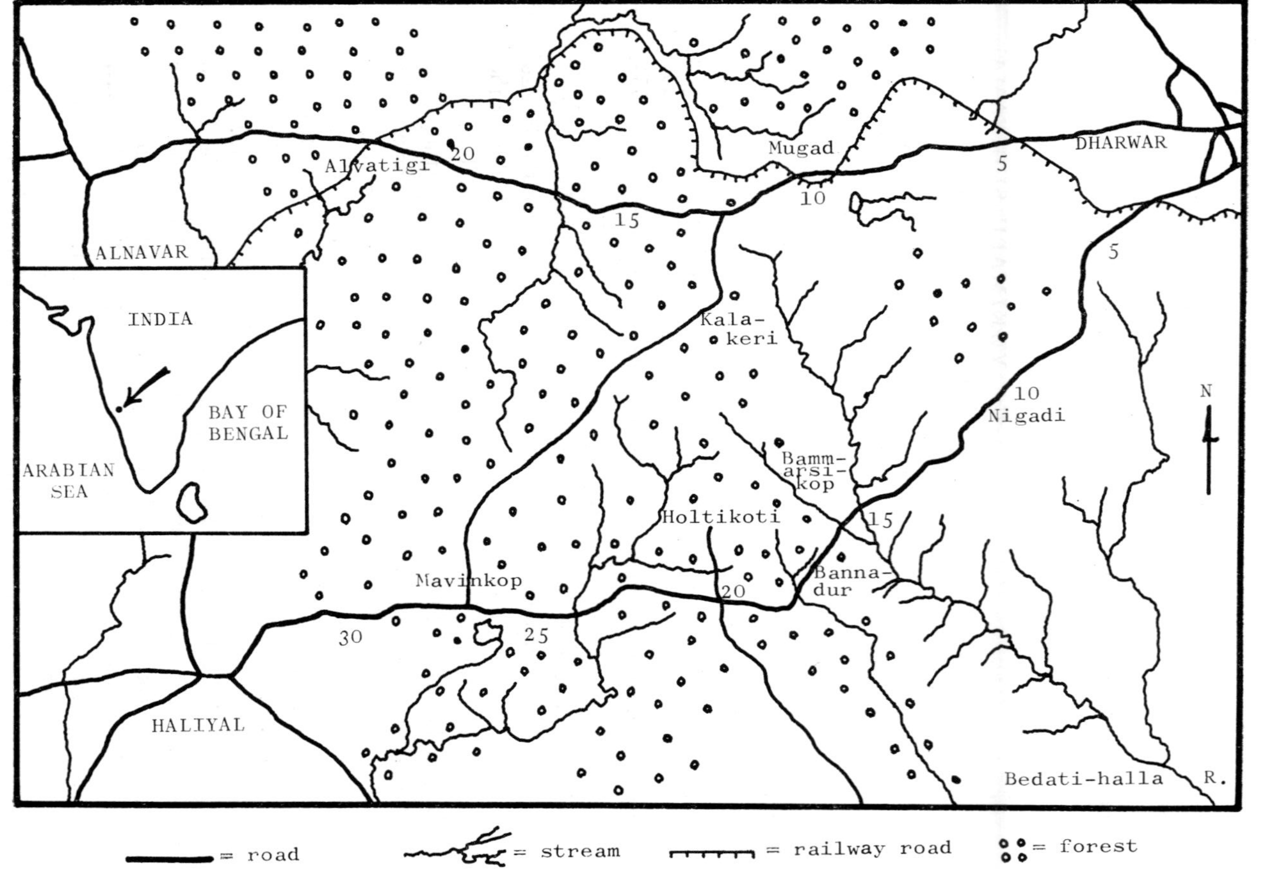

FIG. 1.—Map of Dharwar. ○○/○○ indicates forest; the remainder, except villages, is grassland or cultivated fields. The figures under the road show the distance in kilometers from Dharwar.

PLATE 13.1.—The forest of Dharwar after the rainy season and in the dry season

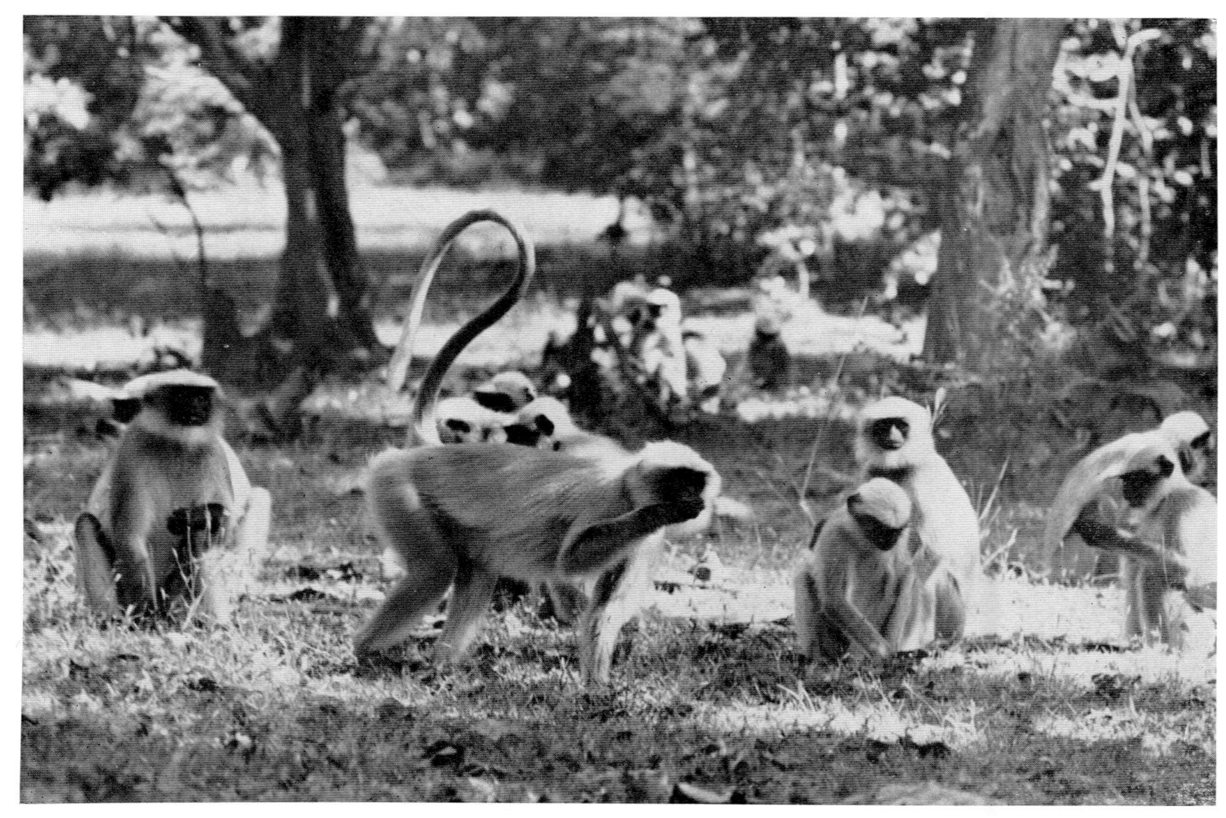

Plate 13.2.—A resting troop of *Presbytis entellus* in the dry deciduous forest of Dharwar

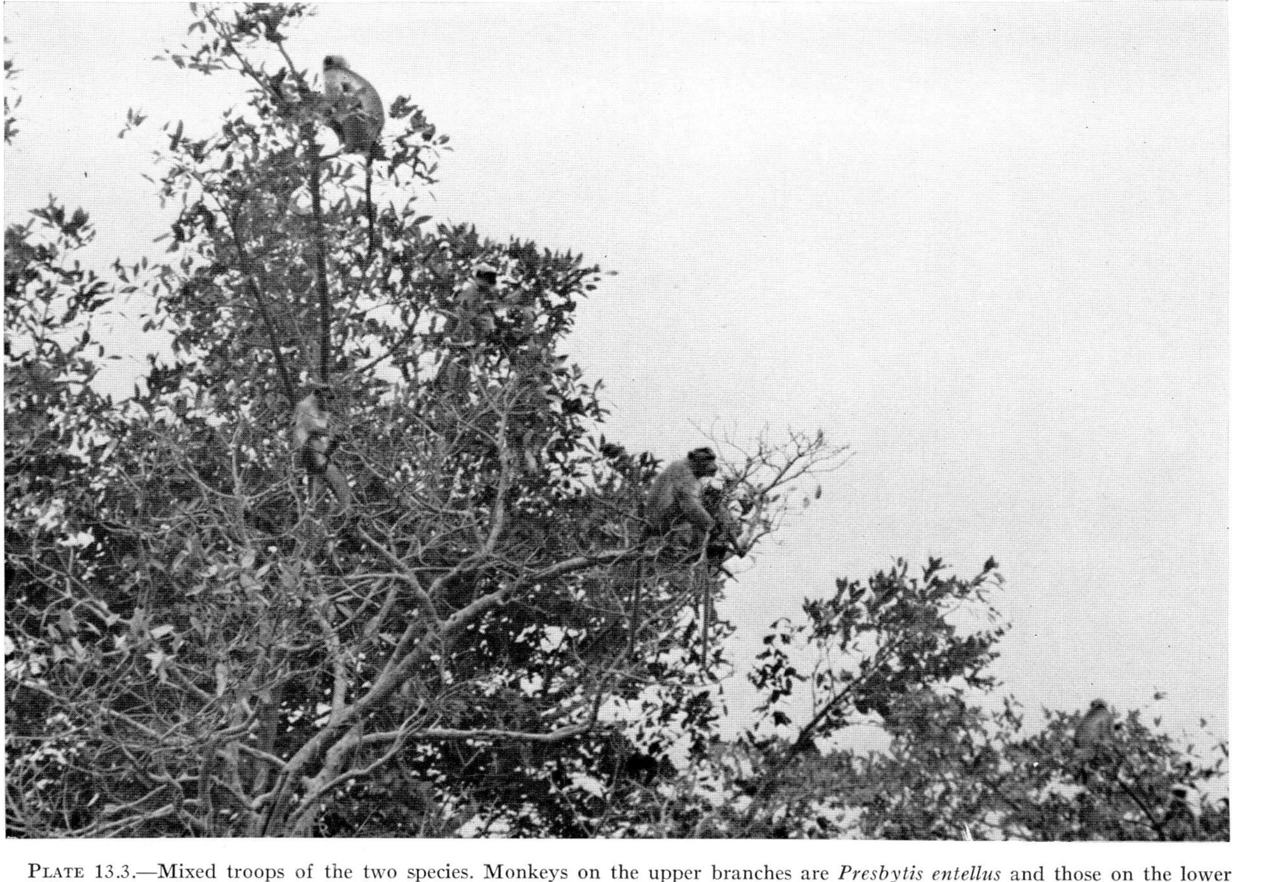

PLATE 13.3.—Mixed troops of the two species. Monkeys on the upper branches are *Presbytis entellus* and those on the lower branches are *Macaca radiata*.

Plate 13.4.—Infant transferring. A young female (*left*) takes a newborn infant. It screams and looks at its mother (*right*), but she does not prevent the kidnaping.

Dharwar, Mysore State, India, to the west of the Western Ghats of the Deccan Peninsula (Fig. 1). The vegetation there is mostly that of the open scrub forest. Between December and April most of the leaves in the forest fall. The highest temperature in the dry season, which is between November and May, is not above 40° C, and the annual rainfall is about 1,000 mm. The main food of the langurs consists of leaves, buds, flowers and the bark of plants. They are not strictly vegetarian; sometimes they eat caterpillars and eggs of insects found on the leaves. About 75 per cent of their day is spent in the trees and the other 25 per cent on the ground.

TROOP COMPOSITION

Forty-four groups along 27.6 km of a road that runs from Dharwar to the west side were recorded (June to September 1961) (Sugiyama 1964). There were two kinds of groups. Thirty-eight groups were bisexual troops (Fig. 2), and the rest were all male groups. Most of the bisexual troops had only one adult male (Fig. 3), but some troops had two, three, or even four adult males. But even the latter had only one large full-grown adult male, the other males being young adults. A few troops had more than five large males but some of these males may not have been permanent members of such troops. The average troop size was 15.1 animals. The average number of adult females in a troop was 8.0 and that of juveniles plus infants was 5.3.

I made a short survey at Raipur, central India, where Jay had studied langur ecology. Sixteen troops and three male groups (one a solitary male)

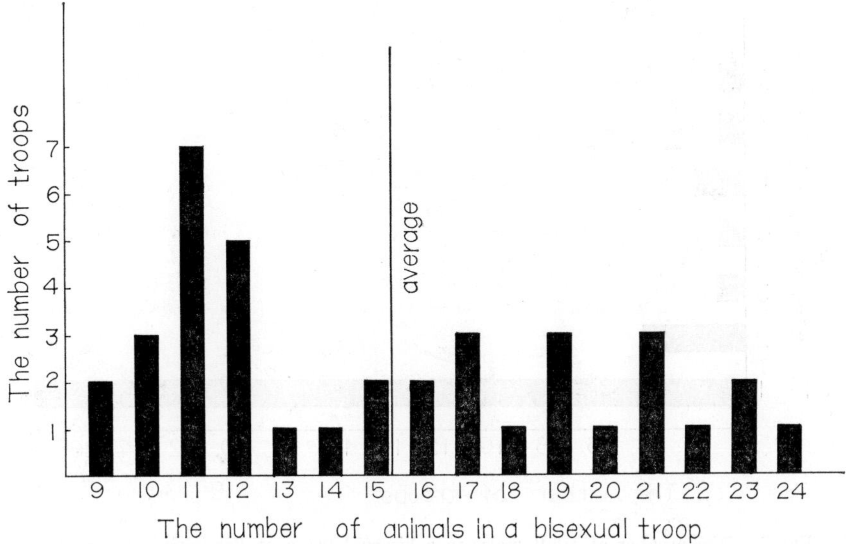

Fig. 2.—Size of the bisexual troop of *Presbytis entellus* (June to September, 1961)

were found. The average troop size was 29.1 animals, and seven troops of sixteen had two or more adult males. If these differences are significant, there must be some differences between the population mechanisms of the langur community of the Dharwar and Raipur areas.

HOME-RANGE AND INTERTROOP RELATIONS

In the second stage of the study, some particular troops 20 to 22 km from Dharwar, along the same road, were selected for more intensive study. All the troops that were selected for this study were followed every day from morning until evening for 73 days, and the exact home ranges of all the troops were identified (Sugiyama, Yoshiba, and Parthasarathy 1965*a*). The home-range size of troops appears to be determined not only by troop size but also by differences in vegetation and the influences of adjacent troops (Fig. 4; Table 1). Each home range can be technically divided into two parts: the area that is used frequently by the troop and the less frequently used parts. Very often home ranges overlap, but the frequently used areas are less affected. When two adjacent troops come close to one another during the course of their daily movement, the leader males of both troops begin to fight. In most cases, such fighting is not severe but is more of a display. The territories average 8.9 hectares. From the data on the forest part of the roadside survey and the above home ranges, the population density of the hanuman langur in Dharwar Forest was about 100 langurs/km.[2]

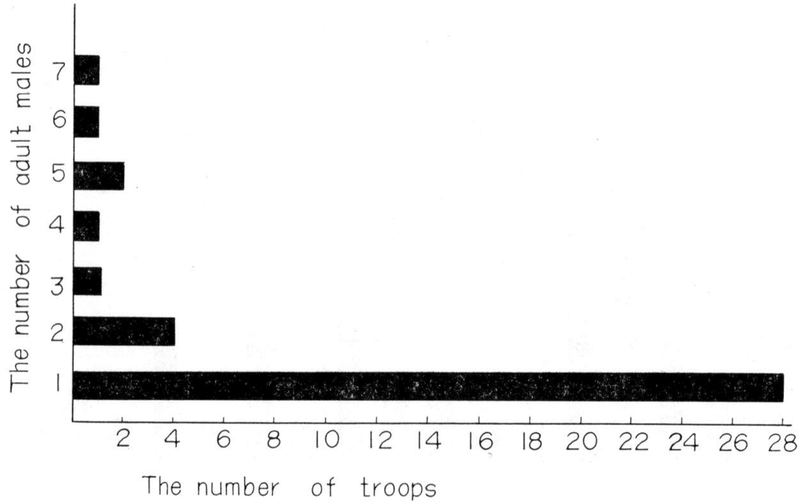

FIG. 3.—The number of adult males in a bisexual troop of *Presbytis entellus* (June to September, 1961).

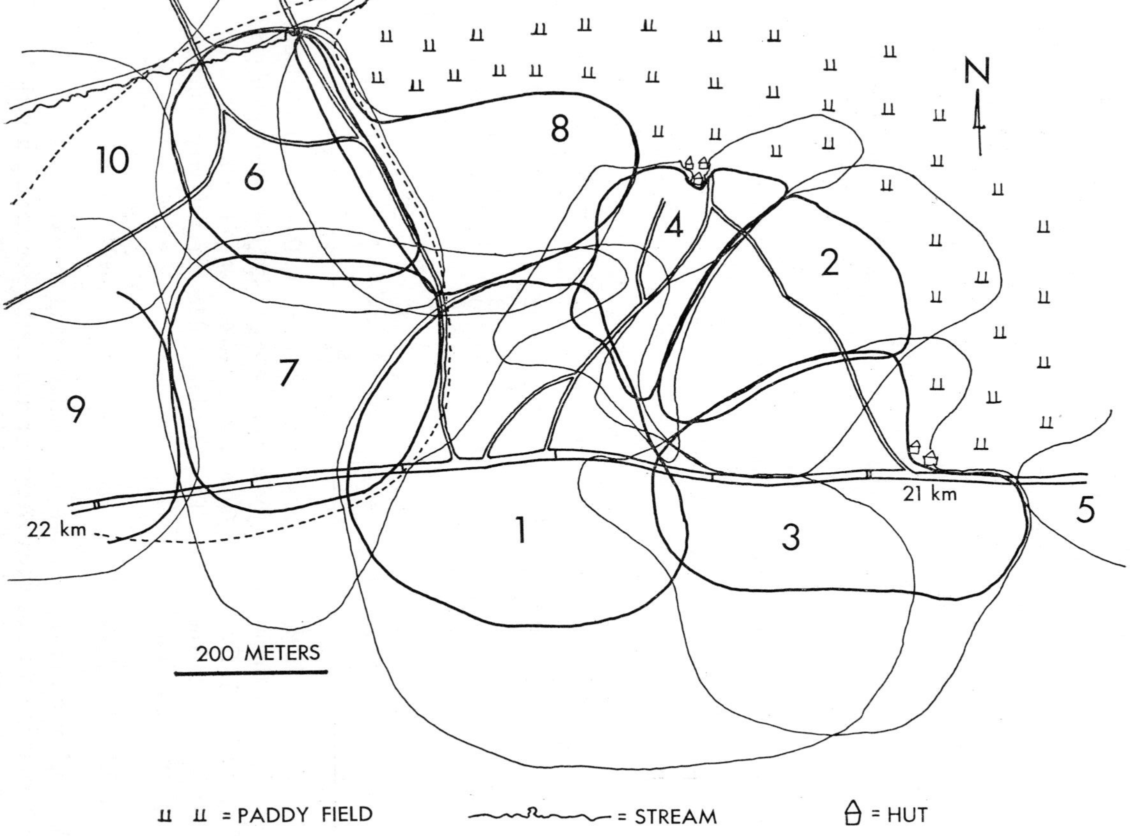

FIG. 4.—The home ranges of some troops of *P. entellus* (October, 1961–March, 1963). The area enclosed by the thicker line is the territory, and that enclosed by the thinner line is the home range of each troop. The number shows the name of the troop. The area enclosed by the broken line is the home range of the male group.

INTERSPECIES RELATION

The same habitat is shared with the bonnet monkeys. But their troops are larger, and their home range is two to four times larger than that of the hanuman langur. In spite of larger troop size, over-all population density of bonnets is lower than that of langurs. Very often intermixed troops of these two species were seen. In spite of this, there was no fighting between them (Plate 13.3). Sometimes troops of both species followed the same routine all through the day, although they seemed to ignore the presence of one another.

Haddow (1952) reported considerable differences in the diets of species of *Colobus* and *Cercopithecus* that inhabit the same areas of Uganda. Bonnets prefer fruits and insects to leaves. By and large, there does not seem to be any

TABLE 1

THE COMPOSITION AND THE AREA OF THE HOME RANGE OF THE EIGHT BISEXUAL TROOPS (NOV., 1961)

Troop No.	Composition (Individuals)					Home Range (Hectares)	
	Full Adult ♂	Young Adult ♂	Full Adult ♀	Juveniles and Infants	Total	Territory	Home Range
I.......	1	2	11	10	24	15.8	31.5
II.......	1		7	9	17	6.5	13.9
III......	1		6	5	12	9.7	18.5
IV......	1		8	1	10	5	10.3
V.......	1		8	11	20		
VI......	1		6	3	10	8	12.5
VII......	1		13	2	16	10.4	19
VIII....	1		11	4	16	7.1	11.9
Average	1	0.3	8.8	5.6	15.6	8.9	16.8

marked difference in the food habits of the two species in this area. In the high forest of the southern hills of India, bonnets live with lion-tailed monkeys (*Macaca silenus*). Even in the same tree two troops, one of each species, do not react much to each other. Their feeding habits are not very different (Sugiyama, unpublished).

NEWBORN INFANTS

Although you can see newborn infants throughout the year, most of those at Dharwar are born between December and April. The gestation period is about 6 or 7 months, judging by my field observations (Sugiyama, Yoshiba, and Parthasarathy 1965*a*) and Hill's laboratory data (1936), and so mating must take place between May and October. The peaks of mating and birth in any one year may differ from place to place (Jay 1963*b*), but the existence of two discrete birth peaks a year at one locale (Prakash 1962, Southwick *et al.* 1961*b*) is not consistent with my observations.

The newborn langur infant has black hair and a pale pink skin. Before the infants begin to walk by themselves, many females that have no infant, and even subadult females, take the infant from their mothers (Plate 13.4). The mothers do not resist, even if the infant is taken by a member of another troop or is only a few hours old and screams violently. This strange phenomenon has also been observed by Jay (1963*b*). In Japanese monkeys and some other species, the mother rarely parts with her infant, although many subadult females attempt kidnaping. When an infant is between 1 and 5 months old, its color changes to that of the juveniles. Three months after birth the infants can play vigorously and respond to other infants. After 6 months they begin to eat leaves, imitating the behavior of their mothers. It probably takes 5 to 6 years for maturation (Sugiyama 1965*b*).

SOCIAL ORGANIZATION

For a study of the social organization of Japanese monkeys in their natural condition, we have adopted three fundamental methods: (1) individual identification; (2) continuous observation for a long period, at least 1 year; and (3) artificial feeding to make observation easier. In many places in India, the third method was not essential, since one could approach the monkeys very closely without disturbing them much.

In contrast to the macaques, the hanuman langurs exihibit neither a strict, functional dominance hierarchy nor a differentiation in their social organization into a central and a peripheral part. Apart from the fact that one large adult male leads the troop, there is no other evident social differentiation.

On the other hand, the male group shows less rigidity in its organization and no functional dominance hierarchy. Theirs is a loose gathering. Sometimes they cohere. At other times the group may split into three or four subdivisions. Sometimes nearly sixty animals were found moving together, and at other times only two were seen. The male groups move over a range that is three or four times larger than that of a bisexual troop, but their moving ranges may overlap that of a troop. Even if a male group has more than five adult males, the single leader male of a bisexual troop usually is able to chase away an entire male group by becoming highly aggressive. On the other hand, the same leader male is less aggressive when he clashes with any adjacent bisexual troop.

SOCIAL CHANGE I: DIVISION OF A TROOP

Many instances of changes in langur society have been observed. The first example relates to the division of a troop (Fig. 5). On August 17, 1961, a solitary male who might have been in a male group came in contact with a troop (Troop 40), which consisted of 1 A ♂ + 6 A ♀ + 5 J ♀ + 2 I = 14.

The solitary male, who was slightly stronger than the leader male of Troop 40, fought and defeated the latter. Later, in spite of many severe injuries suffered by both of them, the solitary male decamped with most of the members of the Troop 40, except the leader male and an adult female. But within a few days three adult females and a juvenile female came back to their former leader and thereafter remained in their original home range (Sugiyama 1964).

SOCIAL CHANGE II: RECONSTRUCTION

The second example deals with the reconstruction of a troop (Fig. 6). There was a large bisexual troop (Troop 30), which consisted of 1 A♂ + 6 J♂ + 9 A♀ + 3 J♀ + 5 I = 24. There was also a male group which consisted of 6 large A♂ + 1 young A♂ = 7. Though their home ranges did not overlap when they met on May 31, 1962, the male group rushed to attack Troop 30 and inflicted severe injuries on its leader male. Soon after this clash, a few

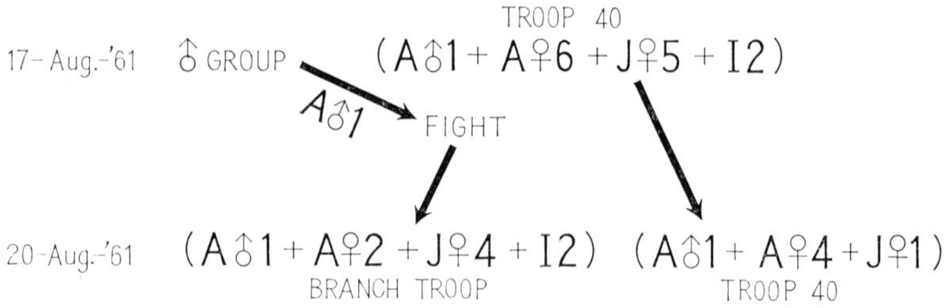

TROOP 40
17-Aug.-'61 ♂ GROUP (A♂1 + A♀6 + J♀5 + I2)
A♂1
FIGHT
20-Aug.-'61 (A♂1 + A♀2 + J♀4 + I2) (A♂1 + A♀4 + J♀1)
BRANCH TROOP
TROOP 40

FIG. 5.—Social change of Troop 40. A♂, A♀ = Adult male(s) and female(s). J = Subadult(s) and juvenile(s) of more than 1 year old. I = Infant(s) of less than 1 year old. Tr. = Troop.

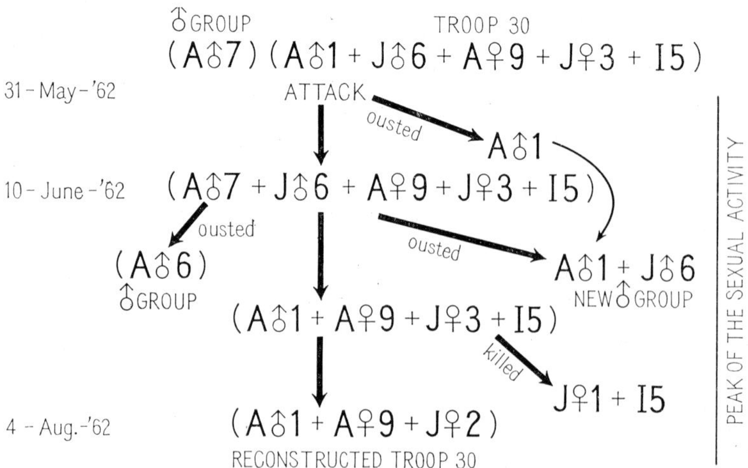

FIG. 6.—Social change in Troop 30

females chose to follow the victorious males of the male group. But the conflict was not over. Next day the male group came back and fought again and again with Troop 30. This type of fighting continued for 10 days, until not only the injured and weakened leader male but also other male juveniles including even a 1-year-old male were ousted from their home range. All the females, including the mother of the ousted 1-year-old male juvenile stayed with the attacking male group. In the following week, six males of the male group were also ousted. Within about 2 months after this event, one juvenile female and all five infants were bitten by the strongest male, the new leader. These unfortunate infants did not get the expected protection from their respective mothers. Finally all that remained were 1 strong A ♂ + 9 A ♀ + 2 J ♀ = 12 in the reconstructed troop. From the beginning of June, sexual behavior involving the new male was observed. The sexual activity soon reached the peak period, and it lasted until the beginning of August. Most of the females delivered their new babies in January and February of the next year (Sugiyama 1965*c*).

SOCIAL CHANGE III: ARTIFICIAL CHANGE

On June 20, 1962, the only adult male of Troop 2 (Fig. 4; Table 1) was removed artificially as an experimental measure to see if there would be any change in the social organization of the troop (Fig. 7). No male from any male group tried to join this "female group" which consisted of 5 A ♀ + 9 J ♀ + 4 I = 18. For several days after the experiment, no change in their daily activity was discernible, and no monkey exhibited fear as a result of the ab-

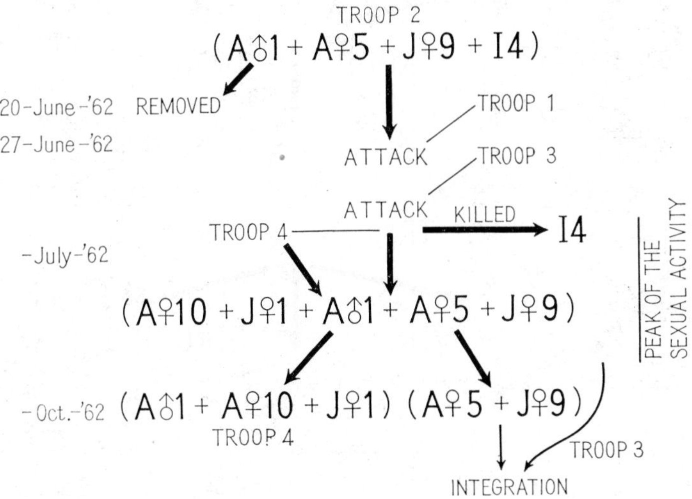

FIG. 7.—Artificial social change in Troop 2

sence of the dominant male overlord who had protected and led the troop. Also, no member deserted the troop. One week after the removal of this male, leader males of adjacent troops recognized that there was no male in Troop 2 and attacked it. The most severe attack was carried out by the adult male of Troop 4, who bit all four infants of Troop 2; the infants succumbed to their injuries. The mothers of these infants literally deserted them during the attack. After this incident, most of the females, including mothers who had lost their infants, copulated with their new male, as had the females of Troop 30 described above. The male of Troop 4 became the leader of Troop 2 and tried to integrate troops 2 and 4, but females of Troop 2 persisted in moving only within their own range, as before, and females of Troop 4 did likewise. Finally the male, failing to integrate these two troops, abandoned the attempt and returned to his original Troop 4. After October another adult male, that of Troop 3, joined Troop 2 and in March, 1963, succeeded in integrating it with his own troop (Sugiyama 1966).

SOCIAL CHANGE IV: CHAIN ACCIDENT[1]

In July, 1962, part of a large male group, which consisted of more than fifty-nine animals living in the scrub forest 20 to 22 km from Dharwar, attacked Troop 5 (Fig 8). The members of the male group were far stronger

[1] Some of the data for this section are taken from the unpublished field notebooks of Dr. S. Kawamura and Mr. K. Yoshiba.

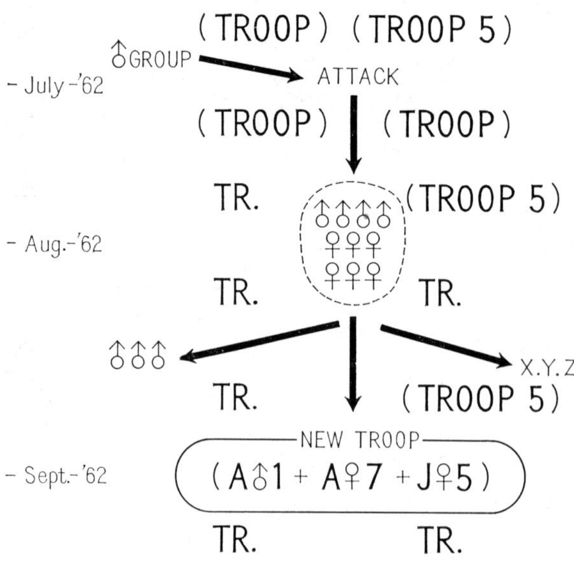

Fig. 8.—Chain social change that occurred in Troop 5 and others

than the leader male of Troop 5. He did not fight, but retreated. When the males moved, some females, including juveniles, joined them. Every day these males attacked not only Troop 5 but also three or four troops which had their home ranges near Troop 5. The attack was repeated on the troops, and females from each troop came to join the males. Finally they established a new troop with a home range south of Troop 5. After the new troop was established, all the males, except one who held leader status in the new troop, deserted it. Mr. Yoshiba, who was observing Troop 5, recognized that ♂ X, ♂ Y, and ♂ Z deserted the troop on or about September 20.

Troop 7 was living about 1 km west of Troop 5. Early in October, ♂ X, ♂ Y, and ♂ Z attacked it and ousted its leader (Fig. 9). For about 1 month these three males governed Troop 7 cooperatively, but early in November, when sexual activity came to its peak, ♂ Y and ♂ Z were ousted from the reconstructed Troop 7. The abdicated leader as well as ♂ Y and ♂ Z attempted to rejoin Troop 7, but they did not succeed. Finally the abdicated leader deserted the original area.

The adjacent troop to the east of Troop 7 was Troop 1, which had three adult males. ♂ Y and ♂ Z attacked Troop 1 and ousted all adult males in early March, 1963 (Fig. 10). When we left India in early April, 1963, two males had killed all newborn infants and were reigning over the reconstructed Troop 1.

DISCUSSION

Excluding one artificial experiment, four out of nine particular troops which we observed for more than 1 year were seized by natural social changes. That is to say, a troop was seized by such social change every 2 years and 3 months on the average. According to my data from the 3-month, roadside survey in the

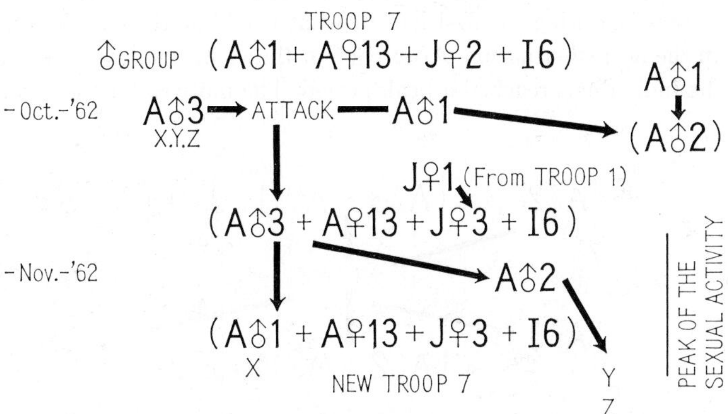

Fɪɢ. 9.—Chain social change that occurred in Troop 7

first stage of the project, a troop was seized by social changes every 3 years, on the average. In most of the troops that I observed, the juvenile males were all less than 2 or 3 years old. If the case of Troop 30 is typical, we can say that these replacements of the leader male lead to the expulsion or death of all males, including juveniles and infants, from the original troop. On this basis, it seems likely that the leader male of most troops that have only male juveniles of less than 2 or 3 years old in Dharwar area has been ruling the troop for less than 3 or 4 years. Whether an abdicated leader who joins a male group has any chance of being reinstalled as the leader of a troop is a problem still to be solved.

In spite of the severe antagonism among males, which we have seen in many cases of social change, a father tolerates his male children. Even after they mature and copulate with the females of the troop, no antagonism was observed between them, and they cooperatively threaten adjacent troops. In the rare cases of multimale troops in Dharwar area, such as in Troop 1, the males may consist of a father and his son. Multimale troops in central India presumably are similar; the principle of social organization in those areas is not very different.

What, then, causes the social difference between the south Indian type and central and north Indian type? There may be a little difference in their environment, adaptation, and population density. There is no census of the population density of the langur in central or north India, but the density in Dharwar is rather high. The high population density causes many males to wander out of the troops and the lack of unoccupied ground in a good habitat makes it difficult for them to establish a new territory. This leads to frequent attacks on the troop by "dissatisfied" males and blocks the increase of troop size.

All attacks that we observed by males against troops, except in the second and the third cases of the chain accident, began between the end of May and August—that is, during the first half of what would be the peak of the mating season in the normal situation. After each of theses incidents, the sexual activities of the individuals reached a peak period. The motives for the formation or

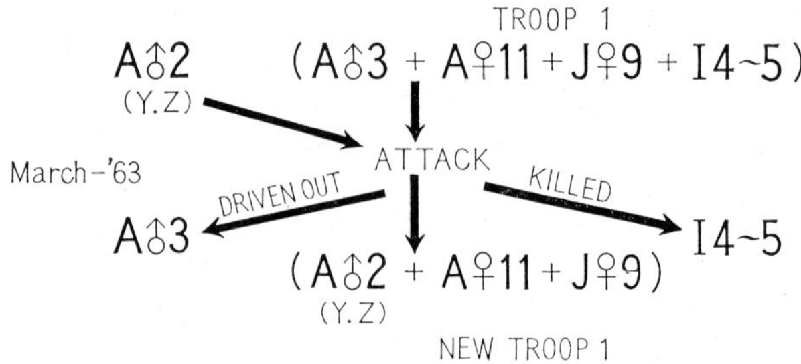

FIG. 10.—Chain social change that occurred in Troop 1

the reconstruction of a troop can be related to the sexual urges of the males of the male group. At the same time these cases show that social stimuli, such as severe fighting among males, change of troop leader, or a severe attack with killing of infants activate the sexual activity of females; in other words, social stimuli play a direct part in activating sexual activity. The artificial experiment was also followed by a peak of sexual activity just after the severe attack by the invading male and, even in Troop 7, active but unseasonable sexual behavior was observed after the incident in November. In our 15 years of observing Japanese monkeys, among which there are few such drastic social stimuli as for the langur, we have never seen active sexual behavior correlated with social changes in the wild.

Why does the new leader male bite all the infants? According to the line of reasoning presented above, the fact that the new leader male bit all the infants (except in Troop 7) and the fact that many of the females who had lost their infants grew excited and copulated with the new male can be correlated. Because a female langur usually delivers an infant every 2 or 3 years unless she loses her infant, loss of the infant has the effect of advancing the estrus of the female. In addition, this astonishing attack may show to the females of the troop the leader's power for organizing them without injury to themselves. Such infant biting is never seen among the macaques, which have a large, well-organized troop composed of many males, many females, and their children.

Now, we can present the following hypothesis on the motive of infant biting, though there is no active evidence that a langur male recognizes his own baby from that of others. That the mother does not bestow any special care on her injured infant, and even deserts it, and that all of the mothers hand over their newborn infants to other females without any hesitation must be related to one of the basic aspects of their psychology. Neither of these behavior patterns is seen in Japanese monkeys. The langur infants have many opportunities to contact most members of their troop, even the leader male, from the time they are a few hours old—a situation which is never seen in the Japanese monkey troop—and this pattern of social behavior must influence the formation of their personality. Further sociopsychological investigations of the difference between macaques and langurs on the formation of personality and the process of socialization during the growth time are needed.

There are many differences in these sociological phenomena between Japanese monkeys and other macaques which have a large and highly organized troop structure. The functional dominance hierarchy among adult males and the spatial differentiation in the central and the peripheral parts of the Japanese monkey troop cause them to coexist in a troop and make the troop stronger and more stable. With Japanese monkeys there is no male group, but rather, there are solitary males who are not members of any troop. The troop is too strong for them to attack. The formation of a new troop after an increase in the population of the original troop was seen only a few times in the 10 years after the study of Japanese monkeys was begun (Sugiyama 1960, Furuya 1960). On the other hand, in the langur troop a rigid dominance hierarchy is not adopted, and antagonism among males is exposed without any regulating mechanism. These differences may help to explain why langurs have a relatively small, unstable group.

The macaque troop is closed as well as strong. In these animals intertroop genetic exchange does not occur, except when a solitary male gains admission to a troop. On the other hand, in the langur society, frequent change of troop leader effects a frequent and considerable genetic interchange.

These characteristics of langur society may be adaptations for the maintenance and the development of the species. Generally speaking, primates dwelling in open country, such as macaques and most baboons, have large, strong, and highly organized groups and forest-dwelling primates, such as gibbons, have a small group with simple organization. Although the hanuman langur in India has become semiterrestrial and is using open ground too, its anatomical character is more arboreal. Sociologically, langur society became distinct from the macaque's way of life, which exploited the advantages of group life as well as the gibbon-like society, and this in turn solved the antagonism among males by creating the one-male-one-female group. In the gibbon society a male and a female with their infants make a group, and there is no sexually dissatisfied nor socially unstable adult male (Carpenter 1940).

Kummer and Kurt (1963) reported one-male groups among hamadryas baboons of the Ethiopian highlands. The difference between the baboon's one-male group and that of the langur is that there is less territoriality and antagonism among one-male groups of baboons, and at dusk many one-male groups come together to sleep. In spite of the fact that they choose the one-male group, they may have a supergroup society. There must be four or five times as many dissatisfied males as group males in their community, because a group male has four or five females in his group, and some males may live out of the group—for example, as male groups or solitary males.

The gorilla also has a small group with few large males. Groups of gorillas, like those of the chimpanzee, are nonterritorial and open to visitors (Kawai and Mizuhara 1962; Schaller 1963). In chimpanzee society the very existence of a group with a fixed membership is doubtful (Azuma and Toyoshima 1961–62, Goodall 1963, Reynolds 1963). Coexistence of the open, small group and the larger society (community) presumably is similar to human society in its early stages and different from the principle of the langur's society.

ACKNOWLEDGMENTS

One of our plans was realized with the help of Dr. Harold Trapido, Virus Research Centre, Poona, India, and with the financial help of a Rockefeller Foundation Grant for Primatological Studies in India (RF60229). Acknowledgments are also due to the Ministry of Scientific Research and Cultural Affairs, Government of India; Department of Forests, Mysore State; Dr. Salim Ali and the staff of Bombay Natural History Society; Dr. C. Anderson and the staff of Virus Research Centre, Poona; Central College, Bangalore and Karnatak University, Dharwar, for their interest in this work.

Particular thanks are due to my colleagues Dr. Syunzo Kawamura, Mr. Kenji Yoshiba, and Dr. M. D. Parthasarathy, who cooperated with me throughout this study in India. I wish to express my hearty thanks to Emeritus Professor Denzaburo Miyadi for his supervision of this work, and to Professor Kinji Imanishi, Dr. Junichiro Itani, the staff of the laboratory of Physical Anthropology, University of Kyoto, and the staff of Japan Monkey Centre, Inuyama, for their kind criticisms and assistance. Acknowledgment is also extended to Dr. S. A. Altmann for his kind help in preparing the manuscript.

REFERENCES

Azuma, S., and Toyoshima, A. 1961–62. Progress report of the survey of Chimpanzees in their natural habitat, Kabogo point area. Tanganyika. *Primates* 3(2):61–70.

Carpenter, C. R. 1940. A field study in Siam of the behavior and social relations of the gibbon. *Comp. Psych. Monogr.* 16(5):1–212.

Frisch, J. E. 1959. Research on primate behavior in Japan. *Am. Anthropol.* 61(4):584–96.

———. 1963. Japan's contribution to modern anthropology. In *Studies in Japanese culture,* ed. J. Roggendorf, pp. 225–44. Tokyo: Sophia University.

Furuya, Y. 1960. An example of fission of a natural troop of Japanese monkeys at Gagyusan. *Primates* 2(2):149–79.

———. 1963. The social life of silvered leaf monkeys. *Ibid.* 3(2):41–60.

Goodall, J. 1963. My life among wild chimpanzees. *National Geog. Mag.* 124(2): 272–308.

Haddow, A. J. 1952. Field and laboratory studies on an African monkey, *Cercopithecus ascanius Schmidti* Matschie. *Proc. Zool. Soc. London* 122(2):297–394.

Hill, W. C. O. 1936. *Ceylon J. Sci.* (20):369–89. (Quoted in Asdell, S. A. *Patterns of mammalian reproduction.* 1946. Ithaca: Cornell University Press.)

Imanishi, K. 1960. Social organization of subhuman primates in their natural habitat. *Current Anthropol.* 1(56):393–407.

Itani, J. 1954. *Takasaki-yama no Saru* (Monkeys of Mt. Takasaki). Tokyo: Kobunsha. (In Japanese.)

———. 1961. The society of Japanese monkey. *Jap. Quart.* 8(4):421–30.

Jay, P. C. 1962. Aspects of maternal behavior among langurs. *Ann. N.Y. Acad. Sci.* 102(2):468–76.

———. 1963*a*. The Indian langur monkey (*Presbytis entellus*). In *Primate social behavior,* ed. C. H. Southwick, pp. 114–230. Princeton: Van Nostrand.

———. 1963*b*. Mother–infant relations in free–ranging langurs. In *Maternal behavior in mammals,* ed. H. L. Rheingold, pp. 282–304. New York: Wiley.

Kawai, M., and Mizuhara, H. 1962. An ecological study on the wild mountain gorilla. *Primates* 2(1):1–42.

Kummer, H., and Kurt, F. 1963. Social units of a free-living population of Hamadryas baboon. *Folia Primatol.* 1(1):4–19.

Nolte, A. 1955. Field observations on the daily routine and social behavior of common Indian monkeys, with special reference to the bonnet monkey. *J. Bomb. Nat. Hist. Soc.* 55(2):177–84.

Prakash, I. 1962. Group organization, sexual behavior and breeding season of certain Indian monkeys. *Jap. J. Ecol.* 12(3):83–86.

Reynolds, V. 1963. An outline of the behaviour and social organisation of forest-living chimpanzees. *Folia Primat.* 1(2):95–102.

Schaller, G. B. 1963. *The mountain gorilla: ecology and behavior.* Chicago: University of Chicago Press.

Southwick, C. H., Beg, M. A., and Siddiqi, M. R. 1961*a*. A population survey of rhesus monkeys in villages, towns and temples of northern India. *Ecology* 42(3):538–47.

———. 1961*b*. A population survey of rhesus monkeys in northern India. II. Transportation routes and forest areas. *Ibid.* 42(4):698–710.

Sugiyama, Y. 1960. On the division of a natural troop of Japanese monkeys at Takasakiyama. *Primates* 2(2):109–48.

———. 1964. Group composition, population density and some sociological observations of hanuman langurs (*Prebytis entellus*). *Ibid.* 5(3–4):7–37.

Sugiyama, Y., Yoshiba, K., and Parthasarathy, M. D. 1965*a*. Home range, mating season, male group and intertroop relations in hanuman langurs (*Presbytis entellus*). *Primates* 6(1):73–106.

———. 1965*b*. Behavioral development and social structure in two troops of hanuman langurs (*Presbytis entellus*). *Ibid.* 6(2):213–47.

———. 1965*c*. On the social change of hanuman langurs (*Presbytis entellus*) in their natural condition. *Ibid.* 6(3–4):381–429.

———. 1966. Artificial social change of a hanuman langur troop (*Presbytis entellus*). *Ibid.* 7(1): (in press).

14

INTERTROOP ENCOUNTERS AMONG CEYLON GRAY LANGURS (*Presbytis entellus*)

SUZANNE RIPLEY

INTRODUCTION

The use of space by individuals and groups is a many-sided problem in the study of behavior. Space has been considered with respect to the relative positioning of individuals vis-à-vis (Hediger 1961; Carpenter 1958; Sommer 1959, 1960; Calhoun 1962) as well as with respect to the location of individuals or social units within a specific geographic area (Howard 1920, Lack 1933, 1934, Nice 1941, Burt 1943, Carpenter 1958, Calhoun 1963, Hinde 1956, Pitelka 1959, Kaufmann 1962). In the study of primate social adaptations it is primarily the latter aspect of the problem of spatial relations that has engaged the attention of fieldworkers. This chapter presents data on intertroop relations and spacing in the Ceylon gray langur, *Presbytis entellus*.

Definitions of territory (Howard 1920, Nice 1941, Burt 1943, Hinde 1956, Bourlière 1956) usually refer to some part of the home range (food source, stores, refuge, nesting site) that the occupant stakes out and defends for a time and within which certain activities are carried out. Burt (1943, p. 350) distinguishes between two types of territory in mammals: one concerning breeding and rearing young and the other concerning food and shelter. The term "defense" is usually used in a narrow sense to refer specifically to fighting, but sometimes it is used in a broad sense (Hinde 1956, pp. 431–32; Carpenter 1958) to refer to any behavior resulting in the exclusion of other conspecifics. Recently Pitelka (1959) directed attention away from an emphasis on the behavioral *mechanisms* of spacing as such, especially defensive fighting, and emphasized instead the *ecological partitioning* of available habitat, through exclusive use, that resulted from these mechanisms. Although Pitelka's shift in emphasis from the behavioral implementation of spacing to its adaptive function is an important contribution, nevertheless the means of

Suzanne Ripley, Eastern Pennsylvania Psychiatric Institute, Philadelphia, Pennsylvania.

effecting this ecological exclusiveness remains an important problem in behavior.

In a recent review, Jay (1965) discusses the application of these concepts of territoriality to the behavior of free-ranging higher primates (monkeys and apes). Jay reports the general occurrence of localized ranges of social units for those taxa in which relatively permanent social groups can be distinguished. Whereas areas of intensive use ("core areas," Kaufmann 1962; "nomadic centers," Tokuda 1964) may be exclusively occupied, there appears to be considerable overlapping of the remainder of the ranges among neighboring groups. Although contacts between neighboring groups sometimes occur, aggressive interactions are extremely rare. For these reasons Jay concludes that territories maintained in the strictly classical sense (by fighting) are not found in the higher primates studied thus far. However, if defense is used in Pitelka's sense, Jay notes that the conclusion is not so clear-cut.

Although defensive fighting is rare, a tendency toward intergroup avoidance is common. This spacing is maintained by patterns of daily movement in which mutual adjustments are made between neighbors by means of visual or auditory location. For example, auditory displays in forest-living species have been reported. They are generally of a broadcast type and are often given at a certain time of day (gibbon, Carpenter 1940; *Callicebus,* Mason 1964; howler, Carpenter 1965), or they may occur when two groups meet (gorilla chest beating, Schaller 1965; macaque branch shaking, Tokuda 1964). This pattern is characteristic of species having well-differentiated social units; in contrast to this, in chimpanzees, among whom social organization is distinctively loose, the loud drumming and vocalizations *attract* widely spaced "groups" (Goodall 1965, Reynolds and Reynolds 1965).

Owing to the paucity of adequate descriptive accounts and conceptual analyses, no definitive statements can be made yet about functions of territoriality in animals generally,[1] or its occurrence in higher primates in particular. However, it is clear at this stage that seen in its broad sense as an ecological partitioning by a variety of behavioral means, territoriality is as much a matter of the relation between neighboring (and potentially competing) resident social units as it is of the relation between any one such resident unit and

[1] Wynne-Edwards' (1962, p. 14) hypothesis (that is, that society is an organization capable of providing conventional competition, including territoriality, which, as a surrogate for undisguised competition for food, is capable of providing feedback for the homeostatic balancing, or controlled expansion, of a species) is cast as a final cause explanation for the existence of society generally, an explanation in which territoriality is interpreted as a powerful convention among many others for the control of population size and quality. In the present chapter we are concerned not with causal explanations for society or territoriality itself but simply with confirming the existence among higher primates of some form of territoriality *by any definition,* and with extending our knowledge of the range of behavioral means employed to maintain it, whatever the ultimate contribution of territoriality to species stability may prove to be. The problem at present is primarily one of conceptual refinement.

a territory which it exploits for whatever purposes. In fact, clues to the determination of the crucial resource, or resources, in respect of which exclusivity of utilization of territory is maintained, lie in the inferred linkage between observed hostile social encounters between units, on the one hand, and observed occurrences of territory utilization or loss, on the other. Therefore, in an attempt to elucidate in part the nature of, and maintaining mechanisms for, the ecological partitioning held to be present in higher primates (Jay 1965), this chapter will focus on the restricted area of intertroop relations among Ceylon gray langurs.

Southwick (1962), following Imanishi (1957), attempted to classify what was known about the patterns of intertroop relations in higher primates. He distinguished two general types: passive and antagonistic. His choice of examples of primates in which groups meet passively (*Cercopithecus ascanius, Gorilla gorilla beringei,* and *Presbytis entellus*) is open to question even in view of our inadequate knowledge of the grouping patterns of these animals. Perhaps the best example of this pattern at present is the nonaggressive close contact of savannah baboon troops at waterholes observed by S. L. Washburn and I. DeVore.[2]

By "antagonistic" Southwick meant "intolerant of the close proximity of other groups" (p. 438). Avoidance and displays characterize interaction between conspecific groups; fighting as a habitual means of spacing is not characteristic of any group of higher primates. Where fighting is found, as in the city-dwelling Achal Tank rhesus groups, Southwick suggested (p. 443) that is is due to the lack of specific means of communication about avoidance. He noted Altmann's (1962) observation that fighting among rhesus groups on Cayo Santiago occurred when the customary monkey chow was not provided and suggested, therefore, that unusual crises, such as food shortages, may precipitate fighting. In addition, Maxim and Buettner-Janusch (1963) reported fighting between baboon groups when an unusually concentrated source of food (garbage dump) became available during the dry season.

In general, in those higher primates that tend toward intergroup intolerance, mechanisms of mutual avoidance are present; and if such groups do meet, displays rather than fighting occur.

It is of considerable interest, therefore, that gray langurs in Ceylon frequently seek other troops and engage them in aggressive encounters. It is further remarkable that these animals are noted for lack of aggression within the troop and that they have the behavioral means of group location and avoidance (vigilance, morning calling), on the one hand, and a display (display jump) which can substitute for aggression in the spacing of troops, on the other.

Gray langur aggressive encounters often take place well inside a home range; they are not the result of accidental meetings; and they are not di-

[2] This has not yet been described in published accounts.

rectly related to competition for food and water sources. Fighting between groups does not appear to be the result of social crowding (Calhoun 1962), nor is it related simply to adult male antagonism, since males can and do live together in both bisexual and unisexual groups. Because the troops do fight, these langurs appear to be territorial in the classical sense, although just what they are defending is not entirely clear.[3] Hinde (1956, pp. 347–48) states, "clearly the nature of the objects or situations defended is of great importance in assessing the biological significance of the territory," but "reliable direct evidence about the functions of territory is scarce." Although the number of instances of intertroop aggression reported here is insufficient to isolate either "apparently advantageous consequences" or "biologically significant consequences in terms of which selection in favor of the behavior can act" (Hinde, p. 348), nevertheless the number of fights recorded is enough to suggest that

TABLE 1

TROOP SIZE

Troop	Total Count	Adult Males
A	19	3
B_1	12	2
B	*c*42	*c*6
C_1	*c*26	*c*4

Sizes for troops A and B_1 given here refer to the constitution of these troops before the shift in April, 1963, of two subadult males from Troop B_1 to Troop A. Although these males appeared to integrate successfully into Troop A, it is not known if the shift remained a permanent change after the study terminated in mid-May, 1963.

these langurs manifest an overt intertroop hostility that conforms to the minimum definition of territoriality and that is apparently rare in Old World monkeys and apes.

POLONNARUWA TROOPS

Observations on four troops of gray langurs were carried out in partly cleared forest near the village of Polonnaruwa, Ceylon, for approximately 375 hours between October 26, 1962, and May 11, 1963. Polonnaruwa is in an area of the dry zone which receives from 50 to 75 inches of seasonal rainfall per year. Most of this falls during the northeast monsoon between November and February. The region is subject to extreme drought during the summer months, especially from June to October. Observations were made on these troops primarily during the wet season when food was abundant (Table 1). Two troops, A and B_1, were well-known; most of the observation hours were

[3] It seems likely to this observer that these langurs are in some way defending the social integrity of the troop, which may be expressed in spatial terms by the concept of troop space. Discussion of the concept may be found in Ripley 1965*a*.

spent on them (Plate 14.1). In two other neighboring troops, B and C_1, only a few key animals were individually identified so that troop locations for range data could be noted at least two times every observation day and so that observations could be made on encounters among all four groups.

Troop Range

As a tentative estimate, 1 square mile of dry-zone forest can support about five to seven troops with an average of twenty-five members each. Range overlap may give each troop a range of ⅛ to ½ square mile. In addition to the four Polonnaruwa troops that occupied a roughly ½-square mile block of partly cleared dry-zone forest (Fig. 1), data on range size were collected from another area in both dry and wet seasons. A square-mile block of arid-zone forest (Bundala) supported seven troops during the driest part of the dry season. Spot checks of locations were made of these seven troops on 34 days of survey between June and October, 1962, and in January (wet season), 1963, for a total of 205 hours. These spot checks give only a general indication of range size for Bundala; important details, including the amount of overlap among ranges and the nature of intertroop encounters, are not available. The

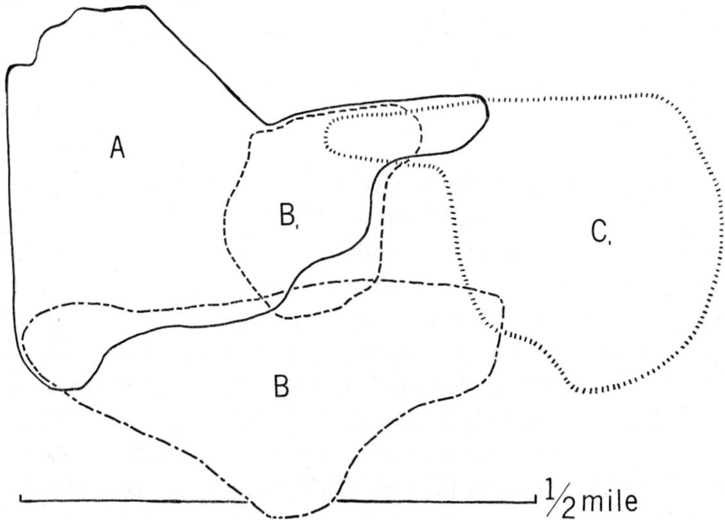

Fig. 1.—Home ranges for the Polonnaruwa troops are based on detailed observations of the movements of troops A (ten permanent members, six infants born, two subadult males joined) and B_1 (ten permanent members, two additional subadult males left to join Troop A) from October, 1962, through May, 1963. Spot locations were made at least twice an observation day on the two other troops. Home-range boundaries are based on composite day range and spot-location charts, and they indicate the areas in which these troops could almost always be found. The exceptions refer to the rare occasions on which the troops left their ranges in the course of intertroop encounters (see Fig. 3).

estimated size of a range, based on similarities in both areas, is assumed for the present to be fairly typical of Ceylon gray langurs.

Gray langurs seem to get along without surface sources of drinking water for several months at a time. The ranges of three of the seven Bundala troops contained no water sources during the driest months (July to October), and there is no evidence that the range boundaries of these troops shifted to include a water source during this time. The Polonnaruwa troops lived beside a large water reservoir but were seen drinking from it only rarely, and these few times were during the wet season when the water edge was close to the trees. The study on the Indian gray langur in Mysore State by the Japan Monkey Center field team confirms the conclusion that the langurs did not drink for many weeks at a time.

Comparison with data on the north Indian langur (Jay 1962) indicates that range size is smaller in Ceylon. Indian troops averaged twenty-five animals also, but their ranges, including core areas, were between 1 and 3 square miles per troop. In Ceylon the smaller troop ranges also include core areas; and since the presence of core areas indicates some choice in habitat utilization, the ranges appear to be large enough that the troop need not use every source of food or refuge within it. In Ceylon, proportionately more of the home range can be considered core area than in Orcha or Kaukori. Jay noted some seasonal shift in the location of core areas but no change in total range size.

Spacing Mechanisms

Morning Whoop

Gray langurs characteristically give loud whooping vocalizations which are audible over long distances. These are given simultaneously by the males sometime after sunrise and before midmorning. These whoops differ in apparent motivation from the whoops given during the day that are part of the display jump. Morning whoops are not accompanied by physical agitation. In the extremely dry areas of Ceylon, where the thick thornscrub forest is from 30 feet to 40 feet high, langurs frequently move to the top of an emergent tamarind (*Tamarindus indicus*) sleeping tree before whooping. Thus, visual as well as vocal means are present for localization. In the moderately dry areas, the morning move to the tops of tall trees (more than 100 feet) was not noted.

The first call usually elicits calls from neighboring groups, as is also true of calls of gibbons, howlers, and the *Callicebus* monkey. The gray langur calls are easily distinguished from those given by the sympatric purple-faced langur, *P. vetulus*. In addition to structural distinctiveness, the purple-faced langur calls are given about ½ hour before sunrise, well before those of the gray langur.

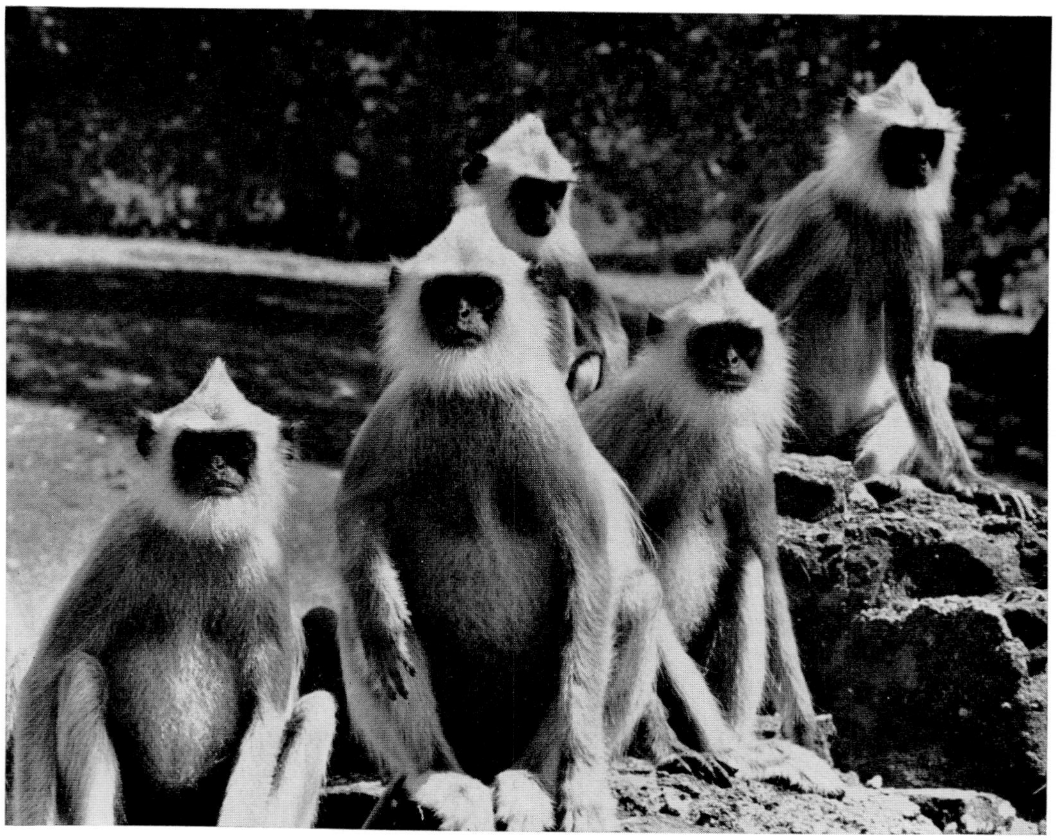

PLATE 14.1.—Five Troop A females intently observe an unusual event

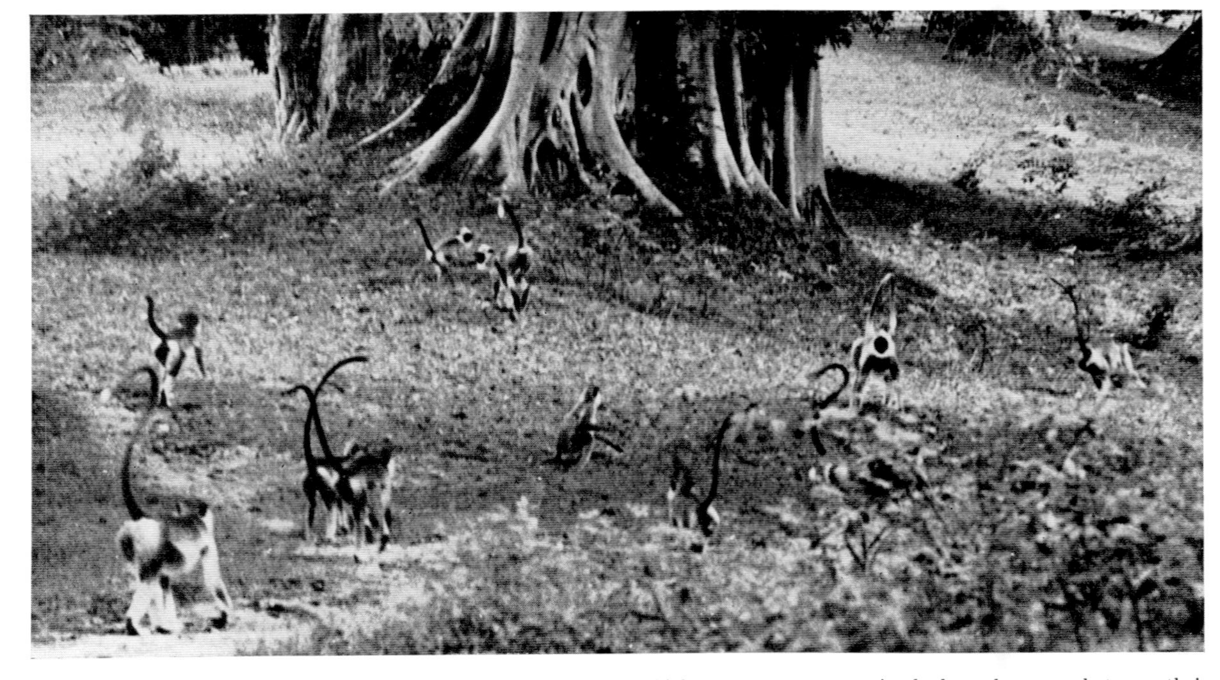

PLATE 14.2.—Troop A runs to an encounter with Troop C_1, which prepares to retreat in the boundary area between their respective ranges.

Display Jump

At any time of the day an auditory stimulus, and especially the sound of a neighboring troop, may elicit a display jump. This is a noisy, impressive spectacle. The males of the troop lead in jumping between branches and trees in a stiff, heavy manner that produces a striking combined visual and auditory effect. Some females without infants join the males. Scarcely touching a support, they launch into long leaps. The males jump onto branches they would otherwise avoid—dry, brittle ones that break under their weight. Even at extreme heights they land on these branches, which give way, sending them crashing down to the next level of branches or to the ground if they are not far from it. During this active show, the males keep up a deep-voiced whooping.

Coordination among displaying animals is remarkable. They never interfere with one another's path during the jumping. They both begin and stop as if at a signal.

These display jumps are certainly heard by neighboring troops. They are an impressive demonstration of strength, agility, and troop unity. In many ways the display jump is a highly mobile, functional counterpart of the individual male macaque's branch shaking or the howler's "vocal battle."

Vigilance

Whereas female gray langurs spend the time when they are not eating in grooming, tending infants, or resting, the males maintain an observational tonus in which they are alive to more distant conditions and events. This behavior corresponds to what Hall (1960) has described as "dominance vigilance" in the chacma baboon. A male langur will climb into a tree from which he has a good view of the surroundings. He will scan the region, keeping his gaze focused in the distance, and give low-level vocalizations. The most common vocalizations are a low grunt given with the mouth closed and a grinding of the teeth. The males seem to be on the lookout for other troops rather than for danger from other sources. Unlike the chacma baboon, however, vigilance in these langurs is not a means of avoiding another troop; rather, it often results in an episode leading to an intertroop fight.

INTERTROOP CONTACTS

During the study period, thirty-one instances of intertroop contact[4] were recorded (Fig. 2). Undoubtedly many more occurred during this time. The

[4] The term "contact" is used here to refer to the entire social event, treated as a unit, within which social interaction takes place. Interaction means specifically the *process* of gestural statement and reply between individuals contained within the contact. A contact is used here to mean a spate of heightened orientation and relevance of acts by a social unit with regard to another social unit. Discussion of the concept and its use is found in Ripley (1965*a*).

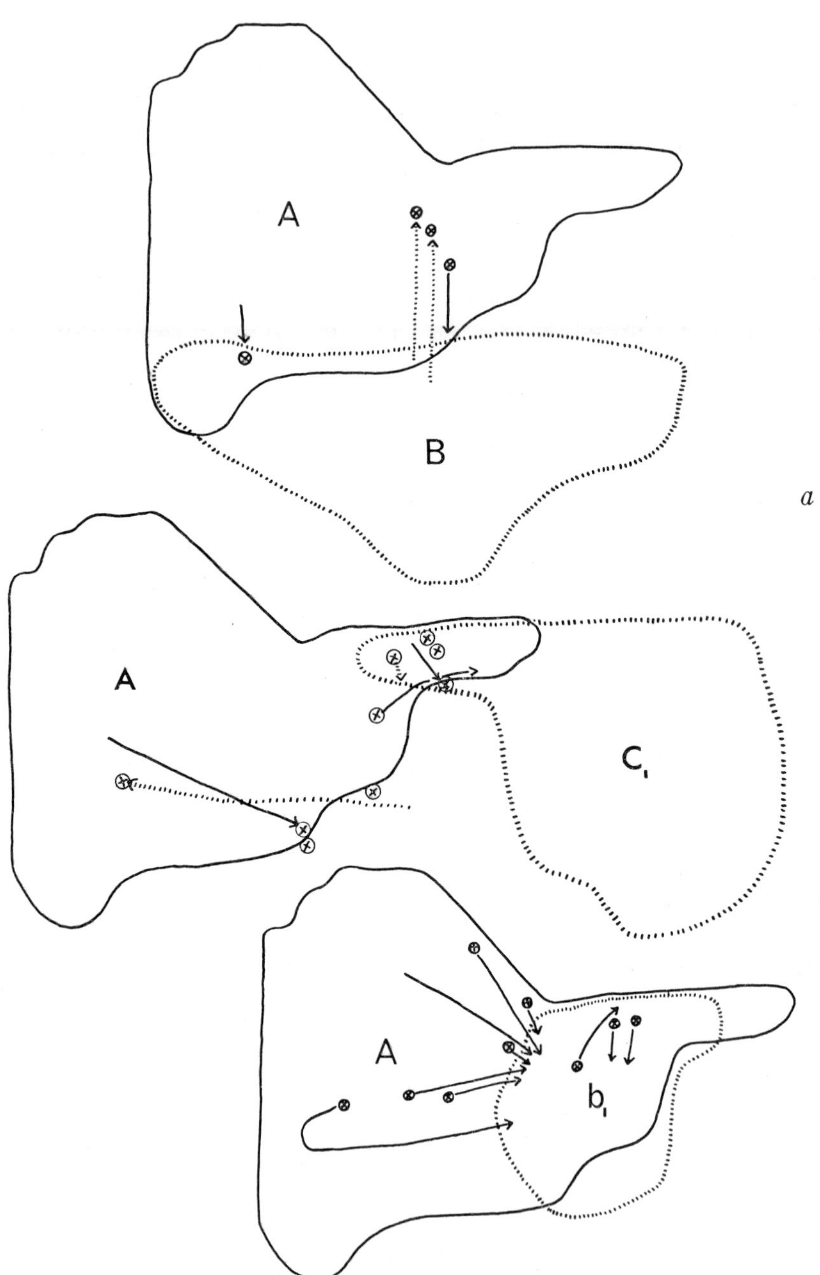

FIGS. 2*a* and 2*b*.—Chases and encounters between troops occur alone or in a series of varying sequences. During some encounters a troop may go far out of its habitual range. Such sallies are rare, and following the encounter the troops return to their normal ranges. The troops never leave their ranges except during an encounter.

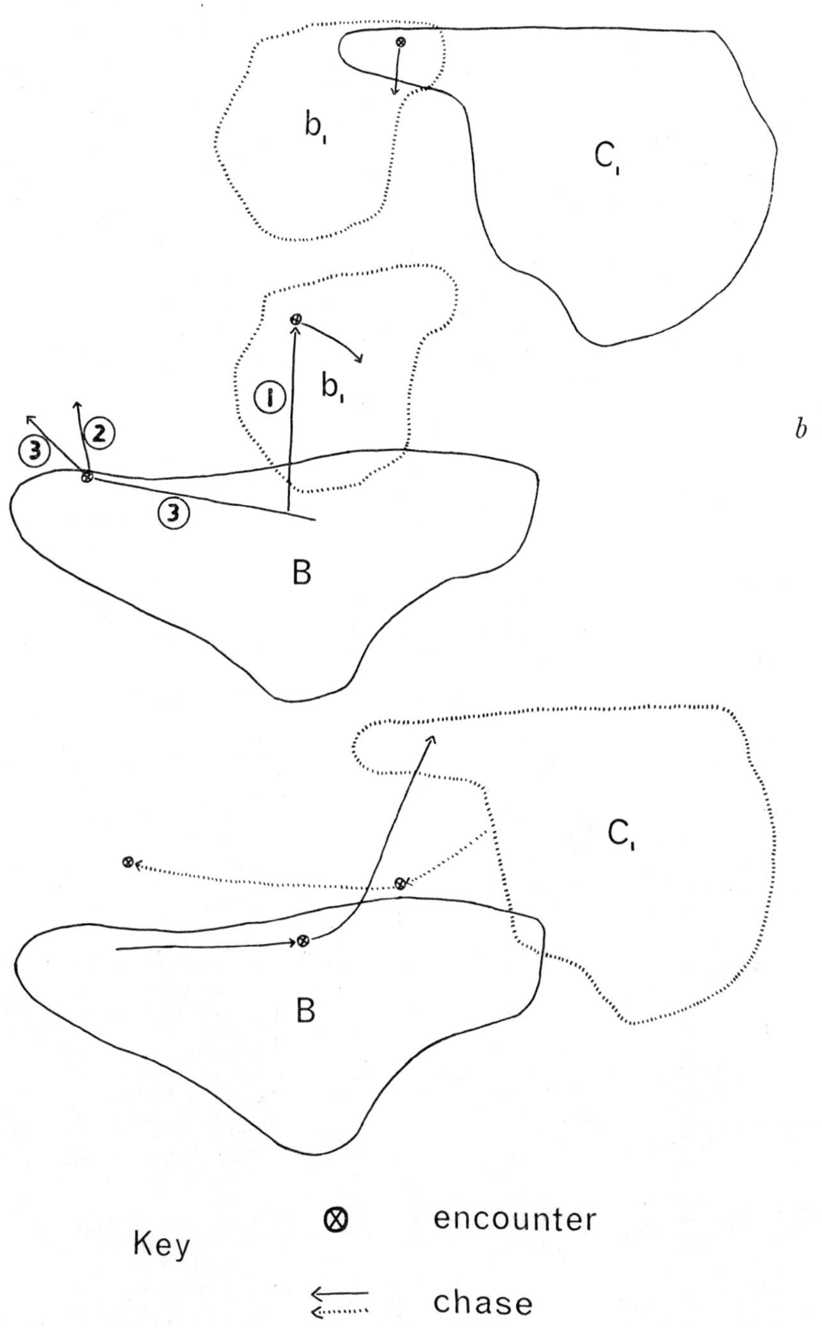

b

data show about one contact per 10 hours of observation time. The order of contacts with dates, time of day, duration, participants, and some indication of the quality of activity taking place are given in Table 2. Most of these contacts occurred in the morning.

Most contacts were antagonistic. They involved chases, displacements, threats, mild wrestling, grabbing, and at times true fighting (including biting). The members of the attacking troop determine the content of the individual interactions which together comprise the contact. The attackers were frequently low-ranking members of their own troop. For example, Gamma, the lowest ranked of three males in Troop A, and to a lesser extent the Beta male, accounted for most of the aggressive encounters in which Troop A was the

TABLE 2

INTERTROOP CONTACTS

Date	Time	Aggressor/Defender	Comments[a]
Oct. 27	0717–0950	C_1 A	Chase
30	0630–0750	C_1 A	Chase, threats
30	0750–0825	B C_1	Displacement
30	0825–1140	A C_1	Chase
30	1230–1300	A C_1	Chase
Nov. 1	0737–0820	A C_1	Displacement, fights
13	1030	A = C_1	No fights, together on border
14	0930–0945	A C_1	Chase
Jan. 17	0930–1030	A C_1	Threats, chase
30	1000	B B_1	Chase
30	1030–1050	A B_1	Attacks, complex interactions
30	1050–1100	B B_1	Chase
Feb. 11	1030	A B_1	Threats
12	0950–1130	A B_1	Displacement
15	1100–1125	A B_1	Chase
16	1115–1130	A = C_1	Threats only
16	1145	C_1 B_1	Threats
19	1600	A B_1	Fiercest fights, males wrestle
21	1220	A = B_1	Threats, no fights
27	1140	A B	Chase
28	0940–1000	A B_1	Chase
Apr. 15	0940	A B_1	Chase
24	0900–0910	B A/B_1	Chase in A's range
30	0810	B A	Enters A's range
30	0815	A B	Chase
30	0900	C_1 B	Chase, fights
30	0930	C_1 A	Chase
30	0950	A B_1	Chase
May 11	1105	A B_1	Chase
11	1040	A ?	Strange troop (other than B, B_1, or C_1) in range of Troop A

[a] Displacement = One individual or troop yields its position to another individual or troop. Chase = A continuous displacement through space. Threat = Gestural indication of a potentially imminent attack. Attack = Unsolicited corporeal incursion by one individual upon another. Fight = An attack that is countered.

attacker. Interestingly enough, Gamma accounted for most of the peaceful contacts with individuals from other troops as well. The generally high social attractiveness (as measured by frequency of grooming and sexual contacts) of low-dominance ranked males agrees with Jay's observations on the Kaukori troop (Ripley 1964, 1965*a*). The Alpha male became involved in chases and at times was an instigator, but he frequently took second place in the attacks. Some females participated in these aggressive encounters, most often those without infants, particularly the two subadult females and three of the eight adults (Plate 14.2).

In one type of encounter an aggressive party goes out from its troop. This generally develops into engagements between males of the defending troop and males of the attack party. Attacking females make short dashes and retreats in the staccato fashion that also characterizes their behavior in intratroop aggressive contacts.

TABLE 3

DOMINANCE RELATIONSHIPS IN SINGLE CONTACTS

DOMINATED	DOMINANT TROOP				TOTALS
	A	B	B_1	C_1	
A		2	0	3	5
B	2		0	1	3
B_1	9	3		1	13
C_1	5	1	0		6
Total	16	6	0	5	

In another type of encounter the whole of each troop is engaged in a shifting constellation of aggressive (chase, displace, tail-pull, bite) and non-aggressive (presenting, posterior grasping, squealing, rarely play, and never grooming) contacts. Often such interactional melees will be separated by quiet interludes during which the animals sit calmly, the males perhaps grunting and grinding their teeth. Then another active episode may begin.

At times the role of aggressor shifts during these episodic encounters. I have indicated such shifts in Table 2 by treating each as a separate event; that two episodes are immediately sequential is evident from the time of occurrence. The apparent unity of action of the attacking troop or party is marked and probably reflects a dependence of the members on the leadership of the attacking male or males. In any one episode, the side which succeeds in displacing or chasing the others is considered dominant (Table 3).

Because Troop A was the principal object of the study, observations of its interaction with other troops are the most frequent. Troop B_1 also has a high rate of interaction owing to its central geographic position. Contacts between these two troops and the two outlying troops, B and C_1, give the picture of intertroop contacts as they relate to troops A and B_1. Of course, contacts between the two outlying troops and their more distant neighbors were not observed; this account reports activities observed in only one restricted area. The frequency of contacts for all troops is certainly higher than indicated here and can be obtained only from a long-term study focused on intertroop conflict. What is significant here (Table 4) is that each troop acts at times as aggressor and at other times as the object of aggression.

Analysis of the contacts reveals the reciprocal dominance relationships of each troop with respect to the others. This material is summarized in Table 4. However, these contacts do not indicate an absolute dominance structure among troops; rather, a troop's position is characterized by a combination of wins, draws, and losses. Troop A appears to be completely dominant over B_1,

TABLE 4

DISTRIBUTION OF CONTACTS BY TROOP

TROOP	TOTAL CONTACTS	ROLE		
		Aggressor	Defender	Neutral
A	25	17	5	3
B_1	14	0	13	1
B	9	6	3	0
C_1	13	5	6	2

with which it commonly interacts; sometimes dominant over C_1, with which it interacts as frequently; and about at a draw with Troop B, with which it interacts infrequently. Troop B is completely dominant over B_1 in all recorded contacts and apparently at a draw with Troops A and C_1. Troop C_1 is dominant over B_1, at a draw with Troop B, and is sometimes dominated by Troop A. Troop B_1 never fares well. Based on the data given, we can indicate a record of relative dominance for the four troops.

Troop A	(+ + =)	+ means *win*
Troop B	(+ = =)	= means *draw*
Troop C_1	(+ = —)	— means *loss*
Troop B_1	(— — —)	

One remarkable feature of these intertroop encounters is that the attacking troop often goes far out of its own habitual range to attack another troop. In the fourth example presented in the following section, Troop C_1 went across the range of its neighbor, Troop B_1, and far into Troop A's range. A troop that occasionally leaves its home range in this fashion does so only in the context of an encounter. Following the chase or fight Troop C_1 returned, albeit leisurely, to its home range. Burt notes (1943, p. 351) that "occasional sallies outside the area, perhaps exploratory in nature, should not be considered as in part [*sic*] of the home range." Despite these incursions into the territory of other troops, no ecological expansion of the home range exclusively exploited by one troop resulted. Burt notes (p. 350) that "it is to be expected that the territory of each and every individual (or group) will be trespassed sooner or later regardless of how vigilant the occupant of the territory might be." Therefore, these few sallies may be legitimately considered apart from the data used to construct the home-range chart. What is peculiar, however, in these langurs is that on occasion some troops are successful in dominating another troop inside the latter's home range; yet the expected ensuing range boundary shifts were not discernible.

This is evidently one example of the ambiguity concerning the functions of territory noted by Hinde (1956, p. 348), and many more data are required in order to analyze the pattern satisfactorily.

A word should be added concerning the range of Troop B_1. From the range chart it appears that the area available to Troop B_1 is almost entirely overlapped by the ranges of its neighbors, especially by Troop A's. During this study, Troop B_1 was in a highly unstable state. From the evidence available to me I am quite certain that it budded off from Troop B in January, 1963. For a time, the new members of Troop B_1 continued to eat and rest in part of Troop B's range, as well as that portion of Troop A's range which they continued to exploit for the remainder of the study. Immediately after the troop split, no aggressive encounters occurred between troops A and B_1; this tolerance appeared to be due entirely to the restraint exercised by Gamma male in Troop A. By the end of January, 1963, hostile encounters began to occur between troops A and B_1. After this time, Troop B_1 confined its maintenance activities (eating, sleeping, social contacts) to its own small range. Although Troop A continued to move over most of Troop B_1's range, it used this region less for eating and never for night sleeping. During this period, Troop B_1 never dominated an intertroop encounter and had the smallest range size.

Further evidence for the instability of Troop B_1 after its formation was the shift in membership of two subadult males from Troop B_1 to Troop A in April, 1963. It is likely, therefore, that the territorial relationships of Troop B_1 with other troops were in the process of formation.

Examples

Some excerpts from the field notes describing the contacts follow:

1. *Feb. 16, 1963, 1115–1130:*

   Troop A and Troop C_1 are near each other at the border of their ranges. Two subadult females from Troop A are sitting on a fence; one of them climbs into a nearby mango tree. A male from Troop C_1 climbs up after her and chases her out. A Troop C_1 female approaches the subadult and grasps her sides in a sort of seated mounting. Males are giving low grunts; not a very tense situation.

This contact was relatively peaceful. No troop can be said to have gotten the better of the other. The interaction took place in an overlapping area of the two ranges in a place that was commonly the scene of intertroop contacts. It is an example of the milder type of activity which occurs when the less-aggressive females are present.

2. *Feb. 19, 1963, 1600:*

Troop A moves into the area of range overlap; Troop B_1 is in the mango tree. Troop A routs them out of the tree, chases, and fights with both the males and females of Troop B_1 in the fiercest fight yet seen. The males actually engage each other physically, wrestle, and pull tails. Troop B_1 is on the defensive and retreats to its range. Troop A does not pursue it further but moves back.

This contact was one of the few that took place in the afternoon and was the only one in which prolonged, serious fighting occurred. It was a fight between two troops that had been closely associated 1 month previously, when Troop B_1 was apparently splitting off from Troop B and was permitted to share Troop A's range for a time. The dominance relationship between the two troops appeared to be stable, and no changes resulted from this severe fight. Later on, at different times, the two subadult males from Troop B_1 joined Troop A. Only subadult males were observed to shift troops successfully.

3. *April 24, 1963, 0800–0810:*

Troop A runs from midrange to the border at top speed. Gamma male and a subadult male formerly with Troop B_1 lead. They meet Troop B_1 at the area of range overlap and chase it out of A's range altogether, giving the grunt with open mouth. Alpha and Beta males from Troop A come up behind, also threatening. They stop short of the encounter, climb a tree, and begin eating. Two females from the routed troop are still eating in this tree. They do not seem upset at the presence of Troop A males. The rest of the routed Troop B_1 sits on the ground in its range.

Gamma male from Troop A rests in a tree not far from Troop B_1. Near him is the juvenile male from Troop B_1 with whom Gamma is often associated when the two troops come together peacefully.

0900:

About five males from Troop B approach along the rampart rapidly; they chase both troops A and B_1. All males mill about, agitated give sporadic display jump with whoops. Troop B males continue to chase and then confront Troop A males inside Troop A's range. Troop B males then run into Troop B_1's range, giving a grunting threat.

This example is interesting because it illustrates one of the rare instances when more than two troops were involved in a single fight. Males of one troop chased those of two other troops simultaneously. Troop B was attacking both the other troops well outside of its own range; nevertheless, in this instance they dominated the combined forces of Troops A and B_1.

4. *April 30, 1963, 0810:*

Beta male and the new subadult male of Troop A run from the middle of their range to the border. They are chasing a group of five males and one female who had been in their range. Then Beta, Gamma, and both the new subadult males chase the group into Troop B_1's range. The strange attack party continues across into Troop C's range. A display whoop is heard from the direction of Troop B. Gamma male from Troop A chases the intruders through Troop B_1's range and to the border of the ranges of troops B_1 and C_1. He then takes a position of good vantage atop a pillar, stares in the direction the intruders have gone, and gives loud grunt threats.

0900:

Troop B and C_1 males fight on the border of the ranges of troops B_1 and C_1. Then Troop C_1 males chase Troop B precipitately into Troop B's range. All rest for a short time. Then Troop B_1 males join the fray by whooping close by. Troop B males then continue their retreat well into Troop A's range, even though Troop C_1 has temporarily stopped chasing them. The rest of Troop B_1 cannot be seen.

0930:

Troop A approaches from the road junction where it had retreated when it was chased out of the corner of its range by Troop B earlier. It is now chased by Troop C_1 males. Troop B_1 is now sighted at the far end of Troop A's range where it has been chased by the hunchback male from Troop C_1. Hunchback now returns at a slow walk to his troop, which rests and watches. Troop C_1 has gone two ranges beyond its own in this maneuver. Hunchback is not now chased or challenged by the males of any of the troops which have been displaced. It is as if the fight were now over and further aggression superfluous.

0950:

Two adult males and two juvenile males from Troop B_1 cross from the woods, across their own range, and continue well into Troop A's range. A party from Troop A consisting of Beta male, a subadult and adult female, one subadult male, and one unidentified animal chase the Troop B_1 party into its own range. They stop the chase at the border to the ranges.

This sequence of contacts was the most complex observed during the study. It involved all four troops, and the action lasted nearly 2 hours. Troop C_1 males moved far out of their range and entered that of troops B_1 and A as aggressors. They were confident and acted with assurance even when they were well beyond their range boundaries. In the course of the 2 hours the entire area was in an uproar, monkeys running in all directions. The normally quiet area was resounding with activity, whoops, and confusion. Since most of the troops were displaced and divided into subgroups, it was difficult to be sure of the whereabouts of any one whole troop. Only the fact that this area was cleared of underbrush made it possible to observe the general course of events.

CONCLUSION

Ceylon gray langurs regularly engage each other in aggressive encounters. Although these contacts are usual occurrences, the cost to the participants is not great since little serious physical injury takes place. In a species in which breeding and births are not confined to one season and in an area where food is so easily obtainable, the variability of range size probably depends upon population density. Thus, the available circumstantial evidence is not sufficient to relate this intertroop aggression directly to either of the types of territory in mammals defined by Burt. Furthermore, range integrity and exclusivity of food resources could probably be maintained by means of troop location and avoidance, an ability which these langurs have in common with many other primates.

One possible explanation for the existence of conflict between langur troops in Ceylon is that fighting occurs when average range size, which is at maximum 3 square miles per troop (average, twenty-five animals), is compressed by population density to nearly the size of the average core area of those troops.

If territoriality is defined either as the defense of an area by fighting or as the exclusive ecological exploitation of some part of a home range, maintained by a variety of behavioral means, then the gray langur of Ceylon is territorial. However, the functions of territoriality in these langurs remains to be explained.

ACKNOWLEDGMENTS

The field study which this chapter reports in part was supported by National Institutes of Health predoctoral fellowship MF-12473 and a research grant supplement M-5542. The writing was done under a program on primate behavior supported by U.S. Public Health Service Grant No. MH 08623.

REFERENCES

Altmann, S. A. 1962. A field study of the sociobiology of rhesus monkeys, *Macaca mulatta. Ann. N.Y. Acad. Sci.*, 102:338–435.

Bourlière, F. 1956. *The natural history of mammals.* 2d ed. rev. New York: Knopf.

Burt, W. H. 1943. Territoriality and home range concepts as applied to mammals. *J. Mammalol.* 24:346–52.

Calhoun, J. B. 1962. A behavioral sink. In *Roots of behavior,* ed. E. L. Bliss, pp. 295–315. New York: Harper.

———. 1963. The social use of space. In *Physiological mammalogy,* eds. W. V. Mayer and R. G. van Gelder. Vol. 1. *Mammalian populations.* New York: Academic Press.

Carpenter, C. R. 1940. A field study in Siam of the behavior and social relations of the gibbon. *Comp. Psychol. Monog.* 16. No. 5.

———. 1958. Territoriality: a review of concepts and problems. In *Behavior and evolution,* ed. A. Roe and G. G. Simpson. New Haven: Yale Univ. Press, pp. 224–50.

———. 1965. The howlers of Barro Colorado Island. In *Primate behavior: field studies of monkeys and apes,* ed. I. DeVore, pp. 250–91. New York: Holt, Rinehart and Winston.

Goodall, J. Chimpanzees of the Gombe Stream Reserve. In *Primate behavior: field studies of monkeys and apes.* ed. I. DeVore, pp. 425–73. New York: Holt, Rinehart and Winston.

Hall, K. R. L. 1960. Social vigilance behavior of the chacma baboon. *Behavior* 16: 261–94.

Hediger, H. 1961. The evolution of territorial behavior. In *Social life of early man.* ed. S. L. Washburn, pp. 34–57. Chicago: Aldine Press, Viking Fund Publ. in Anthropology No. 31.

Hinde, R. A. 1956. The biological significance of the territories of birds. *Ibis* 98:340–69.

Howard, H. E. 1920. *Territory in bird life.* London: J. Murray.

Imanishi, K. 1957. Social behavior of Japanese monkeys, *Macaca fuscata. Psychologia* 1:47–54.

Jay, Phyllis. 1962. The social behavior of the langur monkey, *P. entellus.* University of Chicago doctoral dissertation.

———. 1965. Field studies of monkeys and apes. In *Behavior of nonhuman primates,* ed. A. M. Schrier, H. F. Harlow, and F. Stollnitz, pp. 525–91. New York: Academic Press.

Kaufmann, J. H. 1962. Ecology and social behavior of the coati, *Nasua narica,* on Barro Colorado Island, Panama. *Univ. of Calif. Publ. in Zool.* 60:95–222.

Lack, D. L. 1933. Territory reviewed. *Brit. Birds* 27:179–99.

———. 1934. Territory reviewed. *Ibid.*, pp. 266–67.

Mason, W. A. 1964. A field study of the social organization of the socay monkey, *Callicebus ornatus.* Symposium. Delta Regional Primate Research Center, Nov. 3–6, 1964.

Maxim, P., and Buettner-Janusch, J. 1963. A field study of the Kenya baboon. *Am. J. Phys. Anthropol.* 21:165–81.

Nice, M. M. 1941. The role of territory in bird life. *Am. Midland Naturalist* 26:441–87.

Pitelka, F. A. 1959. Numbers, breeding schedule, and territoriality in the pectoral sandpipers of Northern Alaska. *Condor* 61:233–64.

Reynolds, V., and Reynolds, F. 1965. Chimpanzees in the Bidongo Forest. In *Primate behavior: field studies of monkeys and apes.* ed. I. DeVore, pp. 368–424. New York: Holt, Rinehart and Winston.

Ripley, S. 1965*a*. The ecology and social behavior of the Ceylon gray langur, *Presbytis entellus thersites.* Univ. of California, Berkeley, doctoral dissertation.

———. 1964. "Patterning of aggression in the social behavior of the adult male gray langur, *Presbytis entellus thersites.*" Symposium. Delta Regional Primate Research Center, Covington, Louisiana, Nov. 3–6, 1964.

Schaller, G. B. 1965. The behavior of the mountain gorilla. In *Primate behavior: field studies of monkeys and apes.* ed. I. DeVore, pp. 324–67. New York: Holt, Rinehart and Winston.

Sommer, Robert. 1959. Studies in personal space. *Sociometry* 22:247.

———. 1960. Leadership and group geography. *Ibid.* 25:99–110.

Southwick, C. H. 1962. Patterns of intergroup social behavior in primates, with special reference to rhesus and howler monkeys. *Ann. N.Y. Acad. Sci.* 102:436–54.

Sugiyama, Y. 1964. Group composition, population density, and some sociological observations of Hanuman langurs (*Presbytis entellus*). *Primates* 5(3–4):7–38.

Tokuda, Kisaburo. 1964. Daily movement, range and intertroop relationships of the Japanese monkey. Symposium. Delta Regional Primate Research Center. Nov. 3–6, 1964.

Wynne-Edwards, V. C. 1962. *Animal dispersion in relation to social behavior.* Edinburgh and London: Oliver and Boyd.

DISCUSSION OF SOCIAL DYNAMICS

J. M. WARREN

The chapters in this section of the volume reveal a striking picture of variation, both between species and within species, in the social behavior of monkeys.

For example, groups of langurs are smaller, less stable, and less complexly organized than groups of macaques. Langur mothers readily give up their newborn infants to other female members of the troop; such behavior is almost never seen in *Macaca mulatta* or *M. fuscata*. Dominance relationships seem to be much less important in intratroop interactions among langurs than among macaques, and intragroup aggression is less frequent in langurs than in macaques.

Since langurs are predominantly arboreal and macaques are predominantly terrestrial in their habits, it is interesting but not particularly startling to learn that these types of monkeys differ in their social behavior. The relative freedom from predation enjoyed by the largely arboreal langurs provides an attractive, although unverified, hypothesis to account for the lack of elaborate social organization in this species, and possibly for the more relaxed and nonaggressive individual social interactions among langurs compared with, say, rhesus monkeys.

But how can one explain the wide range of variation in social behavior that characterizes groups of *Presbytis entellus* from different regions of India and Ceylon? The following list of intraspecies differences is representative rather than exhaustive:

Mean Size of Bisexual Groups

Ripley (this volume) and Jay (1963*b*) estimate the average number of langurs in the bisexual groups they observed in Ceylon and in north and cen-

J. M. Warren, Department of Psychology, The Pennsylvania State University, University Park, Pennsylvania.

tral India at twenty-five. Sugiyama observed thirty-eight bisexual groups in south India. The largest had twenty-four members, and the average was fifteen.

Number of Mature Adult Males per Bisexual Group

Almost none of the bisexual groups observed by Sugiyama contained more than one large, fully adult male. Both Ripley (this volume) and Jay (1963*b*), however, report that groups with two or more fully mature males were common in the areas where they studied *P. entellus*.

Quality of Intergroup Conflict

Langur groups in the north of India did not come into contact with one another often. When they did, there was no display of aggression (Jay 1963*b*). Sugiyama found mild fighting between the male leaders of Dharwar langur troops when they happened to meet. Females and young animals did not join in the action. The langurs of Ceylon, in contrast, seem regularly to seek other groups to engage in aggressive displays and mild aggression. Many of the adults, both male and female, participate with the group leaders in these engagements.

Aggression toward Infants by Adult Males

The displacement of an incumbent male leader by a more successfully aggressive intruder was observed in several troops of Dharwar langurs. In all but one instance, the newly ascendant male bit and killed all the infants in the group less than 1 year old. Ripley, however, reported no aggression by adult males toward infants, and Jay (1963*a*) explicitly states she never saw an adult attack an infant.

Defense of Infants by Mothers

Sugiyama noted that mother langurs failed to protect their infants from recently dominant males when they were attacked and that the mothers in fact often deserted the infant during attacks. In the north and central Indian groups observed by Jay (1963*a*), the mothers frequently threatened or struck males that had simply frightened their infants.

At present we can only speculate concerning the relative importance of genetic and ecological factors and the role of social learning in the determination of these variable patterns of social behavior in langurs. The possible adaptative value of the patterns of social behavior peculiar to particular populations of langurs also remains to be determined. Sugiyama found that the male leader of a troop of Dharwar langurs is replaced every 2 or 3 years, and that the new leader typically kills all the infants less than 1 year old. This means that the odds are 1 in 2 or 3 that an infant will fall victim to intraspecies infanticide. It may be, as Sugiyama suggests, that the wholesale induction of estrus in the females of the troop more than compensates, reproductively speaking, for the loss of the infants, but Sugiyama's interpretation is not intuitively obvious. The periodic liquidation of the baby langurs impresses me as being potentially dysgenic in its consequences.

This question could be answered by comparing the growth of population in groups of Dharwar langurs and in groups drawn from populations in which baby biting does not occur, with approximately constant environmental conditions.

The careful documentation of intraspecies and interspecies differences in social behavior among primates provided by the chapters given in this part of the volume serves two very useful purposes: (1) it will discourage superficial speculation regarding the evolution of human social behavior involving descriptions of social behavior in "the" monkey; and (2) it will direct the attention of researchers to important and empirically soluble problems regarding the variation in social behavior, and the causes of this variation, within single species of monkeys.

As one particularly interested in the experimental study of learning, I cannot refrain from making a few final comments on the extraordinary observations reported by Tsumori on delayed-response performance by feral *M. fuscata.* Even though psychologists can no longer accept the claim that delayed response is a measure of symbolic processes in animals (Warren 1965), it is impossible not to be very strongly impressed by the very high percentage of monkeys in the Koshima troop which were able to respond successfully on their first trial. Rhesus monkeys in the laboratory often require 2 or 3 weeks of training, with several trials per day, to acquire delayed-response proficiency. The almost instantaneous learning obtained with Tsumori's ingenious technique for testing monkeys in the field suggests that laboratory studies may provide a very conservative estimate of their capacity for problem solving.

This point is, however, relatively trivial in comparison to Tsumori's demonstration of substantial differences between troops of *M. fuscata* in the speed with which individual monkeys learn the Sand-Digging Test and acquire other novel forms of adaptive behavior, which reinforces previous hints that precultural transmission of learned responses may be studied with considerable profit in groups of monkeys, and also that suitable adaptations of standard learning tasks may prove to be useful dependent variables for assaying the effects of membership in different social groupings upon the adaptive behavior of individual monkeys.

REFERENCES

Jay, Phyllis. 1963*a*. Mother-infant relations in langurs. In *Maternal behavior in mammals,* ed. H. S. Rheingold. New York: Wiley.

———. 1963*b*. The Indian langur monkey (*Presbytis entellus*). In *Primate social behavior,* ed. C. H. Southwick. Princeton: Van Nostrand.

Warren, J. M. 1965. The comparative psychology of learning. *Annual Review of Psychology* 16:95–118.

Part V

COMMUNICATION PROCESSES

15

SOCIAL INTERACTIONS OF THE ADULT MALE AND ADULT FEMALES OF A PATAS MONKEY GROUP

K. R. L. HALL

INTRODUCTION[1, 2]

The patas monkey, *Erythrocebus patas,* is distributed over the savannah and woodland savannah regions of north equatorial Africa from Senegal in the west to Kenya in the east. It shows, in extreme form, physical and social adaptations to living on the ground and finding its food in seasonally arid areas where even baboons may not penetrate. It must be the fastest living primate, being timed traveling alongside a vehicle at 35 mph (Tappen 1965). It is slight and long-limbed in build, the full-grown male weighing about 28 lbs and the adult females 12 to 15 lbs. The sexual dimorphism is thus of the same order as baboons, but baboons average about twice this weight (DeVore and Washburn 1963). The adult male is also conspicuous from the vivid white expanse of his rear view—a feature which probably has an important signaling function for the rest of the group.

The first field study of the species was carried out by Hall (1965) in Uganda in 1963 and 1964. In all groups observed, there was only one adult male, with several adult females and young animals. No other sexually mature males were seen in or near groups, but isolated adult males were occasionally seen far from groups, and on one occasion a bachelor party of four males was located.

Dr. Hall was at the Department of Psychology, University of Bristol, England, until his death on July 14, 1965.

1 This study was supported by the Medical Research Council, London, England.

2 The author wishes to acknowledge the help of Mrs. Barbara Mayer and Mrs. Marilyn Goswell in analyzing the data.

Each group occupied an extensive home range—that of one group numbering thirty-one animals ranging over an area of 51.8 square kilometers (20 square miles). During a 10–12-hour day, a group would travel as little as 700 m when food was abundant, as much as 12,000 m when food was scarce. They very rarely went to water.

In the heat of the day, the group would rest, all the animals together, in one large shade tree, for 2 or 3 hours. At night, each animal went separately up into a tree on the savannah, except for the mother and infant who kept together. Thus a group could be dispersed at night over an area of about 250,000 sq. m.—in marked contrast, as in many other respects, to baboon groups in the same area who usually concentrated in a few large trees by the banks of the Nile. The adaptiveness of this night dispersal in a gregarious, group-living animal is clear if we consider that the only likely night-hunting predator, the leopard, could scarcely catch and kill more than one, or perhaps two, members of the group. The leopard's chances are also reduced by the fact that a group rarely spent successive nights in the same area and by the extreme vigilance of the adult male in reconnoitering a sleeping area at the end of the day.

The unique features of the patas social organization became more and more evident as the field study progressed. The social role of the adult male was clearly and repeatedly that of a watchful guardian. Often he moved far away from his group when surveying a new feeding area, or when the group was disturbed by the too-close approach of the observer. Not only was he exceedingly watchful, but he would, on occasion, perform what appeared to be a diversionary display in which he crashed noisily through bushes, then ran close to the observer and far away from the group. His conspicuous appearance might, in such circumstances, serve two functions—the one to attract the attention of the predator, the other to be easily visible to his group who remained concealed in the long grass.

The quality of the vocalizations of the patas was entirely consistent with this survival pattern of dispersal, concealment, watchfulness, and diversion. On no occasion was the presence of a group given away by vocalization. The adult male never barked when watching the observer or any other source of disturbance, *except* when seeing other patas. I have called the vocal pattern one of adaptive silence. They have many calls, most of them known from our laboratory group, but they are almost all muted and audible to the human ear only at about 100 m or less. These presumably serve to signal intentions, movements, spacing, and so forth, at the minimum loudness necessary within the group when it is in visually restricted circumstances, such as in long grass or thickets. Contrast this again with baboons who are frequently very noisy and make little attempt to conceal themselves.

Against this background of the natural way of life in Uganda, we have to analyze and interpret the social organization. In wild groups, aggressive interactions were infrequent, and no fighting was seen. The adult females sometimes joined in threatening the adult male, usually, it seemed, when he was trying to mate with one of them. On no occasion was he seen to initiate an attack on one of them or on any other member of the group, but he became exceedingly agitated, and presumably potentially very aggressive, if he saw another group or, as happened on one occasion, an extragroup adult male. As the home-range size indicates, spacing between groups was so great that group interactions were very rarely observed.

Sanderson (1957, and personal communication) says that, in West Africa, patas groups were almost always led or "bossed" by an adult female. As we know from baboon studies (Hall 1960), the spatially "leading" position may not imply "leading" in the social sense of the term. In baboon groups, peripheral males are usually to the fore and to the sides of the nucleus of the group formed by adult females and their young and some of the adult males. A similar pattern is seen in macaques. But, in patas groups, from what we have seen so far, there are no peripheral males or females. The animals are either *in* a group or right outside it. Since the adult male patas is often quite far away from his group, even in the course of normal feeding, it would seem to follow that one or more of the adult females may indeed have a special role in looking after or leading the *routine* activities of the other females and the young animals. Therefore, it was not surprising to find from analyzing the field data that an adult female was almost always the first to move away from a day-resting tree, the rest of the group following her, with the adult male quite often being at the rear. It would scarcely be economical, socially, if the patas male had not only to be continually alert to external threat, either from predators or from other patas, but also had to lead the routine activities of the group. The two roles would be incompatible.

Here then we see a beautiful example of socially adaptive behavior in a primate species designed to insure survival in the open grasslands of Africa. Many aspects of the behavior remind one of savannah-living mammals such as hares or antelope. The one-male unit is a breeding unit, and the mating system must be of the harem kind, with the occupant male exceedingly alert to intrusion from other males.

LABORATORY GROUP STUDY

The very nature of these animals in the wild, and the limited time available for the field study (6 months in all, with about 650 hours of observation of the groups), made it impossible to arrive at any more precise analysis of the social interactions within the group. In the Bristol laboratory, a group was set up, modeled on the wild groups in that it consisted of one adult male, two to four adult females, one juvenile (and now young adult) male, and two infants born to the females of the group. This group has just sufficient space to live as a relatively stable unit, provided that it is not interfered with by removing individuals, putting other individuals in, and so on. Such social changes were, however, deliberately induced at times for the purpose of analyzing the resulting tensions in the interactions. Other changes occurred spontaneously, as when a female became sexually receptive and young animals grew up.

From the point of view of studying social interactions and social potentials of a species, laboratory group observations, or experiments, are of little use without the baseline data obtained from the field. Once these normative data are available, even though only preliminary, some valid interpretation of the laboratory social interactions can be attempted. Further, the laboratory data can and should be used as an aid to detailed analysis in another field study, in which it should be possible for the observers to concentrate upon particular social problems. Evidently, the laboratory situation provides little or no scope for the adult male to demonstrate his particular social role in regard to external stimuli, but it does allow of some kind of analysis of his role within the group, particularly in relation to the adult females and the young males.

SOCIAL INTERACTION ANALYSIS

While it is now clearly recognized by field and laboratory group observers that all behavior data and social interactions should be recorded in such a form as to be quantifiable, detailed descriptions are still the basis of analysis, for, until these have been made, it is impossible to know precisely what can or should be quantified when dealing with a hitherto unknown species. Even in the wild patas groups, it was possible to work latterly with prepared score sheets to record social play, grooming, and aggressive interactions, as well as the details of the daily routine. The quantitative profile thus resulting provides a useful basis for intergroup comparisons, as well as for comparing the activities and social organization within one group at different times of year. Comparisons between species can likewise be attempted, as was done with baboons and patas inhabiting the same areas of the Murchison Park.

The laboratory group program has had two main objectives: (1) to work out in detail the social pattern, however limited its context may be, and the changes brought about in this pattern intentionally or spontaneously; and (2) to use the species group as a basis for a social analysis of the subtleties and significances of particular kinds of behavior, such as grooming, aggression, and mating (Boelkins, to be published).

A particular problem on which our recent work has centered concerns the relationships between the adult male and the adult females. These are expressed in a wide variety of social behavior patterns, details of which have already been given elsewhere (Hall, Boelkins, and Goswell 1965).

The animals whose interactions are described in the following sections are

Designation	*Sex and Age Class*
M.	Full-grown male
M.2	Young adult male
M.3.	Juvenile male (born in Bristol)
M.4.	Infant male (born in Bristol)
F.1.	Adult female; mother of M.3
F.2.	Young adult female
F.3.	Adult female; mother of M.4
F.4.	Adult female

The Adult Male

With M.2

The interactions between these two males are quite distinct from those of any other pair of animals in the group.

The series of observations analyzed begin in November, 1963, and end a year later. The end came with a vicious attack on the young male by the adult—the first contact attack that had ever occurred. The young male's blue scrotum and penis had become clearly visible about 14 days prior to this. He had not, however, attempted to mate with any of the females.

On October 27, 1964, M.2 went up to F.3 who was then sexually receptive and sniffed at her rear. M threat-faced M.2 and chased him, with loud lip smacking. M then went after F.3 and mounted her, then chased M.2 with lip smacking. M.2 was sitting well away from him. M.2 put his fingers to his mouth—a usual reaction by him when frightened of M. M returned to F.3 and mounted again, then immediately began again to chase M.2. M.2 screamed loudly, leapt onto pipes high up on the wall where M could not reach him, and clung there until M went away. M again mounted F.3, then sat quietly on lower bars. M.2 groomed F.1 for 15 sec. F.1 (not sexually receptive) pursued F.3 slowly, and F.3 took refuge behind M, who threatened F.1. F.1 avoided M and continued slowly but persistently walking after F.3, who ran away. F.1 continued walking after F.3.

M suddenly, and without any visible provocation, attacked M.2. M.2 now had great difficulty in evading M, and he screamed continuously. M, his mantle fur particularly erect, making him appear very large, continued to pursue M.2, lip smacking noisily and at high frequency. M cornered M.2, struck at him with his hands, and grappled with him on the ground. F.1 rushed up and attacked M from the rear, thus giving M.2 time to escape temporarily. M pursued M.2 very fast indeed, but M.2 was allowed to escape into the adjoining room.

M.2 had the appearance of extreme terror. He had foam about his mouth, defecated several times, trembled violently, and hid under a chair. One hour later, he was still hiding under the chair, and had not touched a banana placed near him. M's aggressiveness ceased immediately M.2 was removed.

It had already been observed, on many occasions, that M's threshold of aggression was greatly lowered as soon as one of the females with him became sexually receptive. When mounting to copulate, he was always ready to turn upon and threaten any other animal that came near him. But most clearly this was shown again and again by the fact that he would now repeatedly strike at his own reflection in two metal doors, whereas normally he was completely habituated to these not very clear visual impressions.

Thus the presence of a receptive female has the effect of building up the aggressiveness in the adult male in proportion to the degree of his interest in that female—as judged by his frequency of mating with her. But only now, for the first time in the many months of M and M.2 being associated in the group, did a full-scale attack on M.2 take place.

From the manifest terror of M.2, it is reasonable to suppose that in the wild he would have fled from the group when such at attack took place, unless he were supported by the counterattacks of the adult females of the group. Such support would seem a highly probable result if the adult male were becoming sexually inadequate, but not otherwise.

At the time of the attack, M.2 was estimated to be 3 years and 10 months old, weighing 20 lbs. He would thus probably not be full grown for about another year or 18 months.

Prior to the attack, the interactions of the two males had been entirely peaceful, except for occasional threat gestures by M if M.2 came too close in the feeding space on the floor. But M.2's behavior had, for about a year previously, become increasingly ambivalent toward M, in that he made threat gestures with his arms in the direction of M, though only when well out of M's reach, while at the same time he uttered a distinctive squealing call, jerking his head sharply from side to side, glancing toward M. There is no "presenting" gesture of submissiveness in patas monkeys, so that this out-of-reach, nervous gesturing may be the nearest equivalent to placatory behavior, although it appears to alternate with gestures that have very low-intensity threat intent—these never being accompanied by movement toward the adult male except of the placatory kind.

The sight of M's genitals appears to attract the attention very strongly both of the young adult male (M.2) and of the juvenile male (M.3). M.2 has, on several occasions, made gestures, when directly below M on the scaffolding, of striking lightly toward, and even touching M.'s testes with his hands. If M responds by the threat face, M.2 at once moves away, uttering the squeal call, cringing, and glancing repeatedly toward and away from M's face. Often these gestures by M.2 are accompanied or preceded by the play face in which the mouth gapes open. Indeed, much of M.2's behavior appears to be that of mock fighting gestures which, however, are almost always made at a distance of several feet from M.

M.2 has occasionally groomed M, but only in an apprehensive manner quite unlike that shown in the "confident" relationship between an adult female and M, or between two adult females. M.2's grooming has almost always been with *one* hand held out at full arm's length to touch and lightly pat the fur on M's back. M.2 has never groomed M from the front, and there is no lip smacking by M.2. Often when the highest-ranking female, F.1, is grooming M, M.2 approaches the pair but sits on the far side of F.1, thus keeping her between him and M. He may then lightly and with one hand groom F.1's back fur.

The only response toward M.2 that is initiated by M is, occasionally, a light touch on the back with one hand. Although M.2's usual response to this gesture is to leap away, it is perfectly clear that there is no threat intended in the gesture. Indeed, it is reminiscent of the light touch of the hand which a baboon male may sometimes give to the back or rear of a young male that has presented to him.

The complexity of the interactions in the patas group is shown on several occasions when M has counterthreatened M.2 and M.3 after either one or both have struck with their hands toward M's testes. On April 27, 1964, when this happened, F.2, with whom M was then mating, joined M.2 and M.3 in threatening M. The three much smaller animals were crouched, heads lowered close to the ground, threat gaping directly at M who stood facing them, noisily lip smacking, but did *not* move toward them. M.2, on this occasion, as also on a few similar occasions, just after the threat interaction had finished, sat on the floor and swept sawdust vigorously aside with a circular gesture of one hand. It seems probable that this gesture, like the waving of the arms when at some distance from M, is indicative of threat arousal, but it is not *directed* at M.

With Adult Females F.1, F.2, F.3, F.4, and the Young, M.3 and M.4

The relationship between M and the adult females depends on several factors, among them being (1) sexual receptivity or otherwise of the female; (2) presence of dependent or partly dependent young with the female; and (3) status of the particular female in relation to the other females in the group.

Sexually receptive female.—F.4, though fully mature and possibly older than any of the others, has never been accepted into any social relationship by the other females; she remains at the farthest point from them in the room, has often been the target of *redirected* aggression, and has never had any grooming relationship with any but the adult male whose protection she seeks whenever possible by going and sitting close to him, but moving quickly away as soon as one of the other females approaches him; he has never interfered with any such removal. F.3, the lowest ranking of the three females, persistently seeks to approach M whenever she is sexually receptive. Although he has repeatedly mated with her, and will temporarily keep the other females from interfering with her when thus engaged, it is difficult for her to keep contact with him because of the close attention of F.1 and F.2 in trying to drive her away from him. By contrast, F.1 can almost always have contact, either in grooming or just sitting close with M, but this contact is usually initiated by F.1 and not by the male.

It is difficult to know precisely what effect female rank has upon birth probability. In the wild groups, most adult females in March and April, 1964, were carrying small infants. Possibly high rank increases the probability of fertile mating if several females were sexually receptive more or less at the same time, but it is more likely that high rank increases the probability of the infant's survival after birth, and possibly, as we shall discuss later, favors the chances of the infant, if female, to attain high rank in the group, or, if male, being retained in the group rather than being ousted on reaching sexual maturity.

Presence of young.—The most significant general finding is that the two mothers of our patas group are extremely protective of their respective young, and this protectiveness is manifested by a fearless aggressiveness toward any other member of the group, including M. There are numerous examples from our data which show that the balance of power is strongly in favor of the mothers, because M never attempts more than low-intensity threat gestures toward an infant or juvenile that encroaches upon his feeding space or his sitting location, and these gestures almost invariably elicit an immediate counterthreat by the mother, in which she is sometimes joined by one of her associate adult females.

Examples:

1.—M.3, apparently attracted by M's genitals when M is above him on the scaffolding, may try to touch M's genitals with his hands. M's usual reaction is the threat face at M.3, possibly with a lunge down at him. F.1 immediately counterthreats M, being usually joined in this by F.2 or F.3, whoever is nearest to the interaction.

2.—M.3, when aged 16 months, chased after M and play struck toward him with his hands several times. M withdrew, then turned and struck back with his hands toward M.3, without touching him. On this occasion, F.2 (not M.3's mother) threat gaped at M, but F.1, who was sexually receptive, took no action.

3.—M was mating with F.1. M.3, her offspring, approached very close to the pair and struck at M from below. M at once retaliated by threat gaping and striking with his hands, first at M.3, then at M.2 who had also approached, then at F.1 who at once counterthreatened.

4.—Infant M.4 occasionally gets separated from his mother, F.3, and gives the "lost" call. F.3 at once shows agitation and usually turns immediately in threat toward M, even though M has made no gesture whatever toward M.4. M reacts to F.3's threats by defending himself, keeping her away with his hands. Once, M.4, when playing, fell from the scaffold to the floor, landing close to M who was sitting there. M.4 squeaked. F.3 ran to him, grabbed him to her belly, and threatened M. On many occasions when F.3 was the only female in the group, with M, M.2 and M.3, if M.4 called out in the course of play between him and M.2 or M.3, she was as likely to direct her protective threat at M as at the other two younger males.

These interactions, consistently of the same pattern, provide abundant evidence that the adult females will not tolerate any interference with their young from the adult male, and that they are ready to offer protective threat at the least sign of agitation by the infants. This protective threat is generalized so that it takes in any animal in the group whether it has had anything to do with the agitation or not.

While it is perfectly clear from group feeding and the spacing of the animals that the adult male takes the center of the feeding space, the females with their young positioning themselves in relation to him according to their own relative rank, the adult male will never assert himself except by the briefest threat, toward either the mothers or their infants without invoking retaliation from them.

Status of a female.—The ranks of the females within our group seem to be established without reference to the adult male: that is, rank is the result of interactions among the females, such as are described below, in which the adult male takes no part. When rank is established, it is the highest-ranking female who initiates all approaches to the adult male by placing herself closest to him, by grooming him more frequently than he is groomed by other females, and by eliciting grooming from him in greater proportion.

The female initiative in this was particularly well shown when F.2 supplanted F.1. F.2 achieved ascendancy, with M taking no part whatever in the interactions, except to watch what was going on. He at once accepted the change of rank by entering into grooming relationships with F.2. The only exception to this is when a lower-ranking female is sexually receptive, but, as stated above, this female is in a distinctly difficult situation, often being harried by the nonreceptive higher-ranking female.

Adult Female Interactions

When a stable relationship has been established, as between F.1 and F.3, F.1, and F.2, the behavior of the subordinate differs significantly from that of the dominant female.

Grooming is invited by the dominant and is received by her in far greater proportion than it is given by her. Grooming may be initiated by the subordinate who approaches the dominant cautiously, glancing repeatedly into her face, and the dominant does *not* look at the subordinate. Although grooming between such a pair is mutual, it is clear that the relationship has to be nicely balanced. While it is possible that in larger groups with much greater space to maneuver, more subtle or even equal relationships between females might occur, none ever existed in our laboratory group. Either a clear dominance relationship was established or the subordinate was so persistently harried that it did not enter into any real social relationship with the dominant at all, just keeping continually out of the latter's way (as was always the case with F.4).

The following are the ways in which females expressed rank: (1) by being the major grooming recipient, and inviting grooming, (2) by position on the floor when feeding, (3) by position at the look-out window, (4) by persistently harrying the subordinate before any other mutual relationship has been established. Harrying consists of walking fairly slowly after the subordinate and causing her over and over again to move away. It consists particularly of walking-toward, without looking the subordinate in the face, that is, with the action of turning the head from left to right and *looking past* the subordinate to either side ("facing away" in the displays of gulls). Striking with the hands may occur if the subordinate does not move out of the way; this seems to be the process whereby female ascendancy is achieved when it is not contested; when it is contested, as in the case of F.1 and F.2, actual fighting face-to-face occurs; it was difficult to see why F.1 "won" on earlier occasions, because the fighting injuries were fairly light, and neither could be said to have gained an advantage as a result. Hence, presumably, in a series of equal encounters, the already dominant female manages to retain her place.

The relationship is also expressed quite subtly in the spacing during feeding on the floor. The dominant female sits close to the pile of seeds that she is eating, and usually does not fully extend her arm in picking them up. The subordinate feeds facing the dominant, and repeatedly glances up watching for signs of threats or movements toward her. The dominant female does not glance about her, except possibly at M if he is there, and she may sit with her back to the subordinate or in whatever position is convenient for her in getting the food.

Tension within the group is always temporarily greatest upon the return of an animal that previously was in the group. It is on such occasions that combined aggression toward the newcomer may occur, but the agility and speed of the victim will usually ensure its avoiding any actual contact, until tension is allayed and a new stable relationship achieved.

The most important interaction between adult females took place when F.2 was returned to the group, which then consisted of M, F.1, F.4, M.2, and M.3. F.1 and F.2 fought fiercely, and though neither sustained severe injuries, F.2 now persisted in harrying F.1 to such an extent that F.1 capitulated to the change and allowed F.2 to become the No. 1 female. However, the course of events, rather than the outcome, is important, because it demonstrates very clearly that the adult male is the focal point around which these aggressive interactions take place, *but that he accepts the outcome,* rather than in any way aiding in it. A summary of the events that started on October 13, 1964, and ended with F.2 being removed from the group on October 19th, is as follows—preceded by observations of an earlier encounter in which F.1 retained her rank:

April 30, 1964:

The group consisted of M, M.2, M.3, F.2, F.4.

1430:

F.1 was reintroduced to the group from which she had been removed a few days previously; F.1 at once reasserts her rank when her own young, M.3, strikes at F.2 —probably as a result of the greatly increased tension in the group—by fighting with F.2, then by ousting F.2 from proximity to M.

1445:

M comes right up to F.1, touches her on the back with his hand, then stands to be groomed; F.1, however, still has to threaten F.2 away, as F.2 tries to get near M; M approaches F.2, but F.1 at once ousts her; F.2 now for the second time attacks and fights with F.1, and M himself strikes at F.1 with his hand but does not engage in any other way with the fighting females; F.1, at the end of a fight lasting 30 seconds, continues to go after F.2, and F.2 gives way.

TABLE 1

DURATION AND FREQUENCY OF GROOMING EPISODES BETWEEN F.2 AND F.1 IN THE THREE 1/2-HOUR PERIODS IMMEDIATELY FOLLOWING RESUMPTION OF GROOMING RELATIONSHIPS AFTER AGGRESSIVE INTERACTIONS

Period	No. of Grooming Episodes in Terms of Groomer		Total Duration (sec) as Groomer		Per Cent Total as Groomer	
	F.2	F.1	F.2	F.1	F.2	F.1
First half hour..	8	2	1,110	75	93.7	6.3
Second half hour	10	4	645	58	91.7	8.3
Third half hour.	5	2	233	111	68.0	32.0
Total........	23	8	1,988	244		
Normal grooming ratio.......					64.0	36.0

1450:

F.2 now changes her behavior completely, trying to approach F.1 without any threat intention, and touches F.1's thigh with her hand at full arm's length.

1455:

F.2 continues now to follow F.1 whenever F.1 moves and starts to groom F.1's back fur briefly.

1505:

F.2 grooms F.1's front (that is, aggression between them now completely reduced) and F.1 relaxes completely as she is thoroughly groomed by F.2 for 270 seconds—a much longer time than the average for these two.

The total of grooming episodes between F.2 and F.1 is of interest, as it is far in excess of anything that normally occurs when relationships within the group are stable (Table 1).

Not only is the absolute amount, in time and frequency of grooming, far greater than the normal during the first and second half hours, but the *proportion* of grooming carried out by F.2 (over 90 per cent) is much in excess of the normal proportion of 64:36 for this pair. These data clearly indicate the significance of grooming among the adult females as the necessary behavior for aggressive tensions to be reduced, and the very high proportion given by the subordinate serves both to indicate submission and thereby to prevent further aggression by the dominant. As Marler (1965) points out, grooming "serves at least partly to pacify relationships between the communicants," and it diverts the recipient to an activity which is incompatible with aggressive behavior. During this period, there were only two other grooming interactions in the group: M.2 tentatively groomed F.1 for 10 seconds, and F.1 started to groom M but for only 5 seconds.

Once the aggressive interactions between F.1 and F.2 had ceased, and the grooming relationship was being reestablished, the aggressive tension in F.2, being no longer directed toward F.1, was now very readily diverted to others in the group and to her own image as reflected in the metal doors inside the room. Thus, F.2 lip smacking noisily, struck at the metal door several times with her hands and tugged at it with both hands, and turned her aggression onto F.4 who, being outside the hierarchy, is normally ignored. The relationship between F.2 and M was also tense. F.2 had several times tried to get near M, but each time was ousted by F.1. F.2 actually threatened M four times, once in the high-intensity crouch-threat posture in which she approached within 3 feet of him. His reaction was either to threat stare and gape at her, or to *look away* from her. F.2 and F.1 later both chased F.4—again an unusual reaction induced by the tension between them.

October 13, 1964: the group consisted of M, M.2, M.3, F.1, F.4.

1520:

F.2 was reintroduced, that is, the reverse procedure to that of April 30, when F.2 was already in the group and F.1 was reintroduced.

F.1 was sitting close to M. F.2 at once goes toward F.1; both utter threat calls, then fight viciously but briefly; the threat interactions between F.1 and F.2 now center around M, as F.1 tries to sit near him but is now each time ousted from this position by F.2, without M taking any action toward either of them; the initiative is repeatedly taken by F.2, and F.1 tries to keep M between herself and F.2's threatening approaches.

1530:

F.2 initiates a second fight with F.1; M lip smacks noisily, but no other action from him; F.2 invites grooming from M, but gets no response, and she briefly grooms him; as soon as she moves away from M, F.1 approaches M, and M briefly grooms her.

1540:

F.2 approaches M, and F.1 at once retreats, and F.2 grooms M briefly; both F.2 and M lip smack noisily as F.1 approaches them, and now, even when F.2 leaves M, F.1 does not venture near him.

1550:

The same interactions continue, namely, F.*2* always ousting F.1 if F.1 goes near M. F.*2* repeatedly harries F.1 by chasing her or walking after her.

1600–20:

One or both females occasionally redirect their aggression to F.4; once, when F.1 is chasing F.4, F.*2* strikes at F.1, but F.1 continues chasing F.4, and F.*2* chases F.1.

1650:

F.*2* still occasionally chases F.1, and once grabs her, making her scream; as F.1, fleeing, passes close to M, he joins with F.*2* in attacking her, and she is allowed to escape into the holding pen in which she is shut for the night.

October 14, *0930:*

The same interactions as the previous day, with F.*2* initiating attacks and occasionally harrying F.1; F.1's son, M.3, leaps on F.*2* who at once turns on him, causing him to squeal, but F.1 makes no attempt to counterattack F.*2*, as she would previously have done.

1015:

F.1 once retaliates when attacked by F.*2*, instead of trying to keep out of her way.

1100:

F.1 creeps up on F.4, who is not looking and grapples her from behind; F.4 breaks loose, and flees. F.1 pursues her, but F.*2* attacks F.1 who flees and is now attacked by both F.*2* and M.*2* (who had not previously joined in any of these interactions).

1400:

F.*2* attacks F.1, and M joins in very aggressively against F.1, and F.1 tries to hide in the corner of the room.

Later the same day, tension is a little reduced, but F.1 will not eat food thrown onto the floor of the room.

October 15, *0930:*

F.*2* alternately prowling after F.1 and keeping close to M.

0945:

F.*2* goes after F.1; F.1, passing near F.4, utters threat calls, and both F.*2* and F.1 chase F.4, followed by M.3, while M threatens generally, that is, stands threat gaping toward any animal that comes near him.

October 16: The group was removed from the room, for cleaning purposes, and then was put back into it together. Except for M, who is kept in a cage in the next room, the situation remains the same—namely, F.*2* sometimes prowling after F.1, but F.1 keeps well out of her way and does not retaliate when F.*2* and F.1's son, M.3, engage in rough play in which M.3 shrieks. F.1 will not eat on the floor with the others and is now repeatedly ousted whenever she goes to the look-out place; even when F.*2* is grooming M.*2*, she at once stops grooming in order to chase F.1 from the look-out; when M.*2* approaches F.1 and attempts to mount her, F.*2* always intervenes and ousts F.1; (M.*2* is not yet sexually mature).

October 17: The group as on October 16; that is, without M, and the situation as between F.1 and F.2 remains the same; the infant M.4 is introduced into the group (the son of F.3, who is not in the group); F.1 lip smacks noisily; F.2 does a noisy display bounce and chases F.4 and is joined in this by F.1, but M.4 is completely untouched, because the aggression at once turns into an attack by F.2 on F.1 in which the latter is badly bitten.

The outcome of these several days of aggressive interaction between the two females is that F.1 is definitely ousted from her No. 1 rank, and there is not a single attempt at placatory or grooming behavior between the two. M was removed in order to observe whether his presence had any major effect on the dominance relations of the females. As was expected from his initially passive role in the aggression of the females, and from his later acceptance of the reversal of rank shown by his joining in attacking F.1, the ranking is determined almost entirely by the females themselves. This was already seen to be the case in the interactions on April 30, when F.1 maintained her rank.

It should be noted that the factors which brought about the change in rank are still not entirely clear. F.1 *retains* her previous rank when she is *introduced to the group from outside.* F.2 overcomes F.1 when *F.2* is introduced from outside. Possibly, the introduced animal may be under greater tension than the ones already in the group, and thus might have an initial advantage, but this has not been the result when other females, such as F.3, have been introduced. Further, the change of rank between F.2 and F.1 was strongly maintained over several days; therefore it is likely to be simply the consequence of the growing up and probably reaching sexual maturity of F.2 that is the crucial factor leading to the change.

The female interaction process was likewise clearly seen when F.3 was introduced to a group consisting of M, M.2, F.1, and M.3. F.3 was clearly subordinate from the start, although at least as large as F.1. F.1 began by repeatedly harassing her in the usual way. F.3 did not fight with F.1; she just tried to keep out of her way. F.3 was sexually receptive during the first 3 days, and M mated with her repeatedly. F.1 frequently tried to oust F.3 from M, but did not attack her when F.3 was with M, except when there was a lull between mating series. Thus, although the dominant rank of F.1 was asserted against F.3 on every possible occasion and F.3 therefore did not alter her rank vis-à-vis F.1 because she was sexually receptive, there is no doubt that low rank does not prevent impregnation (as is also shown by the number of infants with their mothers in the wild groups in March and April, 1964). It does, however, prevent any sustained grooming relationship being established between the subordinate female and adult male, and it *could* indicate that the offspring of the subordinate might have difficulty in asserting itself against the offspring of the dominant—as is suggested by some of the findings of the Japan Monkey Center on *Macaca fuscata,* and by Sade (this volume) on *Macaca mulatta.* But the situation in this respect is by no means simple, as we shall see in discussing socialization.

On the third day after F.3 was put into the group, a fairly stable relationship between F.1 and F.3 was established. F.1 approached F.3, who did not now flinch away, and they sat face to face very close together. F.3 groomed F.1, and F.1 groomed F.3. Seven weeks later, this relationship still held, F.3 being definitely the subordinate, and only retaliating toward F.1 if her infant (M.4), who was also introduced to the group on the fifth day, were threatened by F.1.

In the case of F.2 and F.3, no stable relationship could be established; F.2 continued to harass F.3 even after several days without any grooming interactions occurring.

The provisional conclusions from these long series of observations on the adult female interactions are:

1. Aggressive interactions may sometimes center around the adult male in that one important priority of rank is in getting close to him even when the females are not sexually receptive. The highest-ranking female will keep nearest to him in the feeding space on the floor, but this is *her* choice rather than his, the greater distance of the second-ranking female being maintained because she would be threatened, not by him, but by the dominant female if she went nearer.

2. The limited evidence available from so small a group in so restricted a situation strongly indicates that it is the females who work out their respective ranks among themselves and that the male accepts the outcome of aggressive interactions without making any attempt to interfere. He reacts by threat only when one of them approaches too close to him, and there has not been a single occasion on which he has initiated an attack on one of them.

3. To this limited extent, the laboratory observations, based on heightened tensions within the group brought about by changes of membership as well as on the stable relationships achieved within the group, appear to provide the social detail that confirms and elaborates the outline of the social organization as depicted in the field study. That is to say, the one-male unit of the wild is highly organized around the adult females, the adult male having a clearly defined role as breeder and as watcher for external threat, with all major initiative *within* the group coming from the adult females.

Socialization of the Young

The most obvious characteristic of the patas mother in relation to her infant is, in the laboratory group, her high degree of protectiveness. No reciprocal relationships between females without infants and females with infants—such as occur in baboon groups (Hall 1962)—were observed, and the contrast between the patas and the common langurs as described by Jay (1963) is even more marked. It has to be assumed that the process of socialization characteristic of a species must fit in clearly with the form of social organization observed in the species. Thus, Jay describes how, several hours after parturition, the mother langur will allow an adult female to hold the infant for several minutes, and an infant may be held by as many as eight or ten females and carried as far as 75 feet from its mother in the first 2 days of life. The social unit for the protection and nourishment of the newborn langur does not include the adult male, this being in contrast to the baboon newborn whose mother remains with other adult females and young close to the most dominant adult male in the group. If an adult male langur accidentally frightens an infant, the mother instantly threatens, chases, and often slaps the male. Other adult females near the infant may join the mother and chase the male as far as 25 yards.

In the wild patas groups, when there were many infant-1's (March and April, 1964), the adult male was usually far away from the rest of the group, and mothers with infants were never seen to associate closely with him. Indeed, it follows from our field observations of these animals that the most useful position for the adult male is on the periphery where he can watch for predators and possibly divert them from the mothers and young while they hide in the grass or run away.

In the laboratory group, the fact that the adult male stands outside the mother-infant relationship is also readily apparent by the spacing among the animals and likewise by the readiness of the mother to threaten the male should there be an accidental encounter between that male and an infant. The adult male has never initiated an attack on an infant, but if he does threaten one of them for approaching too close, we have seen that both the mother and the other females will join in strongly threatening him and making him move away. In this, they are quite fearless, and their sometimes close threats to him in confined space have never led to retaliation by him, except in the form of slapping defensively at them to keep them away.

Nowadays, the question of the rank of the mother and its relationship to the rank of her infant is much discussed, and it is sometimes assumed that the offspring of the high-ranking female will itself, through social learning, achieve high rank in the group when it reaches maturity. But the many factors involved in such social interactions must be clearly appreciated. Thus, there is no doubt whatever that M.3, the offspring of F.1, has a very privileged position in the group which is given him by the firm base of his mother's rank. He ventures very close to the adult male on the floor feeding space and learned quickly to be the first to obtain any special food item put into the room, because his mother would immediately threaten any other animal that might dispute his priority. But, as soon as his mother was removed from the group, his privilege was lost, M.2 asserting himself immediately.

The infant of F.3 (M.4), a year younger than M.3, is likewise constrained in his feeding access because of his mother's lower rank, but he is so robust an animal that he enters very vigorously in the play interactions of M.2 and

M.3, and it is by no means clear that he would necessarily emerge as a low-ranking or outcast male in a wild population.

This brings us to a brief consideration of the possible implications of the strongly organized female hierarchy in relation to the males. If the female ranking system is such that it is primarily determined *among themselves* and not by preference or intervention on the part of the adult male, it may follow that the initiative in keeping a male within the group or allowing one to be ousted from the group comes through the female hierarchy. In the case we have considered—namely that of M.2 being attacked severely by M soon after M.2 reached sexual maturity—it seems more than likely that the sexually mature offspring of the highest-ranking female might not be driven out, or if he were driven out, this female and her nearest associate females might well leave the group with him, and thus start another breeding unit.

Many months of intensive field observation of a patas population would be necessary before any conclusions on this important point of social organization could be reached, but we are convinced that exactly this type of problem could be thoroughly worked out in outdoor compound studies of the species, given a sufficient space for the animals to change their groupings without lethal fighting.

The social play of the patas is, as we have described it, exceedingly vigorous and spectacular in the wild, much of it being high-speed chasing—some of it mock-fighting. Again, to what extent these play interactions serve to establish ranks among those of the same age group is quite unknown, and certainly could only be satisfactorily studied in long-term observations of fairly large captive groups.

SOCIAL ORGANIZATION AND SPECIES COMPARISONS

Two of the major categories of behavior that reflect the social organization of a species are the friendly and the aggressive. Although these appear to work against each other, they are both equally adaptive and indeed complementary, so that the frequency, intensity, and quality of the one may correlate or balance with the other. Many of the interactions one sees in the laboratory group of patas are expressed by alternating or fluctuating tendencies. An individual act of aggression, that is, one aggressor against one victim, may be quite uncomplicated. But many of these episodes *are* complicated in that they involve not just the two animals but others as well. In an episode involving several animals, we find that the aggressor will lip smack noisily while looking toward others in the group who may then join with her, also lip smacking, in attacking, for example, the scapegoat of the group, F.4. This lip smacking seems to differ from that which occurs in grooming interactions only in its great rapidity and noisiness, but the expressions which accompany it—the threat face and so forth—are entirely different. Possibly this lip smacking is an exaggerated form of that which signifies a relaxed, friendly relationship, and is thus used, in the aggressive context, as a signal which, combined with head turning, may enlist aid from the aggressor's associates. If this is so, then it is only secondarily a threat-intention signal.

Another distinctive feature of patas aggressive interactions is the gape yawn. In its directed form, this may accompany the threat face in which the aggressor is looking straight at the victim. In its *undirected* form, it can happen repeatedly without any overt threat intentions or expressions, and here it would appear to be simply a tension or arousal indicator. And it should be noted that, whereas aggressive interactions among baboons are often noisily vocal, this is very rarely the case with the patas, who will fight in complete silence or who will threaten intensely, with only a low muffled vocal accompaniment of *huh-huh-huh-*, a repetitive series of grunts.

Similarly, the peculiar "ingratiation" behavior, as we have called it, shown by M.2 toward M, seems to be compounded of submissive gestures, including the squealing vocalization, and very mild, and, one might almost be permitted to say, unintentional threat gestures with the arms, but which are not, however, accompanied by any "real" threat expressions such as the threat face.

These more complex and subtle social exchanges are paralleled, on the friendly side, by grooming and play invitation gestures, such as the play bounce, mouth-to-mouth, and so on, but, interesting though these are, the general nature of the social pattern is seen most clearly in the quality, frequency, and duration of the straightforward exchanges involved in grooming and threat.

We have seen that the grooming between F.1 and F.3, prior to F.1's deranking, is tremendously in favor of F.1 as recipient (about 90 per cent), until the relationship stabilizes with reduction of tension to its normal level between these two of about 68 per cent. We likewise find that the reintroduction of any of the adults with whom the other adults had previously been on grooming terms sets off a much greater amount of grooming than the normal, while a sexually receptive female will not only groom more often with the adult male than normal but will receive a far higher proportion of his attention as groomer —always provided that the female in question is not being harried by a more dominant female. The final point about the grooming relationships is that the amount of time one of the pair spends as groomer of the other is significantly related to the rank relationship between them *in the group as then constituted.*

Thus, M received grooming, on the average, in a proportion of about 80 per cent from the three adult females, F.1, F.2, and F.3. Between the females, F.1 received grooming from F.3 at 80 per cent but only 64 per cent from F.2. M.2 groomed F.1 and F.3 at about 75 per cent, thus indicating, as did his general behavior in the feeding space and other situations, his continuing subordination to them.

The other indications of rank are clearly to be seen in the various forms of behavior that indicate dominance, on the one hand, and subordination, on the other. Spacing is one such indicator; moving toward (or away) is another; posture or attitude is another. Thus the subordinate frequently glances around or at another because he or she has to be on the alert against infringing the social distance.

All these indications of social relationship add up to a complex pattern of communication behavior which, however, is changing in the course of time as the young mature and the elders decline.

What is the general significance of the pattern for the patas as a species? Here, we have to consider the patas in comparison with some of the other species on which extensive field and captivity data are available.

As an animal that is adapted to ground living, we can first and most usefully compare it with the baboons. By and large, it looks as though those species in which social ranks, in terms of various behavioral criteria, are most clearly discernible, are those in whom, according to Marler (1965), tactile signals such as are involved in grooming are especially prominent. Amongst the Old World monkeys, at any rate, it seems likely that the ground-living species are much more clearly organized in terms of aggressive-friendly interchanges than arboreal species. Further, the physical differentiation in terms of size and appearance is much more evident in the ground-living species, for, as DeVore and Washburn (1963) point out, "the answer to the degree of sex differences appears to be that this is the optimum distribution of the biomass of the species" (p. 346). In other words, there are obvious protective advantages in the large size of male baboons, but there is an equally *economic* advantage, so far as food supply is concerned, in the females being small.

Now, the proportion of full-grown males to females in baboon groups varies quite markedly from area to area of Africa, the greatest discrepancy being apparent in the harsh habitat of southwest Africa and in the hamadryas groups of Ethiopia. The hamadryas, living in a harsh environment, have a social organization that compares interestingly with that of the patas, namely the one-male unit system (Kummer and Kurt 1963) in which one adult male may have in his unit as many as nine adult females and their offspring. The patas have the same basic system, and this would seem to be the most economical system that can ensure adequate food, on the one hand, and a one-male breeding unit, on the other. But, further, the patas, unlike the hamadryas, live in quite small groups, each in a very large home range, so that one group rarely meets another. Spacing between groups is maximal—on the monkey scale—and probably spacing between individuals within the groups is also maximal. And, because the patas have evolved alternative adaptations of high-speed locomotion and the peculiar role of the adult male as a watcher and diverter, the adult females have had to achieve a social role that is in degree, if not in kind, rather notably different from that of the females of the other ground-living species.

Although the greatly reduced space of the laboratory environment must inevitably bring out in higher intensity and frequency the social behavior patterns upon which rank is based, this is precisely the value of the captivity situation when the data can be used for comparison with, and elaboration of, field data. When we establish larger enclosures, it will become possible to work out the interactions over a long period not only in one group but also in and between two or more groups. Further field study is planned in a yet unsampled area of the species distribution, namely the savannah regions of the northern Cameroon. From the studies of other widely distributed species, such as the baboons and the langurs, it is reasonable to expect considerable regional variations in the patas populations. But, however great this variation may be, the captive group studies can serve as experimental models whose significance can easily be interpreted by reference to the range of ecological conditions in which the species exists in nature.

REFERENCES

DeVore, I., and Washburn, S. L. 1963. Baboon ecology and human evolution. In *African ecology and human evolution,* ed. F. C. Howell and F. Bourlière. New York: Viking Fund Publications in Anthropology.

Hall, K. R. L. 1960. Social vigilance behaviour of the chacma baboon, *Papio ursinus. Behaviour* 16:261–84.

———. 1962. The sexual, agonistic and derived social behaviour patterns of the wild chacma baboon, *Papio ursinus. Proc. Zool. Soc. Lond.* 139:181–220.

———. 1965. Behaviour and ecology of the wild patas monkey, *Erythrocebus patas,* in Uganda. *J. Zool.* 148:15–87.

Hall, K. R. L., Boelkins, R. C., and Goswell, M. J. 1965. Behaviour of patas monkeys, *Erythrocebus patas,* in captivity, with notes on the natural habitat. *Folia Primatol.* 3:22–49.

Jay, P. 1963. Mother-infant relations in langurs. In *Maternal behavior in mammals,* ed. H. L. Rheingold. New York: Wiley.

Kummer, H., and Kurt, F. 1963. Social units of a free-living population of hamadryas baboons. *Folia Primatol.* 1:4–19.

Marler, P. 1965. Communication in monkeys and apes. In *Primate behavior: field studies of monkeys and apes,* ed. Irven DeVore. New York: Holt, Rinehart and Winston.

Sanderson, L. 1957. *The monkey kingdom.* Garden City, N.Y.: Hanover House.

Tappen, N. C. 1965. Comment on Hall, K. R. L. Ecology and behavior of baboons, patas, and vervet monkeys in Uganda. In *The baboon as an experimental animal.* San Antonio, Texas: S.W. Research Foundation.

16

AUDITORY COMMUNICATION AMONG VERVET MONKEYS (*Cercopithecus aethiops*)

THOMAS T. STRUHSAKER

INTRODUCTION

The primary purpose of this chapter is to present and describe a catalogue of the audible behavior of the vervet monkey (*Cercopithecus aethiops johnstoni*). This catalogue is a list of those sounds of vervets that I was able to distinguish either by ear or from tape recordings and spectrographic analysis, and is not a pretense at a definitive list. It is possible that with further study this list may be increased or decreased. In addition, when available, information on and impressions of the conditions evoking the sounds and the communicative function of the sounds are given.

A brief résumé of vervet social structure is presented here to facilitate the interpretation of the auditory repertoire. Within the Amboseli Reserve of Kenya, East Africa, the vervet monkeys lived in relatively closed and stable social groups with an average size of about twenty-five individuals. For all age classes, the sex ratio was about one-to-one. Each social group defended a well-defined and relatively stable territory against neighboring vervet groups. A linear and relatively stable dominance hierarchy existed among the adults and older immature individuals within each social group. This dominance hierarchy was expressed primarily through priority to food and spatial positions and aggressiveness in agonistic encounters.[1]

Thomas T. Struhsaker, The Rockefeller University and New York Zoölogical Society, New York.

[1] I am greatly indebted to Drs. Stuart Altmann, Irven DeVore, and Peter Marler and Mrs. Jeanne Altmann for their assistance and advice in this study. Thanks are also due to the staffs of the Departments of Zoology at the University College of Nairobi, Kenya; and the University of Alberta, Edmonton, Canada. This investigation was supported in part by Public Health Service Fellowships 1 FL-MH-19, 381-01 and 5 FL-MH-19, 381-02 from the National Institute of Mental Health, U.S. Public Health Service.

METHODS AND MATERIALS

On May 30, 1963, I initiated a field study on the behavior and ecology of the vervet monkey in the Masai-Amboseli Game Reserve, Kenya, located at 2° 45′ S, 37° 15′ E, at an elevation of 3,700 feet. Intensive observations were made through June 12, 1964, with a few brief absences during this time. A final revisit to the study area was made on August 17 and 18, 1964. A minimum of 2,248 hours was spent observing, 157 hours photographing, and 64 hours tape recording the vervets at Amboseli. All the tape recordings were made under field conditions at Amboseli, of vervet monkeys that were relatively undisturbed by man. Whenever possible, data on the conditions evoking the sounds and the associated behavior were recorded.

The tape recordings were made with an Uher 4000 Report-S portable tape recorder at 7½ ips with a frequency response of about 40 to 20,000 cps (manufacturer's specifications), parabolic reflector of about 2½ feet in diameter, and a Shure dynamic microphone—Model 430.

The tape recordings were analyzed with a Kay Electric Sono-Graph, Model 661-A, using the HS shaping switch and wide bandpass filter (300-cps resolution). Pitch (frequency) and temporal measurements of the sonograms were made with a transparent plastic grid overlay.

For tests of significance of the means of the various measurements of the sonograms, the Wilcoxon test for unpaired samples was used for small, non-parametric samples; a modification of the Wilcoxon test for unpaired samples was used for large, non-parametric samples (Alder and Roessler 1964); the Student *t*-test was used for small, normally distributed samples; and a two-tailed *T*-test or Critical Ratio was used for large, normally distributed samples.

Definition of Terms

The following terms are used to describe the various features of the vervet sounds as seen on the sound spectrograms:

Unit: The unit is the basic element of a vervet sound or call, and is represented as a continuous tracing along the temporal (horizontal) axis of the sonogram.

Phrase: The phrase is a group of units that is separated from other similar groups by a time interval greater than any time interval separating the units within a phrase.

Bout: A bout is a grouping of one or more phrases separated from other similar groupings by a time interval greater than that separating any of the phrases within a bout.

Nontonal unit: A nontonal unit is composed of sound that is more or less continuously developed over a wide range of frequencies. This has also been called "noise" by Andrew (1963*b*) and "harsh noises" by Rowell and Hinde (1962).

Tonal unit: A tonal unit is composed of sound characterized by one or more relatively narrow frequency bands and has been referred to as "clear calls" by Rowell and Hinde (1962) and "sound" by Andrew (1963*b*). Units with a harmonic structure are included in this category.

Compound unit: A compound unit is composed of both nontonal and tonal sounds that appear as a continuous tracing on the sonogram.

Mixed unit: Units composed of both tonal and nontonal sounds that are rather superimposed on one another are called mixed units. The tonal and nontonal aspects are more or less separated by differences in frequency.

Distribution of major energy of nontonal sounds: The distribution of the major energy of a nontonal sound is represented and thus determined by the darkest portion of the tracing on the sonogram. This distribution is generally over a smaller range than the frequency range of nontonal sounds. The distribution of major energy of nontonal sounds is described by an upper frequency (highest pitch of major energy) and a lower frequency (lowest pitch of major energy).

Age classes: The following terms are used to represent the estimated age classes: adult ($\geq$ 4 years old), subadult (3–4 years old), juvenile (1.5–3 years old), young juvenile (0.5–1.5 years old), and infants (0–0.5 years old).

Other terms not defined here are considered to be self-explanatory.

The major groupings or subdivisions of the following catalogue represent stimulus situations. Cataloguing sounds according to their stimulus situations has been arbitrarily selected from several other possible methods, such as according to physical properties of the sounds, messages communicated, or responses evoked by them.

Intragroup Agonism

Situations within this category include intragroup agonistic encounters, which ranged from moderate to very high intensity. Such situations usually included threats, counterthreats, chases, and, less commonly, actual physical contact including grabbing, hitting, and biting. It was my impression that these agonistic encounters were most commonly engaged in by adult females and juveniles and less commonly by adult and subadult males.[2]

[2] All measurements of recorded calls are summarized in a table that is not included in this chapter owing to its lengthiness. This table gives, for each recorded call, the age and sex of the vocalizer, per cent of tonal and nontonal sound in the call unit, distribution of major sound energy in lower and upper frequency components, pitch of fundamental frequency and of harmonics, duration of call unit, number of units per phrase, interval between units, duration of phrase, and duration of interval between phrases. For these frequency and time data, the sample size, mean, range, and standard deviation are given.

This table has been deposited as Document Number 8840 with ADI Auxiliary Publications Project, Photoduplication Service, Library of Congress, Washington, D.C. 20540. A copy

*Intragroup Chutter**[3]

The Intragroup Chutter (Plate 16.1) was a low-pitched, nontonal, staccato-like call. This call was sometimes intermingled in the same phrase with the Squeal, Chutter-squeal, Squeal-scream, or Scream calls. The Intragroup Chutter was given only by adult females and juveniles (1–3 years old). Audibly I was unable to distinguish any age or sex differences in this call. Furthermore, the apparent differences between adult females and juveniles in upper frequencies of major energy, durations of units, and intervals between units were not significant at the 0.10, 0.10, and 0.20 levels, respectively.

The Intragroup Chutter seemed most similar to the Intergroup Chutter B. However, in duration of units, the Intragroup Chutter was significantly greater than the Intergroup Chutter B at the 0.01 level. This difference also indicates that the difference in the duration of phrases of the two calls may be significant. The Intergroup Chutter A was the only other call which seemed similar to the Intragroup Chutter. The Intergroup Chutter A and the Intragroup Chutter differed significantly from one another as summarized in Table 1.

The mouth was closed or only partially opened and the teeth were not exposed while the Intragroup Chutter call was given. This call was commonly accompanied by inaudible threat gestures, such as staring, exposing eyelids, head bobbing, and so forth.

The chutter seemed to serve as a solicitation for aid from others and as an aggressive threat against the opponent(s).

*Squeal**

The Squeal (Plate 16.2) was a high-pitched, rather uniform, and tonal call. Harmonics were absent from this call, in contrast to the presence of harmonics in both the Lost Squeal and Weaning Squeal calls. It differed from the Chutter-squeal in lacking the nontonal portion, which in the latter preceded and/or succeeded the tonal sound.

*Chutter-squeal**

The high-pitched, tonal portion of a Chutter-squeal unit (Plate 16.2) was generally preceded and/or followed by a nontonal portion, which spectrographically resembled the Intragroup Chutter. One mixed phrase, which was recorded, resembled the Squeal-scream. The Chutter-squeal is similar and may be ontogenetically related to the Lost and/or Weaning squeals. (See section on Weaning Squeal and Lost Squeal below.)

---

may be secured by citing the Document number and by remitting $5.00 for photoprints, or $2.25 for 35-mm microfilm. Advance payment is required. Make checks or money orders payable to: Chief, Photoduplication Service, Library of Congress.

[3] Tape recordings were made of all calls marked with an asterisk.

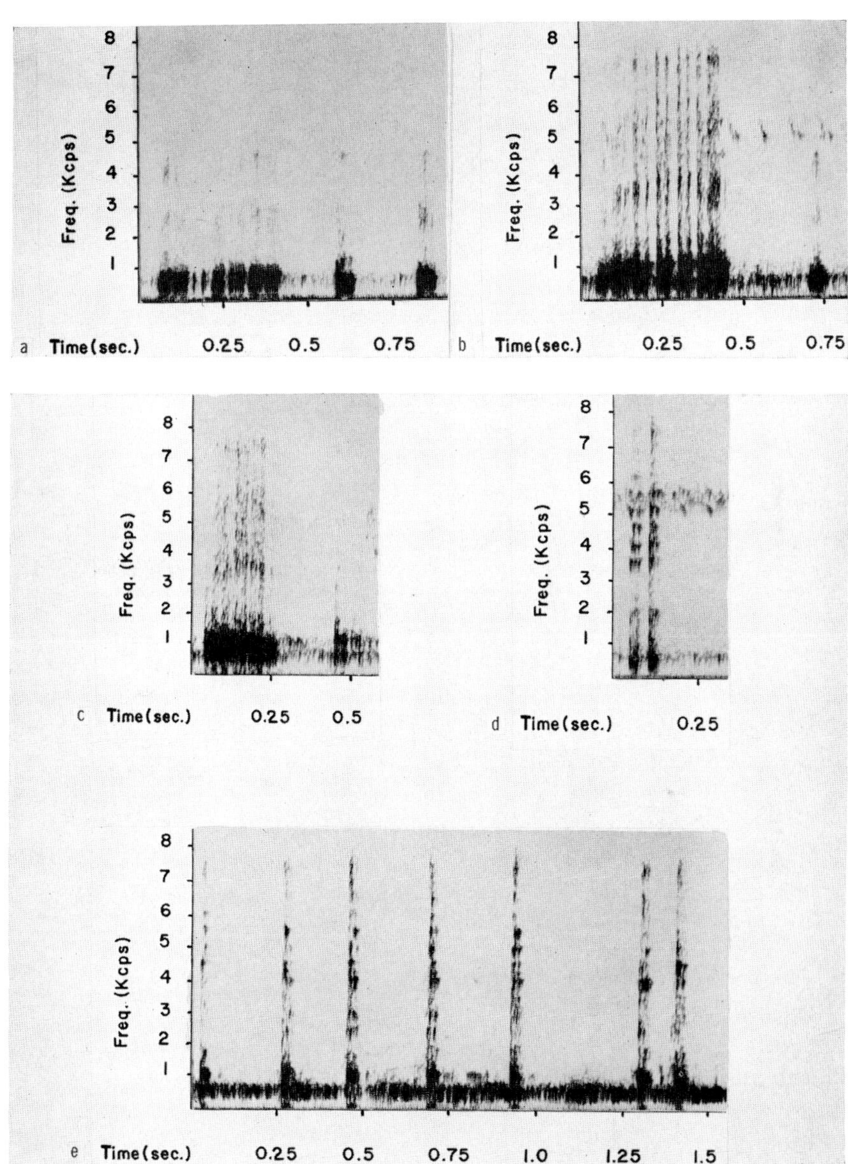

PLATE 16.1.—(*a*) Intragroup Chutter, (*b*) Intergroup Chutter A, (*c*) Intergroup Chutter B, (*d*) Snake Chutter, (*e*) Chutter-Toward-Observer. (Ignore background noise between 5 and 6 Kcps.)

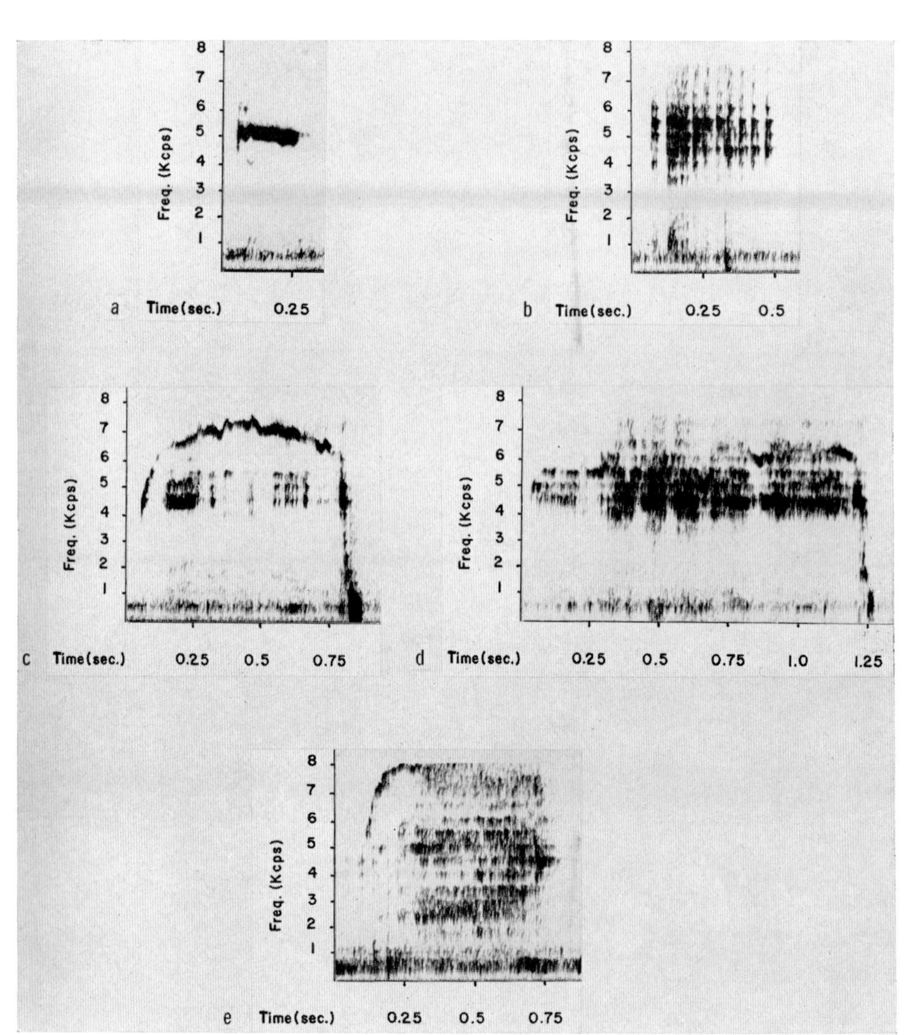

PLATE 16.2.—(*a*) Intragroup Squeal, (*b*) Intragroup Scream, (*c*) Intragroup Chutter-squeal, (*d*) Intragroup Squeal-scream, (*e*) Anti-copulatory Squeal-scream.

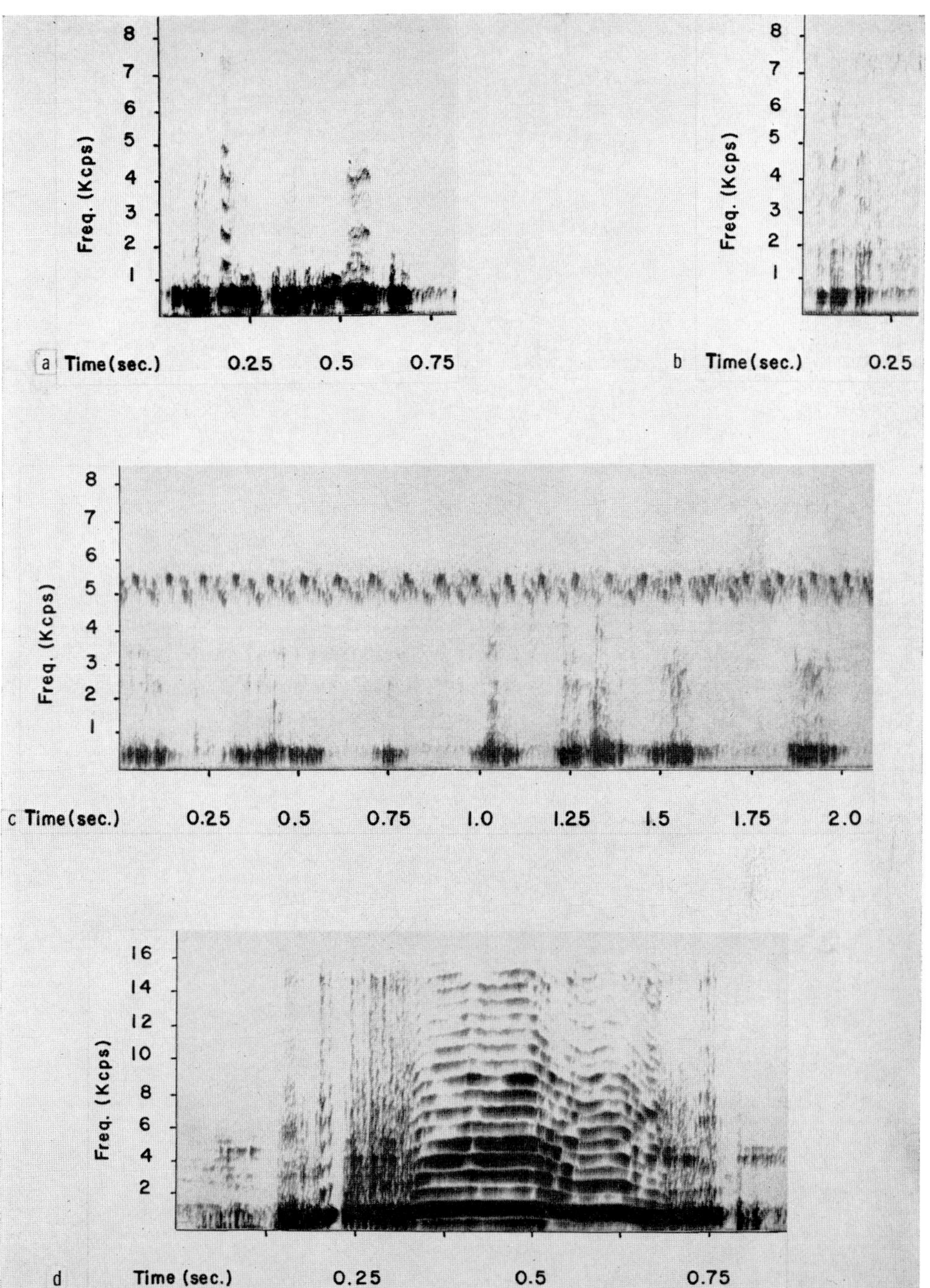

PLATE 16.3.—(*a*) *Woof-waa,* (*b*) Progression Grunt, (*c*) Woof, Woof (ignore background noise between 5 and 6 Kcps.), (*d*) *Waa.*

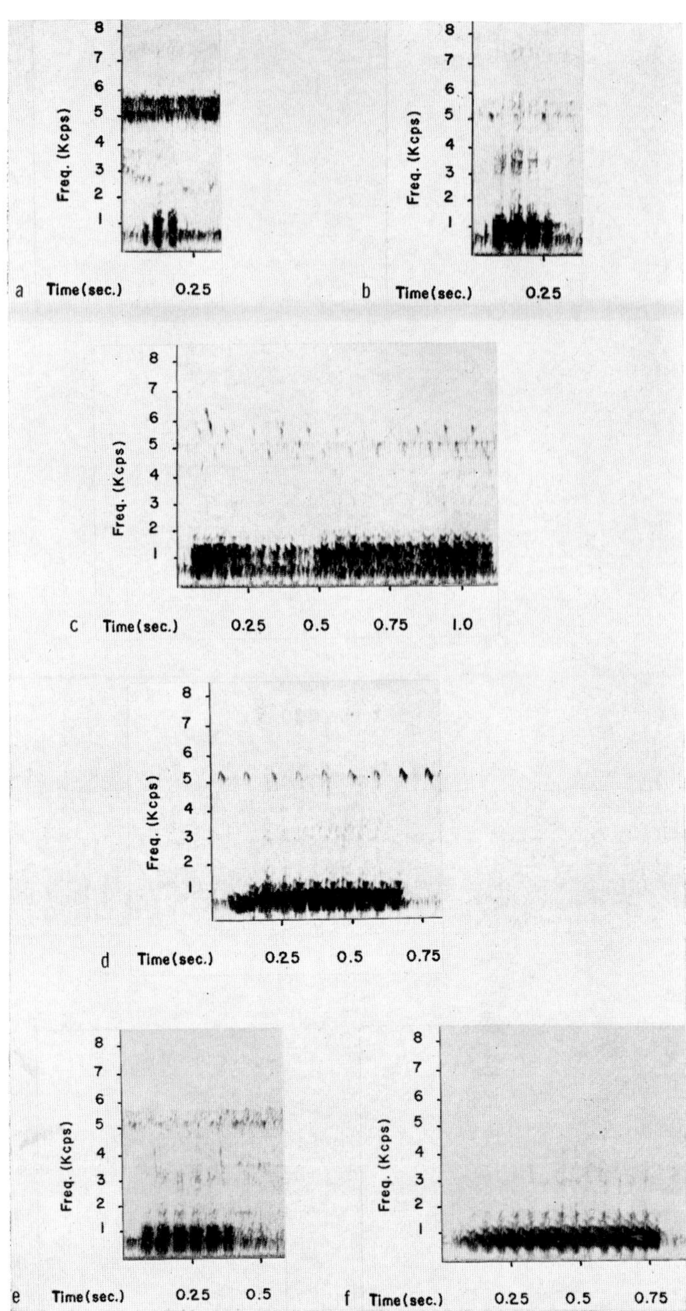

PLATE 16.4.—(*a*) Short *Rraugh,* (*b*) Short *Aarr-rraugh,* (*c*) Long *Rraugh,* (*d*) Long *Aarr-rraugh,* (*e*) Short *Aarr,* (*f*) Long *Aarr.* (Ignore background noise between 5 and 6 Kcps.)

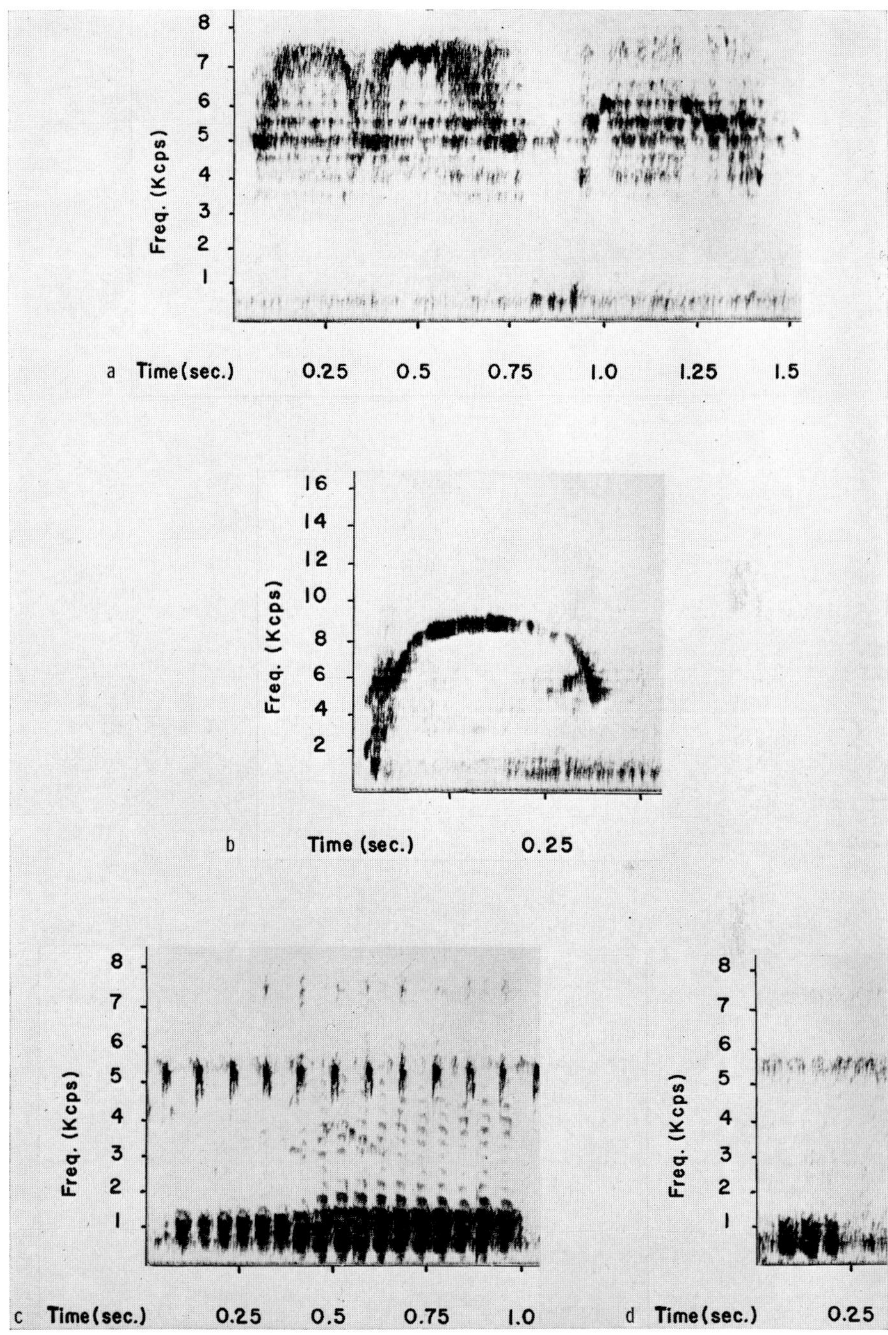

PLATE 16.5.—(*a*) Weaning Scream, (*b*) Weaning Squeal, (*c*) Weaning *rrr,* (*d*) *Eh, eh.* (Ignore background noise between 5 and 6 Kcps.)

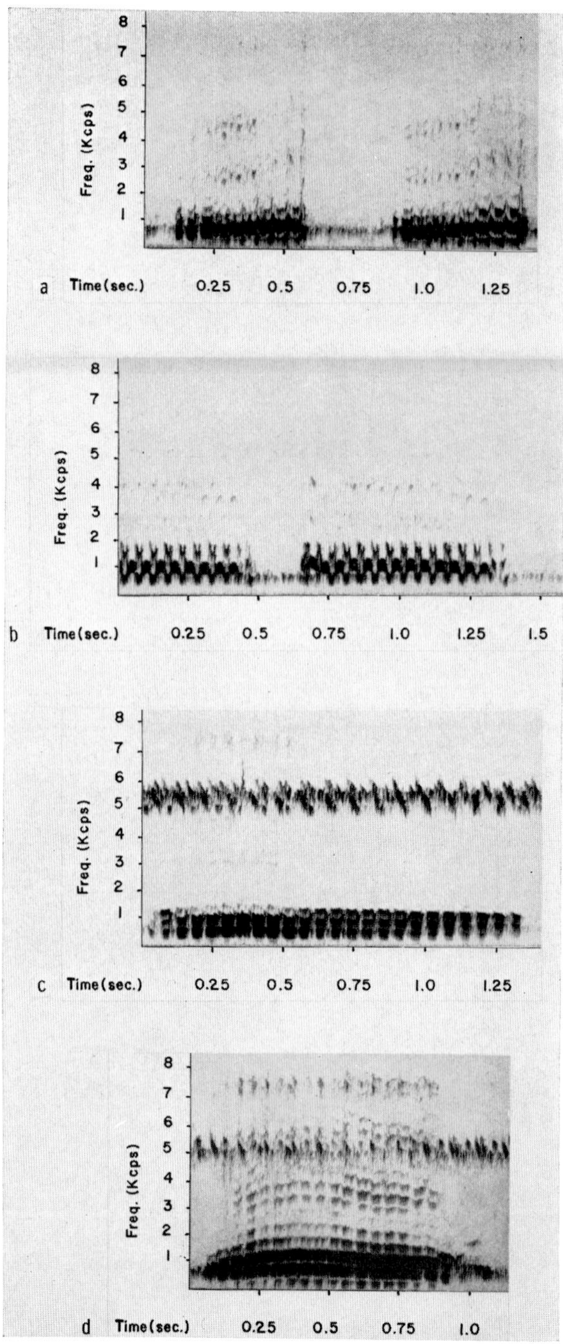

PLATE 16.6.—Lost *rrr* call of (*a*) in$_1$B at 62 days of age, (*b*) in$_1$a at 108 days of age, (*c*) in$_1$DK at 83 days of age, (*d*) in$_1$TK at 93 days of age. (Ignore background noise between 5 and 6 Kcps.)

PLATE 16.7.—In$_1$DK at 93 days of age giving the Lost *rrr* call. (Note the oval shape of his mouth.)

Plate 16.8.—A subadult male giving the Threat-Alarm-Bark

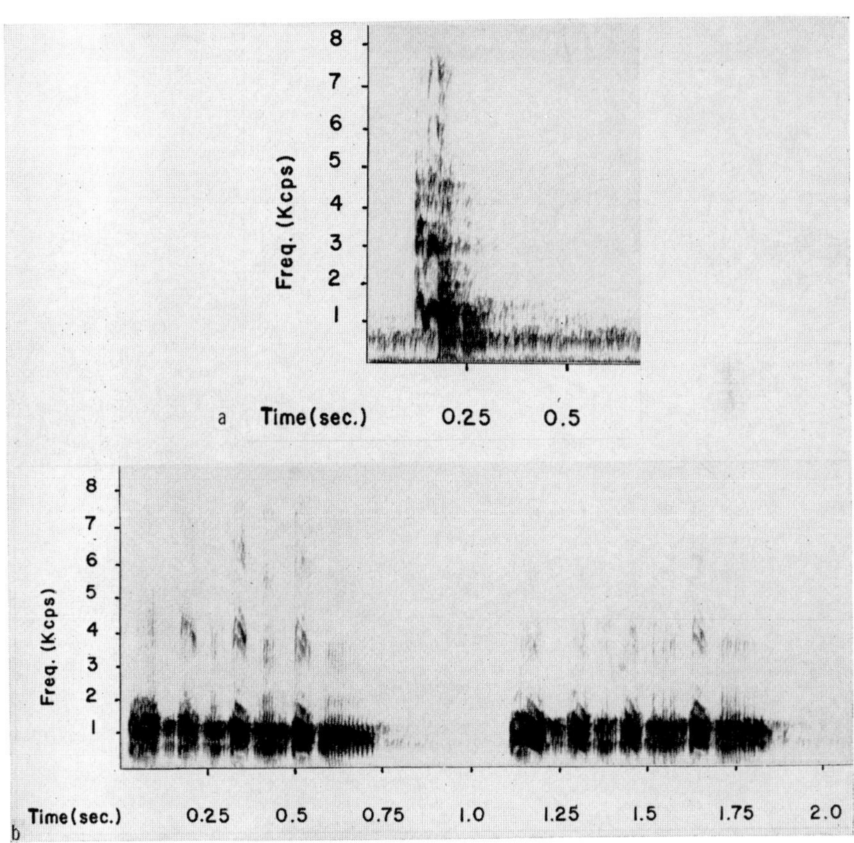

PLATE 16.9.—(*a*) Chirp, (*b*) Threat-Alarm-Bark. (There are two 8-unit phrases in this sonagram. Each phrase begins with an exhalation unit and ends with an inhalation unit. There are 4 exhalation and 4 inhalation units in each of these two phrases; the exhalation and inhalation units alternate with one another.)

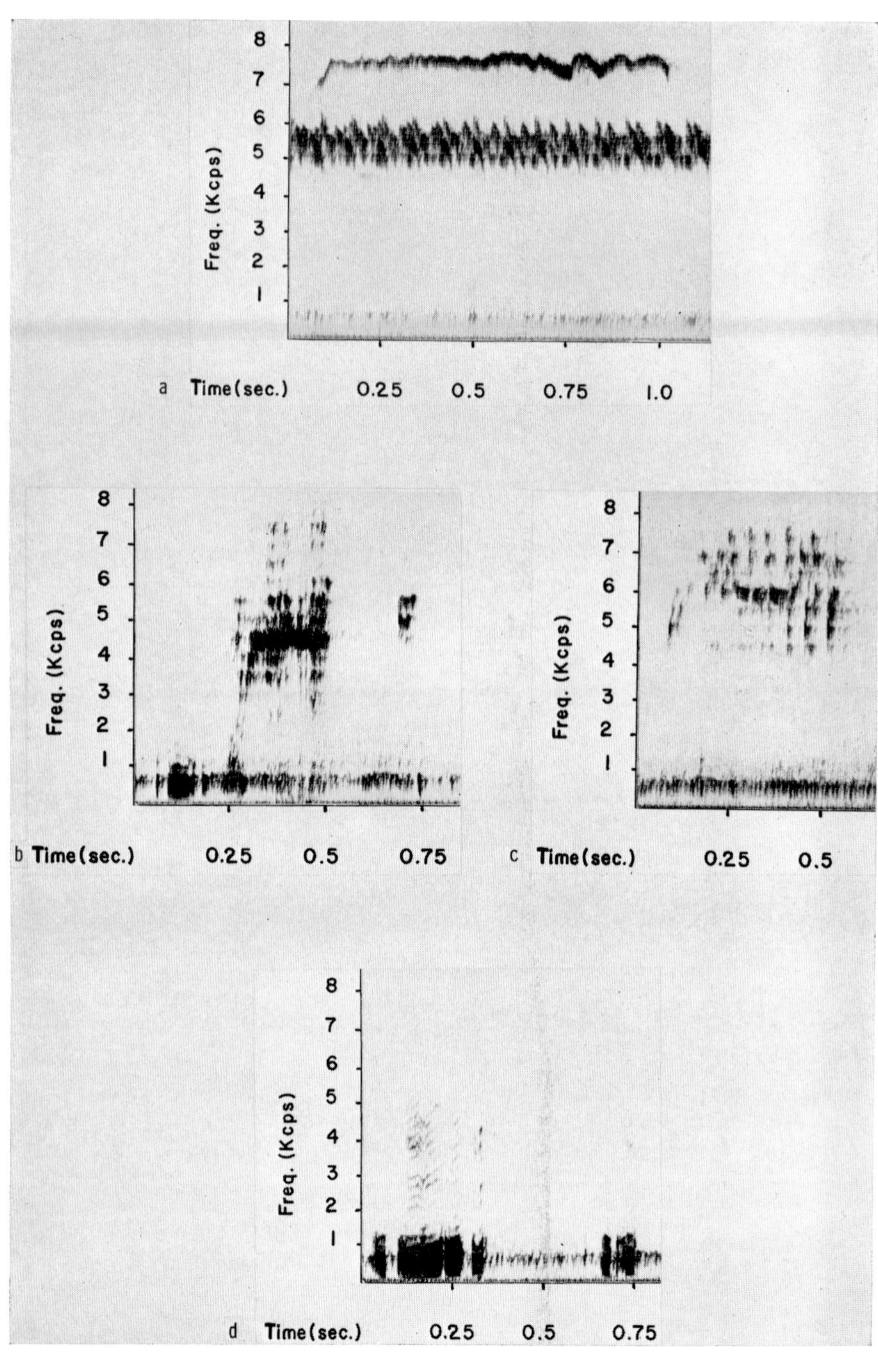

PLATE 16.10.—(*a*) Lost *eee* (ignore background noise between 5 and 6 Kcps.), (*b*) Lost Squeal, (*c*) Lost Scream, (*d*) Lost ***Rrah.***

*Squeal-scream**

This call (Plate 16.2) seemed to be intermediate between the strictly tonal Squeal and strictly nontonal Scream (below). The units were composed of nontonal sounds, but there was a slight suggestion of a tonal band of sound superimposed on the nontonal sounds.

The description of this call is based on only one recorded phrase and should not, therefore, be heavily emphasized as a distinct call.

*Scream**

This Scream (Plate 16.2) was a very shrill-sounding, nontonal call which, spectrographically, appeared to be very similar to the Weaning Scream, Lost Scream, and the infant's Scream toward a "strange" adult male.

The separation of the preceding four calls was based to a large extent on

TABLE 1

RESULTS OF TESTS OF SIGNIFICANCE OF LOST *rrr* CALL

S = Significant; NS = Not Significant; (No.) = Level of Significance; -- = No test made (usually because of inadequate sampling)

INDIVIDUALS (AND AGE) TESTED	CHARACTER TESTED			
	Lower Frequency of Major Energy	Duration of Unit	Number of Units per Phrase	Interval between Units
in$_1$TK (96 days) vs. in$_1$DK (85 days)	DK>TK S (.05)	NS (.10)	--	DK>TK S (.001)
in TK (132 days) vs. in$_1$DK (124 days)	DK>TK S (.02)	NS (.07)	--	NS (.19)
in TK (62 days) vs. in$_1$DK (124 days)	--	NS (.10)	--	DK>TK S (.001)
in TK (96 days) vs. in$_1$TBW (116 days)	--	NS (.20)	--	NS (.10)
in TK (54 days) vs. in$_1$a (54 days)	--	NS (.20)	--	a>TK S (.05)
in$_1$DK (124 days) vs. in$_1$TBW (116 days)	--	NS (.50)	--	DK>TBW S (.001)
in$_1$a (54 days) vs. in$_1$B (62 days)	--	NS (.05)	NS (.76)	--

differences in their spectrographic appearance. Audibly, they were not readily distinguishable. The general impression given by these calls was that of a shrill, high-pitched sound. All four calls were given only by adult females and juveniles (1–3 years old). These calls seemed to be given by individuals who were being attacked or threatened and probably functioned as a defensive threat and as a solicitation for the aid of others in retaliating against the attacking or threatening individual(s). It semed that during these four calls, the mouth position varied from being partially open with the teeth not exposed to a grimace with the unclenched teeth exposed. Differences, if any, in the function of the four calls were not clear, but it is suggested that they may communicate different levels of intensity in agonistic encounters.

Bark

The Bark that was given by subadult and adult males during intragroup agonistic encounters sounded very similar, if not identical, to the Bark given during intergroup agonistic encounters. This Bark seemed to be of a shorter duration than the Threat-Alarm-Bark, which was given toward predators. Although tape recordings were not made of the Bark during intragroup agonisms, it sounded as if each phrase of the Bark consisted of a single exhalation unit, thus differing from the Threat-Alarm-Bark, in which each phrase consisted of at least one and usually several pairs of exhalation and inhalation units.

The intragroup agonistic Bark was rather uncommon, for subadult and adult males rarely engaged in intragroup fights and chases. The male who was giving this call would generally vocalize as he ran toward the juveniles and/or adult females engaged in an agonistic encounter. Although the function of this call is not clearly understood, it seemed to be an aggressive threat that had the effect of disrupting and thus terminating a fight between juveniles and/or adult females.

Attempted Copulation by Subadult or Adult Male with a Nonestrous Female

*Anti-copulatory Squeal-scream**

Spectrographically, this Squeal-scream (Plate 16.2) appeared most like the Squeal-scream that was given by juveniles and females in intragroup agonisms. In addition, a phrase that appeared very similar to an intragroup-agonism-Squeal occurred once with an Anti-copulatory Squeal-scream. Although based on a very small sample of sonograms (five phrases), this call seemed to be extremely variable. Because of the small sample of both types of Squeal-screams, a statistical comparison was not possible. Superficial comparisons of the two types of Squeal-screams and the two types of Squeals can be made from my table in the American Documentation Institute.

The Anti-copulatory Squeal-scream was given by adult and subadult females when an adult or subadult male chased the female to the end of a tree branch or caught her after a chase and then made grabbing movements with his hands toward her or sometimes actually grabbed her hips as if about to mount her sexually. The female usually crouched low, by flexing her arms and legs, faced the male, and then gave this Squeal-scream, with her mouth forming a grimace and her unclenched teeth exposed. The male, after a few seconds of grabbing at the female, commonly moved away from her. I never observed a male mounting a female who was giving a Squeal-scream such as this. Presumably, all such females responding in this manner were not in estrus. This Squeal-scream seemed to inhibit the male's attempts to copulate with the female.

Red, White, and Blue Display

*Woof, Woof**

The *Woof, Woof* call (Plate 16.3) may precede and/or follow the *Waa* and *Woof-waa* calls. There was much variability among individuals with regard to all measurements on the *Woof, Woof* call. In addition to the complete nontonality of the *Woof, Woof* call, the most obvious differences between the *Woof, Woof* and *Waa* calls were in the distribution of major energy of nontonal sounds, unit and phrase durations, and number of units per phrase. The *Woof, Woof* and *Woof-waa* calls are compared in the section on the *Woof-waa* call.

In comparing the *Woof, Woof* call and the Progression Grunt (see footnote number 2, p. 283) of adult female LEG and subadult male GS we see that only the interval between phrases is consistently different between these two calls. In the *Woof, Woof* calls, the minimum intervals between phrases for these two monkeys were 0.042 and 0.126 second, whereas the minimum intervals between phrases of the Progression Grunt were greater than or equal to 2.40 seconds. It is not certain whether it was this difference in intervals between phrases, the difference in behavioral context, or the combination of the two that allowed the monkeys to distinguish these two calls. However, it was my impression that these two calls were readily distinguished by the vervets; that is, the *Woof, Woof* call was never answered with a similar call from another individual, whereas the Progression Grunt was invariably answered with a similar call by another individual.

*Waa**

The *Waa* call (Plate 16.3) seemed to be, primarily, a tonal exhalation of long duration with an occasional nontonal inhalation of short duration. Portions of the exhalation were sometimes nontonal. The development of the harmonic system in the tonal exhalation was most remarkable.

*Woof-waa**

The *Woof-waa* call (Plate 16.3) appeared to be composed of *Woof* units and units that were suggestive of *Waa* units. The *Woof-waa* phrases and units were of shorter duration than the *Waa* phrases and units and were completely non-tonal as opposed to the tonal and compound units of the *Waa* call.

The *Woof-waa* call was significantly greater than the *Woof, Woof* call in duration of units (0.02 level), number of units per phrase (0.001 level), and possibly so in duration of phrases.

Lip Smacking (see *Lip Smacking,* below)

Teeth Chattering (see *Teeth Chattering,* below)

The above five sounds were given during the same set of circumstances. As mentioned earlier, within each relatively closed and stable social group of vervets, there was a linear dominance hierarchy among the adults and older immatures. A dominant individual had priority to food and spatial positions over a subordinate and commonly supplanted a subordinate from the same.

Among the adult and older immature males, those about 2½ years or older, there was a relatively stereotyped display correlated with dominance. In this display a dominant male approached a subordinate male in a "confident" manner (alert, but relaxed in appearance with an absence of extreme muscle tonus and an absence of rapid sideward glances of the head). As the dominant male came to within a few feet of the subordinate, he (the dominant) held his tail vertically erect, thus exposing his red peri-anus, the white medial strip of pelage running from the red peri-anus to the base of the scrotum, and his turquoise or powder-blue scrotum (hereafter referred to as Red, White, and Blue). The dominant male then displayed this Red, White, and Blue to the subordinate male by walking back and forth in front of the subordinate who usually sat in a hunched manner or crouched very low with all four limbs flexed. Sometimes, while displaying the Red, White, and Blue, the dominant vervet walked in a circle around the subordinate, who continually shifted his position, although maintaining a hunched or crouched posture, so as to continually face the dominant male. Throughout this display, the dominant male was silent, and the subordinate male gave any one or all of the above five sounds. After a few moments, either the dominant walked away, the subordinate walked away, or the two walked away in opposite directions.

On some occasions, this display was terminated by a behavior pattern called the "False Chase." In the False Chase, the dominant male galloped away in a slow and confident manner, looking straight ahead, with his tail erect and the Red, White, and Blue exposed, while the subordinate galloped along behind him in a slow, hesitant, and jerky manner, with his limbs partially flexed and usually giving the *Woof, Woof* call. When the dominant stopped galloping so did the subordinate. The subordinate never caught the dominant, even when the dominant stood still—thus the term "False Chase."

Once an adult female gave the *Woof, Woof* call. She gave this call as she looked toward a subadult male who was also giving the *Woof, Woof* call toward a dominant adult male.

While giving the *Woof, Woof* call, the mouth was closed or only very slightly opened. A grimace with unclenched teeth occurred during the *Waa* call. A mouth position intermediate to these two occurred during the *Woof-waa* call.

The above three calls apparently expressed subordination and, conceivably, may have inhibited aggression by the dominant male. It is suggested that these three different calls represent three different levels of intensity, with the *Woof, Woof* being least intense, the *Woof-waa* being intermediate, and the *Waa* being most intense. Presumably, the most intense call was most effective in inhibiting attack by the dominant individual.

Close Proximity of Subordinate and Dominant Monkeys

Relationships involving close proximity among juveniles, adult females, and males (other than the Red, White, and Blue Display) are included in this category.

Wa-waa

This call was, undoubtedly, homologous to the *Waa* call given by subordinate males to whom the Red, White, and Blue was being displayed.

This call was heard only from one particular juvenile female (juvenile female NN, about 2½ years old). While giving this call, her mouth formed a grimace with her unclenched teeth exposed. She gave this *Wa-waa* call when she was approached by the most dominant adult female (DK) in the group. On some of these occasions, this juvenile female was holding the infant of DK. A grooming sequence generally followed the approach by the dominant female and the *Wa-waa* call of the subordinate juvenile female. No aggression by the adult female toward the juvenile female was ever observed after the juvenile had given this call.

It seemed that the *Wa-waa* call expressed subordination and thus may have inhibited aggression by a dominant animal. The inhibition of aggression would permit the close proximity necessary for grooming between subordinate and dominant individuals. It was not understood why this call was, so far as known, restricted to one particular female and evoked only by the approach of another particular female.

Rraugh

*Long Rraugh**.—The long variation of the *Rraugh* call differed from the Short *Rraugh* (Plate 16.4) in having both tonal and nontonal components and in having phrases of longer duration (significant at the 0.02 level). However, there was no significant difference between these two calls in duration of units (.10 level) or in number of units per phrase (.10 level).

*Short Rraugh**.—As implied above, the Short *Rraugh* call had only nontonal sound, and its phrases were of shorter duration than those of the Long *Rraugh* call.

Although these two variations were readily distinguishable spectrographically, such a distiction was not made by the author during the field study. Similarly, I could not distinguish a functional difference between the two variations.

Spectrographically, the *Rraugh* calls appear very much like the *Aarr* and *Aarr-rraugh* calls (see sections on *Aarr* and *Aarr-rraugh* calls below). Audibly, I could readily distinguish the *Rraugh* and *Aarr* calls, but I was unable so to distinguish the *Rraugh* and *Aarr-rraugh* calls. There was also a rather superficial resemblance between the *Rraugh* and the infant Lost *rrr* call, which may reflect on ontogenetic relationship (see Lost *rrr* call, below). The *Rraugh* calls were given predominantly by young juveniles, about 1 to 1½ years old. On one occasion, a subadult female, about 3 years old, gave this call as she approached an adult male whom she subsequently groomed.

The mouth seemed to be closed or only partially open with the teeth not exposed during the *Rraugh* calls.

The juveniles gave the *Rraugh* calls as they approached older, dominant individuals. Such an approach usually led to a grooming interaction in which the juvenile was the groomer. With adult females that were holding their infants, such an approach usually resulted in the juvenile grooming, handling, and holding the infant. The *Rraugh* calls sometimes continued from the approach into the initial stages of the grooming sequence.

It was my impression that the *Rraugh* calls expressed nonaggression or appeasement and thus allowed the subordinate juvenile to approach and groom an older, dominant individual or handle the infant of a dominant female. The exceptional case of the subadult female mentioned above could also be explained in this manner.

Lip Smacking

Lip Smacking, which was the rapid and alternate parting and meeting of the upper and lower lips, was frequently audible at distances of about 10 yards. The teeth were not conspicuously exposed in this behavior pattern. All age and sex classes, except infants, were observed to smack their lips. This behavior pattern was frequently intermingled with other audible patterns such as the *Rraugh* call, *Woof, Woof* call, and Teeth Chattering.

Lip Smacking was performed by both dominant and subordinate members of a grooming pair. Lip Smacking was also given by juveniles as they approached and handled an infant that was clinging to its mother. Furthermore, as indicated above, Lip Smacking was sometimes intermingled with the *Woof,*

Woof calls given by the subordinate male during a Red, White, and Blue Display encounter.

Lip Smacking seemed to express nonaggression, thus permitting the close proximity of subordinate and dominant individuals.

Teeth Chattering

Teeth Chattering was the audible, rapid, and alternate parting and meeting of the teeth of the upper and lower jaws. The teeth were not exposed during this behavior pattern, for the mouth and lips were only very slightly opened. Only subadult and adult males were observed to chatter their teeth.

Teeth Chattering was performed by dominant and subordinate males in grooming situations and sometimes by the subordinate male recipient of a Red, White, and Blue Display. In the latter situation, the Teeth Chattering was intermingled with the *Woof, Woof* call and Lip Smacking.

Teeth Chattering seemed to express nonaggression, thus permitting the close proximity of subordinate and dominant males, such as in grooming encounters.

Initial Phase of Group Progression

*Progression Grunt**

As the name implies, this was a deep, guttural call of short duration (Plate 16.3). Audibly, it most resembled a single *Woof* unit of the *Woof, Woof* call. Since it was not often repeated more than once or twice by a given individual, it was an extremely difficult call to record on tape.

This call was given by all age and sex classes whose members were 4½ months or older. The infants and young juveniles (4½ mos.–1½ years old) seemed to have a Progression Grunt of higher frequency than older monkeys.

The mouth was closed or only very slightly opened during this call.

The initial Progression Grunt phrase seemed to evoke similar Progression Grunts from all individuals in the immediate vicinity of the vocalizer. Subsequent to these latter grunts given by the other individuals, the initial vocalizer would again give a Progression Grunt which would again evoke similar grunts from the same individuals. The impression was that the vervet giving the initial Progression Grunt was answered with Progression Grunts from several others. Following this first exchange, the initiator and followers frequently exchanged another Progression Grunt. Infrequently, the encounter involved more than two exchanges of Progression Grunts. The over-all chorus rarely lasted longer than about 5 seconds.

Almost invariably the occurrence of Progression Grunts was followed by a group progression. Group progressions were regarded as those movements of the entire group that covered a distance of 50 yards or more across open, short-grass country. Such group progressions occurred within about 10 to 15 minutes after the Progression Grunts. It seemed certain that the Progression Grunts indicated the temporal proximity of a group progression and thus facilitated the coherency of the group. It is uncertain what stimulus evoked the initial Progression Grunt in any particular sequence.

Play

Purr

This call was rarely heard, and as its name implies it had a very low amplitude and frequency. Its duration was estimated at about 1 to 2 seconds. Most play encounters were silent. The Purr was heard only when two juveniles (1–1½ years old) wrestled and grappled with one another on the ground. It was not heard in play which involved running and chasing. The Purr was audible at about 6 feet.

The significance of this call was not apparent. However, such a call may somehow enhance the social play bond between individuals and thus facilitate the occurrence of future play encounters.

Approach of a "Strange" Adult Male toward an Infant

Scream

This Scream may be identical with, or at most a variation of, the Lost and/or Weaning screams of infants. Furthermore, it may be the precursor of the Scream that is given by adult females and juveniles. It was a high-pitched and shrill-sounding call. This call was given by infants of about 1 to 4 months of age within one particular group and toward one particular adult male. This male had recently joined the group from a neighboring group. Whenever this "strange" male came to within about 6 to 8 feet of an infant who was not with its mother, the infant gave this Scream. This Scream seemed to evoke the rapid retrieval of the infant by the mother. As the mother approached her infant, the "strange" adult male moved away from the infant. No such response was shown by the infants toward other adult males within the group, even when the infants were in physical contact with them.

Weaning of Infant

*Scream**

As mentioned above, this call sounded very much like the Scream given by an infant toward a "strange" adult male. In addition, this call appeared, spectrographically, to be most similar to the Weaning Squeal, Lost Scream, and the Scream of adult females and juveniles (Plate 16.5). However, because of inadequate sampling, statistical comparisons of these calls were not possible. On one occasion, all three weaning calls (Scream, Squeal, and *rrr*) occurred in the same bout.

The Weaning Scream was given by all infants as they were nipped about the ears and face and pushed away from their mother's nipples by their mother, that is, weaned. It seemed as though the Scream was given during the most intense weaning situations (when the mother seemed most forceful and persistent in her weaning of the infant). The mother showed no obvious response to her infant's Weaning Scream, although she frequently walked away from the infant. The function of this call is not clear.

*Squeal**

The distribution of the major energy of this call was extremely variable between individuals. In addition, one individual had mixed and compound units, whereas two others had only nontonal and compound units. In one extraordinary Squeal phrase, overtones (harmonics) were developed at 23,000 cps. This phrase and one unit of a Lost Squeal (below) were the only examples of ultrasonic calls encountered.

The Weaning Squeal differed from the Weaning Scream in that the Squeal was composed of compound units and the Scream was composed only of nontonal units (Plate 16.5).

The Weaning Squeal appeared similar to and may have been the precursor of the Chutter-squeal given by adult females and juveniles. Because of the extensive overlap and variability of the various measurements, statistical tests of significance between the Weaning Squeal and Chutter-squeal were not made. The Weaning Squeal was also similar to the Lost Squeal (see section on Lost Squeal).

The Weaning Squeal seemed to be given in weaning circumstances that were less intense than those evoking the Weaning Scream. The function of the Weaning Squeal is not understood.

*rrr**

This call appeared and sounded very much like, if not identical to, the Lost *rrr* call (Plate 16.5). The Weaning *rrr* call given by in_1TK (= infant of class one named TK) at 24 days of age and the Lost *rrr* call given by the same individual at 54 days of age seemed to be no different. Because of inadequate sampling, it was only possible to make statistical tests of significance on the units of the Weaning and Lost *rrr* calls. The Weaning *rrr* and Lost *rrr* calls of in_1TK were not significantly different in duration of units (0.30 level) or in intervals between units (0.10 level). The Weaning *rrr* call was, presumably, given in the least intense weaning situations.

The mother's nipples were inaccessible during both the Weaning *rrr* and the Lost *rrr* calls and may thus account for the similarity of the two calls.

In contrast, the other two weaning calls were given under more intense weaning conditions. Such intense weaning conditions included not only the inaccessibility of the mother's nipples, but also the actual physical pain inflicted by the nipping and biting of the infant by the mother.

Separation of Infant and Mother

*Lost rrr**

This call was characteristically composed of phrases with tonal and discrete units (see Table 2 and Plate 16.6), although nontonal units and "subunits" (units that are not distinctly separated) did occur. As seen in Table 1, the duration of units was relatively consistent, regardless of individual and age variation. The Lost *rrr* call was most commonly intermingled with the Lost *eee* call and less commonly with the Lost Squeal and Lost Scream.

TABLE 2

FREQUENCY DISTRIBUTION OF FUNDAMENTAL AND HARMONICS OF LOST *rrr* CALL

"No. in quotations" = an assumed fundamental based on harmonics present
(No. in parentheses) = frequency of occurrence

Name of Individual and Group	Age in Days	Fundamental (cps)	Harmonic(s) (integral multiple of fundamental)
in_1DK (1530 Group)	85	"250" (28) 250 (91)	2f(117), 3f(113), 4f(114), 5f (31)
		500 (13)	2f(13)
	124	"250" (37) 250 (28)	2f(61), 3f(65), 4f(62), 5f(15), 6f(2)
		500 (7)	2f(7)
in_1TBW (1530 Group)	116	250 (14)	2f(14), 3f(14), 4f(14), 5f(14)
in_1TK (1530 Group)	54	"250" (1) 250 (16)	2f(17), 3f(17), 4f(8), 5f(13), 6f(8)
	62	"250" (2) 250 (16)	2f(18), 3f(17), 4f(18), 5f(14)
	93	250 (32)	2f(32), 3f(32), 4f(30), 5f(27), 6f(12), 7f(2)
	132	Tonality suggested, but poorly, if at all, developed	
in_1a (P-Group)	54	"250" (20)	3f(20), 4f(20), 6f(17), 7f(12)
	108	250 (11)	2f(11), 4f(11), 5f(11)
		500 (13)	3f(13)
		750–1000 (24)	None
		1000–1250 (1)	None
in_1B (P-Group)	62	250 (10)	2f(10), 3f(9), 4f(5), 5f(3)
		500 (1)	None

Sonograms of this call were made from tape recordings of five different and identifiable infants. Three of these infants (named in$_1$a, in$_1$DK, in$_1$TK) were each sampled at different ages ranging from about 51 to 132 days of age. In$_1$DK showed no apparent difference in his Lost *rrr* call between 85 and 124 days of age. The Lost *rrr* call of in$_1$TK was recorded at 54, 62, 96, and 132 days of age. The call recorded at 132 days seemed to differ somewhat from those recorded at the other ages. The duration of units at 132 days was not significantly different (0.05, 0.10, 0.20 levels) from those at the other three ages. However, the interval between units at 132 days was significantly greater (0.05 level) from those at 62 days and undoubtedly different from those at 54 and 96 days (which had intervals even shorter than at 62 days). In$_1$a showed marked differences in the appearance of his Lost *rrr* calls as recorded at 51 and 105 days. The variation of in$_1$a's Lost *rrr* call with age was most evident in the different fundamentals and overtones that were developed (see Table 2). In addition, the Lost *rrr* calls given at 105 days by in$_1$a had longer phrases which were composed of only one unit (or several subunits "smeared" together) in contrast to the shorter phrases composed of several discrete units given at 51 days of age.

The infants, in$_1$DK, in$_1$TBW, and in$_1$TK, were members of the same social group (1530 Group). In general, the Lost *rrr* calls of these three infants appeared to be very similar (see Table 2). However, as summarized in Table 1, there were some significant differences among these three infants. Although data adequate for statistical testing are not available, the differences in percentages of nontonal and tonal sound in the Lost *rrr* calls of these three infants may be significant. It remains to be seen if these individual differences are functional as cues for individual recognition among the vervet monkeys.

The Lost *rrr* calls of the above three infants differed noticeably in their spectrographic appearance from the same calls given by infants in$_1$a and in$_1$B, both of the P-Group (see Plate 16.6). In$_1$a and in$_1$B also differed from one another in the appearance of their Lost *rrr* calls. Calls of these two infants, which were recorded on the same day, when in$_1$a was 108 days old and in$_1$B was 62 days old, indicate that the greatest differences between their calls were in the frequency of the fundamentals of the tonal sound, duration of units, and number of units per phrase (see Table 2). However, the latter two characters were not significantly different between in$_1$a at 54 days of age and in$_1$B at 62 days of age (see Table 1). In spite of this individual and intergroup variation, the Lost *rrr* call was very distinct and not readily confused, spectrographically or audibly, with any other call.

The audible similarities suggest that the Lost *rrr* call might develop, ontogenetically, into the *Aarr, Aarr-rraugh,* and/or *Rraugh* calls. The Short *Aarr* differs from the Lost *rrr* call most obviously in that the former is composed strictly of nontonal sound and has phrases of short duration, whereas the latter is composed primarily of tonal and, secondarily, of nontonal sound and has phrases of long duration. Furthermore, the upper frequency of the distribution of major energy of nontonal sound may be higher in the Lost *rrr* than in the Short *Aarr* call. The Lost *rrr* call is composed primarily of tonal sound, whereas the Long *Aarr* call is composed of about equal amounts of tonal and nontonal sound. Otherwise, there is much overlap in the measurements of the Lost *rrr* and Long *Aarr* call.

Although the sample of the Long *Aarr-rraugh* call (below) is too small to allow statistical testing, it seems quite certain that the Lost *rrr* call generally has units of shorter duration and more units per phrase than does the Long *Aarr-rraugh* call.

The Short *Aarr-rraugh* (below) differed from the Lost *rrr* call in that the former was composed only of nontonal sound and relatively short phrases, whereas the latter was primarily composed of tonal sound and relatively long phrases. In addition, the Lost *rrr* call differed from the Short *Aarr-rraugh* call in having units and phrases of significantly longer duration (0.001 level) and a greater number of units per phrase (0.001 level).

The Short *Rraugh* call (see above) was composed of nontonal sound only, whereas the Lost *rrr* call was composed primarily of tonal sound. Furthermore, these two calls differed from one another in that the Lost *rrr* call was significantly greater in number of units per phrase (0.001 level) and duration of phrases (0.001 level). The Lost *rrr* had significantly more units per phrase (0.001 level) than the Long *Rraugh* call. However, these two calls were not significantly different in duration of units (0.10 level) and duration of phrases (0.05 level).

While giving the *rrr* call, the mouth was puckered and thus appeared as an "O" (Plate 16.7). The teeth were not exposed.

*Lost eee**

The phrases of this call were composed of a single, long tonal unit (Plate 16.9). Each phrase rose sharply in frequency at the beginning, leveled off on a plateau with some frequency modulation, and terminated with a sharp descent in frequency. There was no other call that might have been confused with this one, either spectrographically or audibly. This was a relatively stereotyped call. The only noticeable difference between the two infants whose Lost *eee* calls were analyzed spectrographically was that one of these infants had harmonics developed in two of eight phrases. The remaining six had no harmonics, as was true for all such phrases of the other infant. Furthermore, there seemed to be little, if any, variation in this call from an age of 24 days to 139 days.

As mentioned earlier, the Lost *eee* call usually alternated with the Lost *rrr* call. The combination of these two calls provides the three major properties that can enhance sound localization and which, presumably, assisted the mother in locating and retrieving her infant. The abrupt discontinuity of the phrases of these calls accentuates the difference in the time of arrival of the sound at the two ears of the receiver and thus facilitates directional location of the sound. The low-frequency sounds (long wavelengths) of the *rrr* calls probably permit phase contrast, and the high-frequency sounds (short wavelengths) of the *eee* calls might enhance intensity contrast between the two ears of the receiver, thus further facilitating the location of the infant giving the Lost *rrr* and Lost *eee* calls.

The lips were parted, the mouth corners drawn back, and the mouth opened slightly, but the teeth were not exposed during the Lost *eee* call.

The stimulus situations and behavioral significance of this call were, presumably, the same as those for the Lost *rrr* call (see below).

*Lost Squeal**

The Lost Squeal was infrequently heard. On some occasions, this call was intermingled with the Lost *rrr* and Lost *Rrah* calls. One remarkable Lost Squeal unit had an overtone at 20,000 cps. Typically, the Lost Squeal was composed of a tonal portion that was preceded and/or succeeded by a nontonal portion (Plate 16.9). It thus appeared, spectrographically, to be very much like the Weaning Squeal. However, the duration of units of Weaning Squeals was significantly greater than that of Lost Squeals (0.05 level). Furthermore, the percentages of unit time that were tonal and nontonal may have been significantly different in these two calls. In addition, the Lost Squeal was similar to and may develop into the Chutter-squeal of adult females and juveniles. The Lost Squeal had a significantly greater lower frequency of major energy (0.001 level) than did the Chutter-squeal. The Chutter-squeal had a significantly greater duration of units (0.01 level) than did the Lost Squeal. However, these two calls did not differ in their upper frequencies of major energy (0.10 level).

The stimulus situations and behavioral significance of this call were not clearly understood because of its rarity, but they are presumably similar to those described for the other Lost calls (see below).

*Lost Scream**

Although there are only two sonograms of this call, it appears to be spectrographically similar to the Weaning Scream and may also be related to the Scream of adult females and juveniles (Plate 16.9). The Lost Scream was intermingled with the Lost Squeal and Lost *rrr* calls. The stimulus situations and behavioral significance of this call were probably the same as those of its associated calls.

*Lost Rrah**

This call was most reminiscent of the Lost *rrr* call (Plate 16.9). It was infrequently heard, and occasionally units of this call were intermingled in the same phrase with units of the Lost Squeal call. The comparison of the Lost *rrr* call with the Lost *Rrah* call of in$_1$B at about 62 days of age shows that the Lost *rrr* had a greater percentage of tonal sound than did the Lost *Rrah* call. However, when tonal sound was present in the Lost *Rrah* call, a greater number of harmonics (up to 30f, i.e., 30 times the fundamental frequency) were developed than in the Lost *rrr* call (up to 5f). The Lost *Rrah* seemed to be of lower amplitude than the Lost *rrr* call. In addition, the Lost *rrr* had significantly longer units than did the Lost *Rrah* (0.02 level). The Lost *Rrah* had significantly more units per phrase than did the Lost *rrr* (0.05 level). However, they were not significantly different in their lower frequency (0.30 level) and upper frequency (0.30 level) of major energy and duration of phrases (0.10 level).

It was my impression that the Lost *Rrah* call was given in less intense lost situations, for example, when an infant crawled through and over a thicket toward its mother who was a few feet away. It undoubtedly served a function very similar to that of the other lost calls (see below).

All five types of lost calls seemed to occur in similar circumstances and therefore the following remarks apply to all of them. The lost calls were given only when the mother and infant were separated. Sometimes the distance separating the mother and infant was relatively short, such as when a mother was climbing about in a tree 6 feet from her infant. More commonly, the distance was greater than 45 feet. In situations evoking the lost calls, the infants were usually alone. However, occasionally, infants gave the lost calls while ventrally embracing a juvenile female (1–1½ years old). The mothers of the infants giving the lost calls seemed to respond very slowly to these calls, as judged by the relatively long time interval between the onset of the calls and the time at which the mother retrieved the calling infant. An exception to this occurred on one occasion. I was observing a group with whom I had previously spent relatively little time. As a consequence, the members of this group were less tolerant of my close proximity than were the vervets of major study groups. While observing this group, an infant, who was about 10 to 15 feet from me, began giving lost calls. An adult female, presumably the infant's mother, immediately ran toward the infant. She made rapid glances toward me as she retrieved this infant. With the infant clinging to her ventral surface, she ran off, continuing to make rapid glances toward me until she was about 60 feet away.

From the above observations, it seems feasible that the lost calls served to focus the attention of the mother upon her infant when the infant was out of contact with her. The response of the mother to her infant's lost calls was dependent upon the circumstances surrounding the infant. If potential danger was near the infant, the mother rapidly retrieved the infant; however, if no danger was nearby, her retrieving response was much slower.

As mentioned above, it is not certain to what extent individual variation in lost calls facilitated individual recognition. However, it is my impression that only the mother and, on a few occasions, juvenile females retrieved the calling infant. In all cases in which the mother was known to me, I never observed an adult female retrieving a calling infant other than her own. Such observations imply that the calling infant is recognized (visually and/or audibly) and is differentially responded to by the other monkeys.

The similarity between some of the weaning and lost calls, and the fact that an infant sometimes gave the lost calls while ventrally embracing a juvenile female (none of whom had pendulant nipples) suggest that a major part of the stimulus evoking the lost calls is the absence or inaccessibility of nipples and perhaps milk.

Reunion of Infant and Mother

*Eh, eh**

The *Eh, eh* call was of rather low amplitude and short duration (Plate 16.5). It seemed to be most like the Lost *Rrah* call. The *Eh, eh* call was composed only of nontonal sound, whereas the Lost *Rrah* call had about equal amounts of tonal and nontonal sound. In addition, the Lost *Rrah* call had significantly longer units (0.02 level) and intervals between units (0.05 level) than did the *Eh, eh* call.

The *Eh, eh* call was given upon the reunion of the infant and mother as the infant clung to its mother's ventral surface. Prior to the reunion that evoked *Eh, eh* calls and during the separation, the infant usually gave lost calls. Not all reunions evoked the *Eh, eh* calls.

The significance of this call was not at all clear; however, it is suggested that in some manner the call may have enhanced the social bond between the mother and infant.

Ambivalent Situation with Proximity of Foreign Group and Close Proximity of Subordinate and Dominant Individuals

*Aarr-rraugh (Long and Short Types)**

Because of an inadequate sample size, it was not possible to make statistical comparisons of the Long *Aarr-rraugh* call with similar calls. However, a general and superficial comparison of this call with other similar calls will be made. The unit time of the Long *Aarr-rraugh* calls was composed of about 35 per cent nontonal and 65 per cent tonal sounds, whereas the sound of the Short *Aarr-rraugh* calls was 100 per cent nontonal. The Long *Aarr-rraugh* calls had units and phrases of longer duration and fewer units per phrase than did the Short *Aarr-rraugh* calls, as can be seen in Plate 16.4.

The *Aarr-rraugh* calls sounded intermediate to the *Rraugh* and *Aarr* calls. Superficially, these three types of calls were very similar. However, the duration of units and phrases was much longer in the Long *Aarr-rraugh* call than in the Short *Rraugh* call. Although the duration of phrases in the Long *Aarr-*

rraugh and Long *Rraugh* calls seemed to be the same, the Long *Aarr-rraugh* call apparently had units of longer duration and fewer units per phrase than did the Long *Rraugh* call.

The Short *Aarr-rraugh* call was composed only of nontonal sound as opposed to the Long *Rraugh* call, which was composed of a high percentage of tonal sound. The durations of the units and phrases were shorter in the Short *Aarr-rraugh* call than in the Long *Rraugh* call. The Short *Aarr-rraugh* call had significantly more units per phrase (0.001 level) and longer duration of phrases (0.01 level) than the Short *Rraugh* call. The *Aarr-rraugh* calls are compared with the *Aarr* calls in the section on *Aarr* calls.

The *Aarr-rraugh* call was given in ambivalent situations in which either the *Rraugh* or *Aarr* call might have been appropriate. This call was given only by young juvenile females (1–1½ years old) when a foreign group approached the juvenile and its group and the juvenile simultaneously approached and subsequently groomed a dominant adult. The *Aarr-rraugh* call may have indicated the approach and proximity of a foreign group and/or expressed nonaggression and thus permitted the subordinate juvenile to approach and groom the dominant adult. Assuming that the juvenile might be protected from the foreign group by attaining a close proximity to a dominant adult, it is conceivable that the approach of the foreign group evoked the *Aarr* aspect of the call and the juvenile's approach toward a dominant adult evoked, in turn, the *Rraugh* aspect of the call. The significance of the differences between the Long and Short types of *Aarr-rraugh* calls is not understood.

Proximity of Foreign Group

Intergroup Aarr (*Long and Short Types*)*

The sound of the Long *Aarr* phrases varied in composition from 100 per cent nontonal to 100 per cent tonal. The phrases of some Long *Aarr* calls were composed of discrete units, whereas others were composed of nondiscrete units that appeared to be smeared together along the temporal axis of the sonograms. The duration of phrases was relatively long (see Plate 16.4).

The Short *Aarr* calls were composed of relatively short phrases of discrete and nontonal units (Plate 16.4). There was relatively little variation in this call among five individuals. One individual had units of longer duration and had fewer units per phrase than the other four. However, these two differences had the effect of cancelling one another relative to the duration of phrases.

The Long and Short *Aarr* calls of adult female LEG can be compared (see footnote number 2, p. 283). Although the sample is too small for statistical testing, the differences are probably quite real. The Short *Aarr* call was composed strictly of nontonal sound, whereas the Long *Aarr* was composed of nearly equal amounts of tonal and nontonal sound. Furthermore, the duration of units and phrases were shorter and, for some, there were more units per phrase in the Short *Aarr* call than in the Long *Aarr* call. However, there were fewer units per phrase in the Short *Aarr* call than in the Long *Aarr* call of juvenile male RS (2–2½ years old). The most obvious and consistent difference between the two *Aarr* calls was that the Long *Aarr* call had phrases of longer duration (mean of 0.71 second, $n = 5$) than the Short *Aarr* call (mean of 0.19 second, $n = 20$).

The *Aarr* calls seemed to be most like the *Aarr-rraugh* and *Rraugh* calls. The Short *Aarr* call had a significantly longer duration of units (0.001 level) than the Short *Aarr-rraugh* call. However, the Short *Aarr-rraugh* call had significantly longer intervals between units (0.001 level) than the Short *Aarr* call. The Short *Aarr* call was composed strictly of nontonal sound in contast to the Long *Aarr-rraugh* call, which had a large proportion of tonal sound. In addition, the Short *Aarr* call had units and phrases of much shorter duration and more units per phrase than did the Long *Aarr-rraugh* call.

The Long *Aarr* call differed appreciably from the Short *Aarr-rraugh* call in duration of phrases. Although only one phrase of the Long *Aarr-rraugh* call was available for comparison, its phrase duration and other measurable properties were well within the range of comparable properties of the Long *Aarr* calls. However, based on the different stimulus situations evoking these two calls and my impression of their audible differences, it seems best at this time to treat the Long *Aarr-rraugh* call as distinct from the Long *Aarr* call.

The Short *Aarr* call had significantly more units per phrase than the Short *Rraugh* call (0.01 level) and longer duration of phrases (0.05 level). Furthermore, the Short *Rraugh* call apparently had longer intervals between units than did the Short *Aarr* call. The Short *Aarr* call was obviously different from the Long *Rraugh* call in that the former was composed only of nontonal units and the latter was composed of both nontonal and compound units. The Long *Rraugh* call had significantly longer duration of phrases than the Short *Aarr* call (0.05 level), but they did not differ significantly in duration of units (0.10 level).

The Long *Aarr* call differed from the Short *Rraugh* call in that the former had phrases of longer duration than the latter. The measurements of the Long *Aarr* and the Long *Rraugh* calls overlap and owing to the relatively small sample, do not allow statistical comparisons. However, based on my impression of their audible and stimulus situation differences, I believe that the two calls should be treated as distinct unless more extensive information proves otherwise.

Both the long and short types of intergroup *Aarr* calls were commonly given by juveniles and adult females and less frequently by subadult males.

The mouth was slightly open and puckered and the teeth were not visible while the *Aarr* call was given.

These calls were given only when a foreign group was relatively near the vocalizer. Such calls seemed to indicate the proximity and/or approach of a foreign group and were thus given only in the vicinity of a territoral boundary.

Wawooo

The *Wawooo* call was given only by infants. It was heard from infants as young as 118 days. This call was apparently analogous and possibly homologous to the *Aarr* calls, being given in identical situations and apparently communicating the same message. This call was suggestive of a call intermediate to the Progression Grunt and Lost *rrr* call.

Intergroup Grunt

On a very few occasions when two males of different groups were within ten feet of one another, they exchanged low-frequency and low-amplitude grunts several times in rapid succession. These grunts were most reminiscent of the Progression Grunts. The function of and response to these Intergroup Grunts are not understood.

Intergroup Agonism

*Intergroup Chutter A**

In comparing Intergroup Chutter A calls of two particular adult females (named TK and LEG), I found that TK had significantly longer units than did LEG (0.01 level) and that LEG had significantly longer intervals between units than TK (0.01 level). However, they did not differ significantly from one another in the upper frequency of the major energy (0.50 level), number of units per phrase (0.10 level), and duration of phrases (0.10 level). The Intergroup Chutter calls A and B were sometimes intermingled in the same bout.

The Intergroup Chutter A was compared statistically with the Intergroup Chutter B, Intragroup Chutter, Short *Aarr*, Snake Chutter, and Chutter-Toward-Observer. The results of these tests of significance are summarized in Table 3. In some cases, because of inadequate sampling, it was not possible to make statistical tests of significance. However, it was my impression that the Intergroup Chutter A differed further from the Chutter-Toward-Observer in having more units per phrase and shorter intervals between units (see Plate 16.1).

This call was given by adult females, an infant about 217 days old, and probably all juveniles.

The mouth of the vocalizer was closed or partially open and the teeth were not exposed.

*Intergroup Chutter B**

The phrases of this call were sometimes composed of indistinct units which appeared to be smeared together along the temporal axis of the sonograms. However, many of the phrases did have distinctly separate units. As mentioned above, the major energy of these calls was restricted to the lower frequencies and thus a shorter range of frequencies than it was in the Intergroup Chutter

A. The other measurements of these two calls were almost the same. This category includes calls which spectrographically resemble one another, but which audibly sounded quite variable, ranging from the chutter of an adult female to a wooflike call of a 2- to 3-year-old male to a barklike call of an adult male. This apparent auditory variation might be attributed to age and sex differences in call production.

The Intergroup Chutter B appeared and sounded most like the Intragroup Chutter (see section on Intragroup Chutter and Plate 16.1).

TABLE 3

RESULTS OF STATISTICAL TESTS OF SIGNIFICANCE WHEN COMPARING THE INTERGROUP CHUTTER A WITH THE FIVE CALLS MOST SIMILAR TO IT

	CHARACTER OF CALL TESTED	NAMES OF FIVE CALLS COMPARED WITH THE INTERGROUP CHUTTER A				
		Intergroup Chutter B	Short *Aarr*	Intragroup Chutter	Snake Chutter	Chutter-Toward-Observer
INTERGROUP CHUTTER A	Upper frequency of major energy	A> (.001)	A> (.001)	A> (.001)	A> (.001)	---
	Duration of unit	---	---	A> (.05)	---	---
	Number of units per phrase	---	---	A> (.01)	A> (.001)	---
	Interval between units	---	A> (.001)	NS (.10)	---	---
	Duration of phrase	---	---	A> (.01)	A> (.001)	---
	Lower frequency of major energy	---	---	---	---	A< (0.1)

A< = Intergroup Chutter A significantly less, A> = Intergroup Chutter A significantly greater, NS = not significantly different, --- = no test made, (Nos. in parentheses) = level of significance.

This call was given by adult males and females, and a 2- to 3-year-old juvenile male. The mouth position was essentially the same as that for the Intergroup Chutter A.

Both Intergroup Chutters A and B were given during relatively intense intergroup agonistic encounters. Such encounters usually involved chasing and sometimes grappling, grabbing, hitting, and biting. In addition to indicating that an intergroup agonistic encounter was in progress, these two calls may have solicited the aid of other group members in assisting the vocalizer in its encounter with a foreign group. Furthermore, these calls might have been aggressive threats toward the foreign group.

Bark

As mentioned earlier, this call sounded identical to the Bark that was given during intragroup agonistic encounters. This call was given only by subadult and adult males during intense intergroup agonistic encounters. Although the function of this call was not clearly understood, it seems likely that in addition to communicating the occurrence of an intense intergroup agonistic encounter, this call also functioned as an aggressive threat and may have solicited the aid of others in the attack of the vocalizer against the opponent(s) of the foreign group.

Chirp

This call sounded identical to the Chip given toward major predators (see below). It was given only by adult females and juveniles during rather intense intergroup agonistic encounters. As with the Bark, the Chirp undoubtedly indicated to the other vervets the occurrence of intense intergroup agonistic encounters, functioned as an aggressive threat, and may have solicited the aid of others in the agonistic encounter between the two groups.

Differences in the stimulus situations and functions of the preceding four calls were not readily apparent.

Proximity of Human Observer (the Author)

*Chutter-Toward-Observer**

The range of the distribution of major energy and the duration of the phrases of this call were quite variable, but owing to the small sample, statistical treatment of these inter-call differences was not possible. The other measurements were relatively uniform.

The Chutter-Toward-Observer call was audibly very similar to the Snake Chutter (see section on Snake Chutter) and to the Intergroup Chutter A (see Table 3, section on Intragroup Chutter A, and Plate 16.1).

This call was only heard from three particular individuals. It was given by an unidentified adult male once, several times by an older subadult male (about 4 years old, named LBD), and several times by a juvenile female (about 2 years old, named NN). During this call, the mouth of the vocalizer formed a grimace with the unclenched teeth exposed.

The stimulus evoking this call in the unidentified male was not determined. The juvenile female gave this call whenever I came within about 4 feet of her. She sat upright and stared toward me as she gave this call. Once, as I approached an infant, she ran toward me, exposed her eyelids, stared, and gave this call. The older subadult male gave this chutter when I was relatively close to him and to either one of two particular infants. On some occasions, this male approached me as I stood near one of these infants; then he stood quadrupedally and stared toward me as he gave this chutter from about 4 feet away. The distance between the infant and myself in such situations ranged from about 7 to 15 feet.

Although the communicative function of this call is not clearly understood, it seems most likely that it was a moderate- to low-intensity threat, directed toward me. Furthermore, it was not clear why so few individuals gave this call. It may be pertinent to mention that the juvenile female NN ranked rather low in the apparent dominance hierarchy and was the only individual observed to give the *Wa-waa* call. The older subadult male (LBD) was the second-most dominant male in his group (P-group).

Proximity of Potential Snake Predator

*Snake Chutter**

This low-amplitude call was most like the Intergroup Chutter A (see Table 3 and Plate 16.1) and Chutter-Toward-Observer. The Chutter-Toward-Observer call was significantly greater than the Snake Chutter call in the lower frequency of major energy (0.01 level), duration of units (0.001 level), and interval between units (0.05 level). It also seems likely that the difference in the duration of phrases of these two calls was significant.

The mouth was formed into a grimace with the unclenched teeth exposed during this call. The Snake Chutter was given primarily, if not exclusively, by adult females and juveniles.

On one occasion, an adult male gave a chutter that sounded like a Snake Chutter and might have been directed toward a snake. However, it was my impression that another, undetermined stimulus was involved. Regardless of this particular case, it can be said that the subadult and adult males rarely, if ever, gave the Snake Chutter.

The Snake Chutter was directed toward two species of snakes: the Egyptian cobra (*Naja haje*) and the puff adder (*Bitis arietans*). While giving this call, several vervets clustered together in very close formation about 5 feet away from the snake, staring toward it. In such a formation, and while continuing to emit this call, the vervets moved along beside the snake as it crawled through the thickets. This response was not evoked by at least three other unidentified species of snakes that were seen near the vervets. It seems likely that the Egyptian cobra and puff adder were potential predators upon the vervets or at least potential sources of harm and the latter three species were not. The response to these latter three snake species was variable. Once, two juveniles grabbed at and closely approached an unidentified snake (about 8–10 inches long) in a manner suggestive of exploratory behavior.

On a second occasion, as an adult female sat in a tree, another unidentified snake (green in color and about 3 feet long) climbed to within about 4 inches of her. Suddenly the adult female leapt to another branch about 3 feet away and sat. She uttered no sound. The snake dropped to the ground.

On a third occasion, a thin, lime-green snake, about 2 feet long, climbed past (within about 3 feet) two juvenile females (1 year old) and an infant male (95 days old). One of the juvenile females gave a single, low chutter which resembled a Snake Chutter. All three briefly looked toward the snake and then resumed their previous activities.

Python sebae was seen in the Amboseli Reserve and within the home range of the vervets that were studied most intently. However, it was never seen in association with vervets. *P. sebae* is very likely a predator on vervets (see Isemonger 1962) and may evoke a response similar to that evoked by the Egyptian cobra and puff adder.

The Snake Chutter, in conjunction with the associated behavior, seemed to warn other vervets of and to indicate to them the location of a potential snake predator. The low amplitude of this call seemed to be adaptive in that it effectively warned other vervets, who might be within the relatively short striking distance of the snake, without attracting the attention of other predators to the vervets, as might occur with a call of higher amplitude.

Proximity of Minor Mammalian Predator

Uh!

The *Uh!* call was typically of low amplitude, short duration, and seldom, if ever, repeated. It seemed to be an audible exhalation. The appearance of the vocalizer's mouth during this call was not noted. Presumably, all adults and juveniles were capable of giving this call.

The *Uh!* call was apparently evoked upon initially seeing close by a spotted hyena (*Crocuta crocuta*), silver-backed jackal (*Canis mesomelas*), cheetah (*Acinonyx jubatus*), hunting dog (*Lycaon pictus*), or, irregularly, Masai tribesmen. These species are believed to be minor mammalian predators because they did not evoke a high-intensity alarm response, which included rapid flight into trees or thickets and Threat-Alarm-Barks and alarm Chirps. An exception to this occurred once when the vervets gave a high-intensity alarm response to a female cheetah and her cub that were standing about 75 yards from the vervets. The *Uh!* call could be evoked by any one of these species from as far as about 300 yards. Immediately subsequent to this call, almost all members of the group looked toward the stimulus and then resumed their previous activities or, if the potential predator was approaching them, they moved away. The *Uh!* call may be considered as a low-intensity alarm call which specifically indicated the relative proximity of any of the above-mentioned minor, mammalian predators.

Sudden Movement of Minor Predator (Mammalian and Avian)

Nyow!

The *Nyow!* call was of short duration and not generally repeated. Although not observed in each age-sex class, the *Nyow!* call was probably given by all adults and juveniles. The *Nyow!* call of subadult and adult males seemed to be of lower frequency (pitch) than the similar calls of females and younger males. The *Nyow!* call seemed to have several acoustical variations; however, in the absence of tape recordings and sonograms, this problem cannot be elucidated.

When either a baboon (*Papio* sp. [*cynocephalus*]), bush cat (*Felis lybica*), or spotted hyena (*Crocuta c.*) approached, apparently unseen, to within about 20 yards of the vervets and then suddenly and rapidly bolted away from the area (usually because it had just seen me), the *Nyow!* call was given by the vervets. On one occasion, a warthog (*Phacochoerus aethiopicus*), which similarly approached a group of vervets unseen and then suddenly bolted away upon perceiving me, also evoked the *Nyow!* call from them. This was probably a "false alarm." Subsequent to the *Nyow!* call, essentially all of the vervets looked toward the "foreign" animal as it bolted away. From such observations, it seems reasonable to assume that the stimulus evoking the *Nyow!* call was not so much the species of the intruder as it was the sudden movement of an animal that was near the vervets and whose approach had not been seen by them.

In addition to the above stimuli, the Verreaux's Eagle-Owl (*Bubo lacteus*), one of the largest owls in East Africa, evoked the *Nyow!* call when flying near the vervets. However, the vervets demonstrated no such response when this owl was perched. In fact, vervets climbed within 2 to 3 feet of this owl when it was perched, without showing any response to it. Furthermore, vervets moved across open, short-grass areas, 75 to 100 yards wide, toward the tree in which a clearly visible and audible Verreaux's Eagle-Owl was perched. In a situation such as this, the monkeys would be extremely prone to avian predation, and yet no response to the owl was seen. However, as soon as the owl took flight in the vicinity of the vervets, the *Nyow!* call was given, and the majority of the monkeys looked toward the flying bird. In several encounters between these two species, only once did I observe an owl dive toward vervets (a presumed attack, which the vervets avoided by fleeing into a thicket). Thus the relationship between these two species is not clear. Possibly a more intense predator-prey relationship exists between these two species at night when the owl is presumably a more effective predator. Thus the element of surprise, which seems to be the basis of the previous stimulus situations, is apparently absent when the Verreaux's Eagle-Owl is the stimulus evoking the *Nyow!* call.

The *Nyow!* call seemed to indicate that any of the above-mentioned species was in rapid motion, and thus it directed the attention of the vervets toward the moving stimulus. With the exception of the warthog, all of the other stimulus species were potential predators upon the vervets. It is suggested that the *Nyow!* call was an alarm of moderate intensity, which may be evoked by the sudden movement of any previously unseen animal (particularly by a minor predator) and by the flight of a nearby Verreaux's Eagle-Owl.

Initial Perception of Major Avian Predator

Rraup

The *Rraup* call was of short duration and rarely, if ever, repeated, being given only once by one individual. Although not well established, the *Rraup* call was probably given by all adult females and juveniles more than about 131 days old. Subadult and adult males were not heard to give this call. The homologue of this call among subadult and adult males may be the Threat-Alarm-Bark (see below).

The *Rraup* call was regularly evoked upon the initial perception, either while in flight or perched, of the Martial Eagle (*Polemaetus bellicosus*) or Crowned Hawk-Eagle (*Stephanoaetus coronatus*). Immediately subsequent to the *Rraup* call, the vervets ran or dropped into the dense thickets from the tree branches and the open, short-grass areas. The *Rraup* call and rapid retreat into the thickets were commonly followed by Threat-Alarm-Barks and Chirps (see below), which were given while looking toward the bird. This was especially so if the bird made an attack upon the vervets. The *Rraup* call was given toward Martial Eagles and Crowned Hawk-Eagles who were as much as 300 yards away. With rather strong circumstantial evidence, such as unsuccessful attacks, one occasion of a Crowned Hawk-Eagle feeding on a vervet that the eagle had presumably killed, and the intense response which these birds evoked from the vervets, it is suggested that the Martial Eagle and Crowned Hawk-Eagle were major predators upon vervets.

On four occasions, the *Rraup* call was evoked by an immature Steppe or Tawny Eagle (*Aquila nipalensis* or *A. rapax*),[4] White-Back Vulture (*Pseudogyps africanus*), a Harrier-Hawk (*Polyboroides typus*), and a Gabar Goshawk (*Micronisus gabar*). These were apparently "false alarms," because in response to the *Rraup* call the vervets ran only part way toward the dense thickets, stopped, looked toward the bird concerned, and then resumed their previous activities. Furthermore, the vervets showed no response to these four avian species in the many other encounters that they had with them.

It seems, then, that the *Rraup* call was given upon initial perception of a major avian predator, evoking a rapid retreat by the vervets into the dense thickets and away from the more open areas. Therefore, it may be termed a "high-intensity alarm call," specific for avian predators.

[4] These two are probably conspecific, according to a personal communication from Mr. John Williams of the Coryndon Museum, Nairobi, Kenya.

Proximity of Major Mammalian and Avian Predators

*Threat-Alarm-Bark**

The Threat-Alarm-Bark was composed of abrupt phrases of low frequency and high amplitude. These features made this call audible and localizable over a great distance and thus effectively indicated the position of predators to other vervets. Any phrase of two or more units consisted of a series of alternating audible exhalations and inhalations (Plate 16.10). In a total of 195

TABLE 4

NATURE OF TONAL EXHALATION UNITS OF THREAT-ALARM-BARK

(Nos. in parentheses) = frequency of occurrence

Sequential Position of Tonal Unit in Phrase		1st (11)	3d (3)	5th (2)		7th (3)
Fundamental in cps		250 (11)	250 (3)	250 (1)	500 (1)	250 (3)
Harmonic(s) expressed as an integral multiple of the fundamental	2f	(11)	(3)	(1)	(1)	(2)
	3f	(10)	(3)	(1)	(1)	(3)
	4f	(11)	(3)	(1)	(1)	(3)
	5f	(11)	(2)	(1)	(1)	(3)
	6f	(8)	(2)		(1)	(2)
	7f	(5)	(1)			(1)
	8f	(1)				
	9f					
	10f					
	11f					
	12f					
	13f					(1)
	14f					(1)
	15f					(1)

units analyzed, of which 104 were exhalation units and 91 were inhalation units, tonality occurred in only 19 units, all of which were exhalations. The nature of these tonal exhalation units is summarized in Table 4. Such tonal units occurred in only two out of eight individuals recorded on tape. The Threat-Alarm-Bark was, therefore, predominantly a nontonal call.

With some individuals, the exhalation units were of longer duration than the inhalation units, whereas with others the reverse was true, or the two types of units were of nearly equal duration. The one character which was consistently different between the exhalation and inhalation units was the upper frequency of major energy. The exhalation units had a higher upper frequency of major energy than did the inhalation units. This difference was significant (0.001, 0.01, 0.01 levels) within each of the three individuals (named SG, DP, and LEN) for whom adequate samples were available.

The sequential arrangement of the exhalation and inhalation units was analyzed in forty phrases. In all cases with two or more units, the phrase was initiated with an exhalation unit and terminated with an inhalation unit. Furthermore, the majority of phrases with two or more units were composed of an alternation of exhalations and inhalations. Exceptions to this latter feature occurred in five cases (12.5 per cent). In four of these, the first two units were both exhalations subsequent to which the typical alternating pattern of exhalations and inhalations occurred. The interval separating the first two exhalations in these four cases appeared to be of longer duration than the usual interval between units. The fifth exceptional phrase was composed of only three audible exhalation units, which were separated from one another by very long intervals.

In addition, there were six phrases which were composed of one exhalation unit only. No phrases with only single inhalation units occurred. In four phrases, the exhalation and inhalation units occurred in rapid sequence with essentially no interval between succeeding exhalations and inhalations, thus giving units which were composed of one exhalation subunit and one inhalation subunit. In these four cases, the exhalation subunit always preceded the inhalation subunit. In one additional case, a chutter-like unit occurred between two Threat-Alarm-Bark phrases.

It was my impression that when the stimulus (predator) evoking the Threat-Alarm-Bark was visible and relatively close to the vocalizer, the amplitude of the call increased, the number of units per phrase increased, and the interval between phrases decreased.

The Threat-Alarm-Bark was given only by subadult (about 3 years old) and adult males. During this call, the mouth of the vocalizer was opened widely and briefly (Plate 16.8). While the mouth was open, the unclenched teeth were exposed. The position of the mouth during the Threat-Alarm-Bark differed from a grimace in that during the former the mouth was opened wider and for a shorter period than during the latter.

*Chirp**

The Chirp was composed of abrupt units of low frequency and high amplitude, thereby facilitating the audibility and localization of this call over long distances and thus enhancing its effectiveness as a warning call. The Chirp phrases apparently were composed of exhalations only (Plate 16.10). No audible inhalation was detected. The Chirp call, as is indicated by the name, gave the impression of a short and sharp-sounding call.

The Chirp call was given only by adult females and juveniles. The position of the mouth during this call was essentially the same as that during the Threat-Alarm-Bark.

Both the Threat-Alarm-Bark and the Chirp were evoked by the appearance and relative proximity of Martial Eagles and Crowned Hawk-Eagles, lions (*Panthera leo*), leopards (*P. pardus*), serval cats (*Felis serval*), and on one occasion by a cheetah.[5] All of these species, with the possible exception of lion and cheetah, were judged to be major predators on vervets. These two calls were evoked by most of the above species from a distance as great as about ¼ mile. During a typical encounter, many Threat-Alarm-Barks and Chirps were heard, creating a most impressive display. The longest of these demonstrations lasted for about 4½ hours and was directed toward an immature Crowned Hawk-Eagle as it fed on a vervet, which the eagle had presumably killed.

As indicated earlier, the subadult and adult males' Threat-Alarm-Barks, which are given at the onset of any particular encounter with a major avian predator, are probably homologous to the *Rraup* call of adult females and juveniles. Similarly, this seemed to be the call which subadult and adult males gave upon first perceiving a major mammalian predator. I was unable to distinguish from the Threat-Alarm-Bark whether the predator was avian or mammalian. The initial Threat-Alarm-Bark seemed to function primarily as an alarm, whereas the subsequent Threat-Alarm-Barks, although continuing to function as an alarm, seemed to act primarily as a threat directed toward the predator. On several occasions, adult males give the Threat-Alarm-Bark as they made incipient lunges or incomplete charges toward the predator.

The Chirp call was given by an adult female or juvenile either upon initially perceiving a major mammalian predator or subsequent to the *Rraup* alarm call. The vervets ran into the trees and away from the thickets and open areas when the Chirp was the initial alarm heard. This call seemed to function primarily as an alarm and secondarily as a threat. The latter would certainly be true when the predator was a bird, for adult female and juvenile vervets gave the *Rraup* call as an initial alarm call toward major avian predators.

In summary of the responses of vervets to major predators, subadult and adult males gave Threat-Alarm-Barks, both when they first perceived avian or mammalian major predators and subsequently. In contrast, adult females and juveniles gave the *Rraup* call when they first noticed a major avian predator, and then Chirp calls. If, however, the major predator was a mammal, adult females and juveniles, but not subadult or adult males, gave Chirp calls, both when the predator was first noticed and subsequently. There-

[5] In other localities, where vervets were molested by people, the Threat-Alarm-Bark was sometimes directed toward the author.

fore only the adult females and juveniles gave different responses to major avian and mammalian predators and consequently only they evoked different retreating responses in the other vervets, that is, retreat to the thickets for major avian predators or retreat to the tree branches for major mammalian predators.

Interference with Respiration

Coughing and Sneezing

Both of these sounds seemed to be of a lower frequency (deeper pitch) in subadult and adult males than in adult females, juveniles, and infants. Coughs and sneezes were apparently given by members of all age classes. In all age classes, sneezing, with its higher frequency, was readily distinguishable acoustically from coughing.

Coughing and sneezing seemed to have no function other than the obvious, namely, the elimination of respiratory interference and communicating that a particular individual was coughing or sneezing.

The coughing and sneezing of vervets was distinctive and, with the possible exception of baboons, not confused with that of other species at Amboseli. Coughing and sneezing might, therefore, facilitate the location of vervets by their predators.

Moynihan (in preparation) reports that coughing and sneezing are of social significance in *Callicebus moloch* and other New World monkeys.

Indigestion

Vomiting

Vomiting was only faintly audible. The actual sound seemed to be produced by a combination of air and food being expelled into the mouth. The regurgitated food was either completely retained within the mouth or partially expelled. In either case, the vomiter usually re-ingested most of the regurgitated material, frequently licking the expelled vomit from its hands, feet, and the ground. The actual vomiting commonly occurred in spasms over a period of several minutes. The vomiter was always in a sitting position, and, during each individual spasm of vomiting, the over-all muscle tonus of the vomiter appeared to increase. Vomiting was seen in all age classes with the exception of infants (6 months or younger).

Although data on the frequency of vomiting are not available, my impression is that some vomiting occurred at least once each day within a group of seventeen vervets.

It was presumed that vomiting was evoked by some form of indigestion. Presumably, the vomiting relieved the indigestion.

Juveniles (about 1–1½ years old) or infants sometimes approached the vomiter and placed their muzzles near the vomit on the ground and/or near the vomit on the lips and muzzle of the vomiter. Presumably, this latter behavior was exploratory in nature and was evoked by the sound and associated behavior of the vomiter. The significance of this exploratory behavior is not understood.

DISCUSSION

At least thirty-six physically and/or audibly distinct sounds were recognizable among the vervet monkeys (see Table 5). Twenty-one different stimulus situations evoked sounds from the vervets. At least twenty-one to twenty-three different messages were communicated and eighteen to twenty-two different responses evoked by these sounds (see Table 6).

TABLE 5

LIST OF THIRTY-SIX SPECTROGRAPHICALLY AND/OR AUDIBLY DISTINCT SOUNDS

(The inclusion of two or more sounds together indicates the possibility of identicalness.)

1. Intragroup Chutter
2. Intragroup Squeal
3. Intragroup Chutter-squeal, Weaning Squeal
4. Intragroup Squeal-scream, Anti-copulatory Squeal-scream, and Squeal
5. Intragroup Scream, infant Scream toward "strange" adult male, Weaning and Lost screams
6. Intragroup Bark, Intergroup Bark
7. *Woof*, *Woof*
8. *Waa*
9. *Woof-waa*
10. *Wa-waa*
11. Long *Rraugh*, Long *Aarr-rraugh*, Long *Aarr*
12. Short *Rraugh*
13. Lip Smacking
14. Teeth Chattering
15. Progression Grunt, Intergroup Grunt
16. Purr
17. Lost *rrr* and Weaning *rrr*
18. Lost *eee*
19. Lost Squeal
20. Lost *Rrah*
21. *Eh, eh*
22. Short *Aarr-rraugh*
23. Short *Aarr*
24. *Wawoo*
25. Intergroup Chutter A
26. Intergroup Chutter B
27. Chirp
28. Chutter-Toward-Observer
29. Snake Chutter
30. *Uh!*
31. *Nyow!*
32. *Rraup*
33. Threat-Alarm-Bark
34. Coughing
35. Sneezing
36. Vomiting

The size of the auditory repertoire of vervet monkeys is compared with that of several other primate species in Table 7. Some of the readily apparent differences in repertoire size are undoubtedly due to species differences. However, a certain portion of these differences may be due to differences in conditions under which the animals were studied, differences in the extent or duration of the study, and differences of opinion among the various observers in their definition of what constitutes a distinct sound or call. For example, all of the vocal repertoires that were established by Andrew for various species of primates under laboratory conditions are smaller than repertoires estab-

TABLE 6

SUMMARY OF STIMULUS SITUATIONS EVOKING SOUNDS, AGE AND SEX OF VOCALIZERS, MESSAGES COMMUNICATED, AND RESPONSES EVOKED BY VERVET SOUNDS

A = adult, SA = subadult, J = juvenile, yJ = young juvenile, in_1 = infant

Stimulus Situation	Name of Sound	Age and Sex of Vocalizer	Message Which Was Probably Communicated	Apparent or Probable Response Evoked in Other Animals
1. Intragroup agonism	Chutter	all juveniles, & SA & A females	solicitation for aid and aggressive threat	solicits aid and evokes flight
	Squeal Chutter-squeal Squeal-scream Scream	all juveniles, & SA & A females	solicitation for aid and defensive threat	solicits aid and inhibits attack
	Bark	SA & A males	aggressive threat	disrupts fighting
2. Attempted copulation by SA or A male with anestrous female	Anti-copulatory Squeal-scream and Squeal	SA & A female	indicates that female is anestrus	inhibits copulatory attempts by male
3. Red, white, and blue display	*Woof*, *Woof*	A, SA, & J male and A female	expresses subordination	inhibits attack
	Waa *Woof-waa* Lip Smacking Teeth Chattering	SA & A male	expresses subordination	inhibits attack
4. Close proximity of subordinate and dominant individuals	*Wa-waa*	J female	expresses subordination	inhibits attack
	Long *Rraugh* Short *Rraugh*	SA female & yJs	expresses nonaggression	permits subordinate to approach dominant
	Lip Smacking	all but infants	expresses nonaggression	permits close proximity of subordinate & dominant animals
	Teeth Chattering	SA & A males		

TABLE 6—*Continued*

Stimulus Situation	Name of Sound	Age and Sex of Vocalizer	Message Which Was Probably Communicated	Apparent or Probable Response Evoked in Other Animals
5. Initial phase of group progression	Progression Grunt	none younger than 139 days	indicates temporal proximity of group progression	facilitates coordination of group progression and/or group coherency
6. Play	Purr	yJs	not apparent	may enhance play bond
7. Approach of "strange" A male to in_1	Scream	in_1	indicates approach of "strange" A male	evokes rapid retrieval of infant by mother & rapid retreat of "strange" male
8. Weaning of infant	Scream Squeal *rrr*	in_1	not apparent	not apparent
9. Separation of infant and mother	Lost *rrr* Lost *eee* Lost Squeal Lost Scream Lost *Rrah*	in_1	not apparent	attracts attention of mother and evokes eventual retrieval of infant by mother
10. Reunion of infant and mother	*Eh, eh*	in_1	not apparent	may facilitate social bond between mother and infant
11. Ambivalent situation–(proximity of foreign group & close proximity of subordinate and dominant)	Long *Aarr-rraugh* Short *Aarr-rraugh*	yJs	indicates approach &/or proximity of foreign group &/or nonaggression	others look toward foreign group & permits vocalizer to approach dominant individual

TABLE 6—*Continued*

Stimulus Situation	Name of Sound	Age and Sex of Vocalizer	Message Which Was Probably Communicated	Apparent or Probable Response Evoked in Other Animals
12. Proximity of foreign group	Long *Aarr* Short *Aarr*	SA male, A female, Js, & yJs	indicates approach &/or proximity of foreign group	others look toward foreign group
	Wawooo	in_1		
	Intergroup Grunt	males	not apparent	evokes Intergroup Grunt
13. Intergroup agonism	Intergroup Chutter A	A female, in_1 & probably Js	indicates intergroup chase &/or fight; possibly solicitation for aid; aggressive threat	others look toward foreign group and may solicit aid
	Intergroup Chutter B	A & SA male, A female		
	Bark	A & SA male	indicate intergroup chase &/or fight; possibly solicitation for aid; aggressive threat	others look toward foreign group and may solicit aid and evokes flight
	Chirp	A & SA females & Js & yJs		
14. Near proximity of human observer	Chutter-Toward-Observer	A & SA male, J female	moderate- to low-intensity threat toward observer	not apparent
15. Proximity of snake predator	Snake Chutter	females and Js & yJs	high-intensity warning of snake predator	others look toward snake
16. Proximity of minor mammalian predator	*Uh!*	probably all As & Js & yJs	low-intensity warning of proximity of minor mammalian predator	others look toward predator
17. Sudden movement of minor predator (mammalian & avian)	*Nyow!*	probably all As & Js & yJs	moderate-intensity warning of sudden movement of minor predator (mammalian & avian)	others look toward predator & sometimes run toward trees

TABLE 6—*Continued*

Stimulus Situation	Name of Sound	Age and Sex of Vocalizer	Message Which Was Probably Communicated	Apparent or Probable Response Evoked in Other Animals
18. Initial perception of major avian predator	*Rraup*	probably all A females & Js to as young as 132 days	high intensity warning of proximity &/or approach of major avian predator	others run into thickets and away from open areas and treetops
19. Proximity of major predator (mammalian & avian)	Threat-Alarm-Bark	A & SA males	high intensity warning of & aggressive threat toward major predator (mammalian & avian)	others look toward predator and run to appropriate cover; predator ignores threat or gives aggressive threat in return
	Chirp	females & Js & yJs	high intensity warning of major mammalian predator & possibly aggressive threat toward major predator (mammalian & avian)	others run into trees; predator's response not apparent
20. Interference with respiration	Coughing Sneezing	all	indicates the obvious	none
21. Indigestion	Vomiting	all except in_1	indicates the obvious	others sometimes approach and investigate vomiter's mouth and the vomit

lished by others for the same species under field conditions (Table 7). This difference could, however, be due to either of the other two factors mentioned above. Differences of opinion as to what constitutes a distinct sound or call are not readily resolved. As Altmann (1962, 1965) has suggested, distinct behavior patterns and thus distinct sounds or calls can be determined empirically; that is, "One divides up the continuum of action wherever the animals do." However, in itself, this approach, of splitting the behavior wherever the animals do, is not completely adequate for the establishment of a repertoire of com-

TABLE 7

COMPARISON OF SIZE OF AUDITORY REPERTOIRES

Species	Reference	Number of Sounds
Cercopithecus aethiops	Struhsaker (this chapter)	≥ 36
C. aethiops, *C. nigroviridis*, *C. neglectus*, *C. mitis*, *C. nictitans*, *C. cephus*, *C. mona*, *C. diana*	Andrew (1963*a*, *b*)	5–8
Erythrocebus patas	Hall, Boelkins, & Goswell (1965)	11 (including teeth gnashing & infant calls)
Macaca fuscata	Itani (1963)	37 (including estrus & infant calls)
Macaca mulatta	Rowell & Hinde (1962) Altmann (1962) Andrew (1963*a*, *b*)	20–30 7–17 6–8
baboon (*Papio sp.*)	Hall & DeVore (1965) Andrew (1963*a*, *b*)	15 (including lip smacking) 4–5
Presbytis entellus	Jay (1962)	≥ 10
Alouatta palliata	Carpenter (1934), Collias & Southwick (1952), Altmann (1959)	20
Aotus trivirgatus	Moynihan (1964) Andrew (1963*a*, *b*)	9 or 10 4
Gorilla gorilla	Schaller (1963)	23 (including chest beating)
Pan troglodytes	Goodall (1965) Reynolds (1965) Andrew (1962; 1963*a*, *b*)	25 (including drumming & lip smacking) 13 (including drumming) 6–7
Hylobates lar	Carpenter (1940)	11 (including lip smacking & teeth snapping)

municative behavior patterns. Such an approach does not permit the distinction between the variations of one behavior pattern all of which have the same communicative function and a graded or continuous system of behavior patterns, in which there are an infinite number of communicative functions. It is suggested that repertoires of communicative behavior patterns be based upon both the natural units of behavior as suggested by Altmann and upon the communicative function of the behavior patterns as manifested by the response(s) which they evoke in other animals.

Homologous and/or analogous calls of several primate species are compared with vervet calls in Table 8. Until more information is available on the form, ontogeny, and function of various behavior patterns among the primates and the effects of ecological variables on these patterns, it will remain extremely difficult to distinguish homologues from analogues. The comparison in Table 8 is based primarily on the form or acoustical nature of the calls and the apparent function of or response evoked by these calls. Although this comparison is premature, it seems likely that calls which are similar in form and function in species of the same subfamily are homologues. Difficulties are most apparent in comparisons of acoustically similar calls that occur in many, dissimilar species; for example, the shrieks, screams, screeches, and so forth.

Andrew (1962; 1963*a, b*) has presented some data on auditory communication in several species of *Cercopithecus*. These data are based on captive animals. Andrew (1963*a, b*) has listed five calls which he heard from *C. aethiops*. I was unable to recognize his "Moo" or "Cluck" calls spectrographically or in terms of stimulus situations evoking them. Andrew's "arrr" call may be somewhat similar to the *Aarr* calls which I described. However, the energy of his "arrr" call occurred at higher frequencies (4 to 8 Kc) than the major energy in the *Aarr* calls (0.25 to 1.0 Kc) which I described. "Arrr calls are also given with high grins in response to attack of superiors (*C. aethiops*)" (Andrew 1963*b*). This stimulus situation was radically different from that evoking the *Aarr* call among the vervets that I observed. Andrew (1963*a*) refers to "grunts" which were ". . . given during friendly mutual grooming and when approaching a social fellow." These "grunts" may be the same as the *Woof, Woof* call or Progression Grunt which I described, but again, the behavioral context was very different. Andrews "shriek" is probably the same as the Squeal, Chutter-squeal, Squeal-scream, or Scream calls that I recorded. Such differences as apparently exist between our data may be attributable to geographical variation within the species. Unfortunately, Andrew does not indicate the source of his animals.

The distribution among the age-sex classes of the thirty-six distinct sounds is summarized in Table 9. It is not certain whether the differences in repertoire size between the various age-sex classes were significant. However, it is interesting to note that, significant or not, the difference between the smallest repertoire (infant class with 12) and the largest (young juvenile class with

TABLE 8

Comparison of Possible Homologous and/or Analogous Sounds

The numbers listed with each sound in column one correspond to the numbers of these same sounds as listed in Table 6

Cercopithecus aethiops (Struhsaker, this chapter)	*Cercopithecus aethiops* (Andrew 1963*a*, *b*)	*Erythrocebus patas* (Hall *et al.* 1965)	*Macaca fuscata* (1. Itani 1963; 2. Andrew 1963*a*, *b*)	*Macaca mulatta* (1. Rowell & Hinde 1962; 2. Altmann 1962; 3. Andrew 1963*a*, *b*)	Baboon (*Papio* sp) (1. Hall & DeVore 1965; 2. Andrew 1963*a*, *b*)	*Presbytis entellus* (Jay 1962)	*Alouatta palliata* (1. Carpenter 1934, 2. Altmann 1959)	*Aotus trivirgatus* (Moynihan 1964)	*Gorilla gorilla* (Schaller 1963)	*Pan troglodytes* (1. Reynolds 1965; 2. Andrew 1962, 1963*a*, *b*; 3. Goodall 1965)	*Hylobates lar* (Carpenter 1940)
1. Chutter	...	*huh-huh*	...	...	...	...	...	...	...	...	...
1. Squeal Chutter-squeal Squeal-scream Scream	shriek[1]	shriek *gek-gek-gek*	1. B-1 2. shriek	1. screech 2. *eee* (36) 3. shriek	1. screeching 2. shriek	squeal-scream continuum	...	screams	screams	1. squeal 2. shriek 3. screams	...
3. Lip Smacking	...	lip smacking	1. rhythmic lip movement	1. lip smacking 2. lip smacking (39)[2]	1. lip smacking	...	...	...	...	3. lip smacking	lip smacking
3. Teeth chattering	...	teeth gnashing	...	1. gnashing 2. gnashes teeth (6)	...	...	...	...	...	...	teeth snapping
5. Progression Grunt	...	...	1. A-7, A-10	2. *hŭ*, *hŭ*, *hŭ* (48)	1. grunting	...	1. type 2 2. grunt, type H	...	abrupt grunts	1. grunts 3. *hoo*	type IX
6. Purr	...	...	...	...	...	...	...	...	chuckle	3. *aach-e-aach*	type VII

[1] Also present in *Cercopithecus mitis*, *C. nigroviridis*, *Macaca maura*, *Papio hamadryas*, *Theropithecus gelada* (Andrew 1963) and *Galago crassicaudatus*, *Lemur fulvus* (Andrew 1962).

[2] Also in *Lemur fulvus* infants (Andrew 1962).

TABLE 8—*Continued*

Cercopithecus aethiops (Struhsaker, this chapter)	*Cercopithecus aethiops* (Andrew 1963*a*, *b*)	*Erythrocebus patas* (Hall *et al.* 1965)	*Macaca fuscata* (1. Itani 1963; 2. Andrew 1963*a*, *b*)	*Macaca mulatta* (1. Rowell & Hinde 1962; 2. Altmann 1962; 3. Andrew 1963*a*, *b*)	Baboon (*Papio* sp) (1. Hall & DeVore 1965; 2. Andrew 1963*a*, *b*)	*Presbytis entellus* (Jay 1962)	*Alouatta palliata* (1. Carpenter 1934, 2. Altmann 1959)	*Aotus trivirgatus* (Moynihan 1964)	*Gorilla gorilla* (Schaller 1963)	*Pan troglodytes* (1. Reynolds 1965; 2. Andrew 1962, 1963*a*, *b*; 3. Goodall 1965)	*Hylobates lar* (Carpenter 1940)
8. weaning: Scream Squeal *rrr*	....	squeal	1. F-6	1. gecker &/or rising clicks 2. *ĭk, ĭk, ĭk* (24)	1. ick-ooer	....	....	....	....	....	....
9. Lost *rrr* Lost *eee* Lost Squeal Lost Scream Lost *Rrah*	....	....	1. F-2, F-3, &/or F-4	1. long, high plaintive call 2. *kōō* (56) 3. shriek[2]	....	....	1. type 5	infant squeaks & screams	whine, screech	2. shriek	....
10. *Eh, eh*	....	....	....	....	....	....	1. type 6	....	....	....	....
1. Bark 13. Bark	....	bark	1. C-1 &/or C-5	1. bark &/or roar 2. *!hō!* (29)	1. roaring	....	....	gruff grunts	....	....	....
16. *Uh!*	....	chirrup	1. A-4	....	....	....	....	....	....	....	....
19. Threat-Alarm-Bark	....	bark	1. D-1	....	1. Two-phase bark	alarm bark	2. type C_1	....	roar	1. waa barks 3. high-pitched bark, panting barks	type III
19. Chirp	....	yak	1. D-2 &/or D-3	1. shrill bark 2. *!ka!* (55)	....	alarm chirp	2. type D_1	....	bark	1. waa barks	type III

18) was relatively small. This is in contrast with what Kawabe found in Japanese monkeys as reported by Itani (1963). Apparently Kawabe found that *Macaca fuscata* infants had developed a larger (26) number of sounds than did the juveniles (13) and adults (about 13).

Behavior patterns of communicative significance are sometimes divided into discrete or continuous patterns (Hockett 1960, Marler 1961). The discrete patterns are those which are separate and readily distinguishable from other patterns, whereas the continuous patterns are those which more or less blend continuously into other patterns. The majority of vervet sounds seems to be of the discrete type. However, communicative signals of the continuous type might be represented by the Squeal, Chutter-squeal, Squeal-scream, and Scream group of calls; the *Rraugh, Aarr,* and *Aarr-raugh* group of calls; and the Weaning and/or Lost Squeal and Scream calls. Although variability

TABLE 9

AUDITORY REPERTOIRE SIZE OF AGE SEX CLASSES

Age-Sex Class	Adult Male	Subadult Male	Adult Female	Subadult Female	Juvenile Male	Juvenile Female	Young Juveniles	Infants
Age in years	4+	3–4	4+	3–4	1.5–3	1.5–3	0.5–1.5	0–0.5
Size of auditory repertoire...	13	14	17	13	15	16	18	12

in the physical nature of the call is more apparent in those of the continuous type, the measurements readily indicate that variability also exists in calls of the discrete type. One expects to find a certain amount of intraspecific variability in every biological phenomenon which is relatively successful from an evolutionary standpoint. However, in addition to structural variation (syntactic), the continuous systems also have functional variation (pragmatic). That is, within a continuous system there are an infinite number of functions.

Marler (1965), in reviewing research on primate auditory communication, indicates that the continuous or graded system of auditory communication is more prevalent in the baboons, macaques, and chimpanzees, whereas the *Cercopithecus* and *Colobus* species (with the possible exception of the olive colobus [*Procolobus verus*]) are typified by a relatively discrete system of auditory communication. As mentioned above, the majority of vervet sounds are apparently of the discrete type and thus in agreement with this thesis. Marler (1965) further implies that a grading of calls reduces the size of the repertoire: "It will be clear that several of the estimates in [Marler's] Table 2 of the size of the repertoire of sound types defined as suggested are too large. Nowhere is this more evident than in the vocal behavior of the rhesus monkey. . . . Rowell (1962) has been able to show that these nine sounds (agonistic sounds) actually constitute one system, linked by a continuous series of intermediates." As pointed out above, however, in a graded or continuous system there is an infinite number of patterns and functions. As a consequence the repertoire is not reduced by a graded system, but is in fact increased.

Marler has also suggested an exclusively syntactical approach during the initial establishment of a behavioral repertoire: "Nevertheless, in dealing with communication signals, the explicit consideration of function should probably be postponed until the basic typological categories are set up, to avoid the dangers of circular reasoning," I am suggesting that instead of an exclusively syntactical approach a combined syntactical and pragmatic approach be assumed. This approach places equal emphasis on the description of the call, behavioral context in which the call is given, and the apparent response which the call evokes in other individuals. Such an approach would permit not only a consideration of structural variation (syntactical) but also of functional variation (pragmatic) and would thus allow distinction between a variable discrete system and a continuous system of communication.

SUMMARY

A catalogue of the auditory behavior of the vervet monkey, as established under field conditions, is described. Included in this description are various spectrographic measurements of the calls that were recorded on tape. Information on and impressions of the conditions evoking the sounds and the communicative function of the sounds are given. The auditory repertoires of the different age-sex classes among vervets are compared. Furthermore, auditory repertoires of several primate species are compared with the vervet's repertoire.

REFERENCES

Alder, H. L., and Roessler, E. B. 1964. *Introduction to probability and statistics.* 3d ed. San Francisco: W. H. Freeman & Co.

Altmann, S. A. 1959. Field observations on a howling monkey society. *J. Mammalogy* 40:317–30.

———. 1962. A field study of the sociobiology of rhesus monkeys, *Macaca mulatta. Ann. N.Y. Acad. Sci.* 102 (Art. 2): 338–435.

———. 1965. Sociobiology of rhesus monkeys. II: Stochastics of social communication. *J. Theoret. Biol.* 8(3):490–522.

Andrew, R. J. 1962. The situations that evoke vocalization in primates. *Ann. N.Y. Acad. Sci.* 102 (Art. 2):296–315.

———. 1963*a*. Trends apparent in the evolution of vocalization in the Old World monkeys and apes. *Symp. Zool. Soc. Lond.* No. 10, pp. 89–101.

———. 1963*b*. The origin and evolution of the calls and facial expressions of the primates. *Behaviour* 20 (Pt 1–2):1–109.

Carpenter, C. R. 1934. A field study of the behavior and social relations of howling monkeys. *Comp. Psychol. Monogr.* 10(2):1–168.

———. 1940. A field study in Siam of the behavior and social relations of the gibbon (*Hylobates lar*). *Comp. Psychol. Monogr.* 16(5):1–212.

Collias, N. E., and Southwick, C. 1952. A field study of population density and social organization in howling monkeys. *Proc. Amer. Philos. Soc.* 96:143–56.

Goodall, J. 1965. Chimpanzees in the Gombe Stream Reserve. In *Primate behavior: field studies of monkeys and apes,* ed. I. DeVore, New York: Holt, Rinehart and Winston.

Hall, K. R. L., Boelkins, R. C., and Goswell, M. J. 1965. Behaviour of Patas monkeys, *Erythrocebus patas,* in captivity, with notes on the natural habitat. *Folia Primatol.* 3:22–49.

Hall, K. R. L., and DeVore, I. 1965. Baboon social behavior. In *Primate behavior: field studies of monkeys and apes,* ed. I. DeVore, New York: Holt, Rinehart and Winston.

Hockett, C. F. 1960. Logical considerations in the study of animal communication. In *Animal sounds and communication,* ed. W. E. Lanyon and W. N. Tavolga, pp. 392–430. Washington, D.C.: American Institute of Biological Sciences.

Isemonger, R. M. 1962. "Snakes of Africa, south, central, and east." Johannesburg: Thomas Nelson and Sons Ltd.

Itani, Junichiro. 1963. Vocal communication of the wild Japanese monkey. *Primates* 4(2):11–66.

Jay, P. C. 1962. The social behavior of the Langur monkey. University of Chicago doctoral dissertation.

Marler, P. 1961. The logical analysis of animal communication. *J. Theoret. Biol.* 1:295–317.

———. 1965. Communication in monkeys and apes. In *Primate behavior: field studies of monkeys and apes,* ed. I. DeVore, New York: Holt, Rinehart and Winston.

Moynihan, M. 1964. Some behavior patterns of platyrrhine monkeys. I. The night monkey (*Aotus trivirgatus*). *Smithsonian Miscellaneous Publications* 146(5):84 pp.

Reynolds, V., and Reynolds, F. 1965. Chimpanzees in the Budongo Forest. In *Primate behavior: field studies of monkeys and apes,* ed. I. DeVore, New York: Holt, Rinehart and Winston.

Rowell, T. E. 1962. Agonistic noises of the rhesus monkey (*Macaca mulatta*). *Symp. Zool. Soc. Lond.* No. 8, 91–96.

Rowell, T. E., and Hinde, R. A. 1962. Vocal communication by the rhesus monkey (*Macaca mulatta*). *Proc. Zool. Soc. Lond.* 138 (Pt. 2):279–94.

Schaller, G. B. 1963. *The mountain gorilla: ecology and behavior.* Chicago, Illinois: The University of Chicago Press.

17

THE STRUCTURE OF PRIMATE SOCIAL COMMUNICATION

STUART A. ALTMANN

INTRODUCTION

Social communication among primates is a biological phenomenon and, like any other biological process, presents us with several major questions: (1) What is its structure? (2) How does it function? (3) What is its underlying causation? (4) How does it develop ontogenetically? (5) What is its adaptive significance? (6) How did it evolve? As with other biological processes, descriptions of what we may call the "anatomy" or "structure" of communication are bound to be a primary focus of research attention during this early stage in work on primate behavior. Indeed, our ability adequately to answer the remaining questions will depend upon an understanding of the structure of communication. The special nature of such a structure lies in this: it is a structure not of objects but of events.

THE NATURE OF SOCIAL COMMUNICATION

In 1939, Crawford wrote: "The term *communication* lacks precise definition in its application to animal behavior, for it may be stretched to include almost any sort of anticipatory movement which may signalize an activity to another individual, or it may be limited strictly to vocalizations or gestures which clearly direct or predicate."

Despite Crawford's comments, several authors have attempted to clarify what is meant by *communication.* Révész (1944) wrote: ". . . we may define *communication* as a type of behavior between living creatures of the same or different species, characterized by mutuality, rooted in biological heredity, and constituting one of the general manifestations of life." Unfortunately, Révész's definition is far too vague to be of much use.

Stuart A. Altmann, Yerkes Regional Primate Research Center, Emory University, Atlanta.

Haldane (1953) wrote: ". . . by a communication from animal X to animal Y I mean an action by X involving a moderate expenditure of energy, which evokes a change in the behaviour of Y involving much larger quantities of energy." Now, any mother who has shouted repeatedly at her child in order to get it to do some simple task will not agree that she has expended less energy than has the child.

Perhaps what Haldane was trying to convey can be seen in Hockett's definition (1958): "Communication is those acts by which one organism triggers another." By "triggering," Hockett means that the energy of the response is not the energy of the stimulus (Hockett 1960*a*). Certainly, this is true of virtually all cases of animal communication. My major objection to Hockett's definition, as will become clear below, is that his definition restricts the process of communication to the communicating act.

"Let us start," wrote Birch in 1952, "with a broad definition of 'communication' as the effect of the behavior of one organism upon the behavior of another organism." Whereas Hockett's definition restricts communication to signals, Birch's restricts it to the response to these signals. "The mere transmission and reception of a physical signal does not constitute communication. . . . Communication is not the response itself, but is essentially the *relationship* set up by the transmission of stimuli and the evocation of responses" (Cherry 1961). In short, social communication is a process by which the behavior of an individual affects the behavior of others.

This approach leads readily to our distinction between a community and a society. Members of a community are part of a common communication network. But while a predator and its prey may communicate, and are thus members of the same biological community, they are not members of the same society. A society consists of conspecific, intercommunicating individuals that are bounded by frontiers of far less frequent communication (Altmann 1962*b*).

Let us try to clarify the idea that the behavior of one individual affects the behavior of another. To start with, we must have a plurality both of *messages* (communicative behavior patterns) and of responses. "The message, to convey information, must represent a choice from among possible messages . . ." (Wiener 1948). Similarly, "a sign, if it is perceived by the recipient, has the potential for selecting responses in him" (Cherry 1961). Yet, what does it mean to say that these behavioral responses are affected? At any one moment, the behavior of an individual is not a determined process. Rather, each of the responses of which the individual is capable has a certain likelihood of occurring, depending upon numerous conditions, past and present. If the behavior of another individual produces any change in these likelihoods, then communication has taken place. Thus, when we say that, in a communication process, the behavior of one individual affects the behavior of another, we mean that it changes the probability distribution of the behavior of the other.

This means that in dealing with social communication we are dealing with the contingencies among sequences of events. The sequences may be thought of as a system that is in a particular state at any moment. The states represent the residual influence on the animal's behavior of contingencies, behavioral and otherwise, that antedate the present stimulus. For each state, there is a distribution of probabilities of the next event. Mathematical systems of the type described here are referred to as *Markov processes.* To date, changes in such probability distributions that result from social messages have been published for only one species of nonhuman primate, the rhesus macaque (Altmann 1965).

While the present review was in preparation, I came upon a paper by N. I. Zhinkin (1963) that includes ideas of striking similarity to some of those presented here. He writes (p. 133):

> Much, very much indeed, depends not only on facts but on the *approach* to their collection and treatment. The approach used in the present article is that of constructive algebra. Sounds made by monkeys are regarded as sequences of elements which have been defined in a certain way, and rules for the construction of these sequences are explained. . . . Nor shall we be faced with the question whether, to put it bluntly, a monkey is capable of "conversing" like a human. It would be more interesting to find out how vocal communication between monkeys themselves is carried on. If the sequences of sounds are not random, then they are capable of control to the extent to which entropy is destroyed.

Much relevant material in Zhinkin's paper could not, at this time, be incorporated into this review; the reader is therefore referred to Zhinkin's own work.

THE STRUCTURE OF SOCIAL COMMUNICATION

Fortunately, much of the framework for an analysis of the structure of social communication systems is now available. Of particular influence in my own thinking on this subject have been works of a mathematician, Claude Shannon (1948); a philosopher, Charles Morris (1955); and a linguist, Charles Hockett (1960*a*). In a paper presented at the 1959 meeting of the A.A.A.S., I attempted to show that all of the crucial properties (or similar properties) of human language that Charles Morris singled out are also found in the social communication systems of other anthropoid primates (Altmann 1962*b*). Hockett's much more thorough treatment of the properties of human language (Hockett 1958, 1959, 1960*b*, and especially 1960*a*) as well as the abundance of recent studies of communication in nonhuman primates makes a reappraisal of the properties of primate communication desirable.

Hockett's brilliant analysis of the universal design features of human language is of great heuristic value in the study of animal communication. The beauty of this approach is not that it provides a set of neat categories, or pigeonholes, for behaviors, but that it spells out the implications of membership in these categories, both in their relations to one another and in their more general, ecological and adaptive context.

Hockett's analysis of the design features of human language ranges widely. It deals, at various points, with the channel of communication, the transmitter, the receiver, the messages themselves, and even the ecology and genetics of the whole system. He approaches his subject from the standpoint of semantics, syntactics, and pragmatics. Beyond that, he has attempted to relate his analysis to animal communication in general. Basically, however, Hockett's analysis concerns the universal design features of human language. That analysis will form the backbone of what follows. Because my subject matter is the social communication of all primates and ultimately of all animals, I will feel free to refine and to enlarge upon Hockett's list of design features.

In the long run, this broad, biological approach to communication, which Sebeok (1965) refers to as "zoosemiotics," should be useful to the linguist, both in helping to demarcate what human language is *not,* and because the evolution of human language may have involved not only the development of new properties (Hockett and Ascher 1964), but also perhaps the loss or suppression of old ones. Furthermore, humans frequently communicate through nonlinguistic cues (Sebeok, Hayes, and Bateson 1964). Such paralinguistic and kinesic phenomena may have different properties than language per se (for example, they are often analogical rather than digital messages). In addition, such nonlinguistic processes may markedly affect our interpretations of linguistic messages.

Unfortunately for our purposes, the study of social communication in primates is in its infancy. The data that are available are almost all given in a simple, verbal form that does not lend itself readily to the type of analysis that we shall discuss here. As Marler (1965) has pointed out,

> even the descriptive material is scant and concerns only a hand-full of the better-studied species. Furthermore, the descriptions are mostly concerned with signalling behavior and little attention has been given to responses the signals evoke. In such circumstances, the judgement about what actually constitutes signalling behavior is a subjective one and may need revision in the future. Thus much of what follows is speculative and should perhaps be taken as a guide to the kinds of data we would like to have rather than a review of what has already been accomplished.

Let us now consider design features of communication systems in animals, with particular reference to primates. For the convenience of the reader, the following properties are cross-referenced to the corrsponding ones in Hockett (1960*a,* 1963), Charles Morris (1955), and Altmann (1962*b*). A recent paper by Bastian (1965), which is relevant to this discussion, was not available when this chapter was prepared.

Channel (1 in Hockett 1960*a*)

According to Hockett, the signals used in any language consist, without residue, of sounds that are produced by the vocal cords and received by the ears. Yet, even for a consideration of human social behavior, this restriction of the domain of discussion to acoustical signals, while perhaps convenient for the linguist, leaves unanswered all questions about the nature and quantity of information that is transmitted via other channels or about the basic ecological differences between the various channels. Indeed, some of these nonacoustical messages have a direct influence on the way in which we interpret spoken language and hence would seem to be of direct relevance to the linguists' task.

Primates use a wide variety of olfactory, tactile, auditory, and visual cues in their social communication, taking advantage of ecological features of each of these channels. For example, loud vocalizations are commonly used in long-range communication by species of primates that live in dense forests, such as the black-and-white colobus, *Colobus "guereza"* (Ullrich 1961), and several other colobine primates (Hill and Booth 1957); gibbons, *Hylobates lar* (Carpenter 1940); lemurs, especially *Indri indri* (Petter 1962); and howlers, *Alouatta palliata* (Altmann 1959, Carpenter 1934, Collias and Southwick 1952).

The howler monkey's howl—one out of about twenty vocal patterns in these animals (Altmann 1959)—is one of the most spectacular vocalizations in the animal kingdom. Special laryngeal sacs as well as an enlarged, cuplike hyoid bone apparently serve as resonating chambers, although the mechanism by which the very intense sounds are produced is not yet fully understood (Keleman and Sade 1960, Starck and Schneider 1960). These howls apparently are a proclamation of an occupied area; to whatever extent other groups of howlers avoid areas from which such calls come, they function to maintain the territory of the group (Southwick 1962).

There is an interesting adaptive aspect to the structure of the male howls (Altmann 1966). They are remarkably loud and deep; the fundamental tone, as well as much of the rest of the acoustical energy of the howls, is concentrated into the bottom of the sound spectrum (Plate 17.1). The physics of sound is such that very low-pitched vocalizations will not cast acoustical shadows unless they encounter unusually large objects. Consequently, there will be relatively little absorption of such sounds, and they will penetrate deeply through the trees. In a dense tropical forest where long-range communication by visual displays is often impossible because of intervening foliage, such vocalizations are highly adapted to the difficulties of communication between distant groups. Probably for the same reason, the drumming of chimpanzees can sometimes be heard when the accompanying vocalizations are out of range (Reynolds and Reynolds 1965).

Terrestrial or semiterrestrial primates usually rely more upon visual communication than on any other channel. For example, Japanese macaques, *Macaca fuscata,* according to Itani (1963) "communicate more by pantomime than by vocalizations." Hall, Boelkins, and Goswell (1965) report that patas monkeys (*Erythrocebus patas*) "are mainly nonvocal, communicating chiefly by visual, and secondarily by tactile, cues. The function of olfactory cues in the social behaviour of these animals is not known." Writing of the Indian langur, *Presbytis entellus,* Jay (1962) says: "Gestures have a much larger role in the communication of emotional states than do vocalizations. It is possible to understand and to follow a long, complex interaction among many adults by watching a motion picture of the action without the benefit of a sound track, but a sound track without the film picture is almost meaningless." This last, however, may be more a statement about the nature of the observer than of the subjects.

The reader is referred to recent papers by Andrew (1963, 1964), Bolwig (1959*a,* 1964), and Marler (1965) for reviews of the literature on the ways in which various channels of communication are utilized by primates and the ecological significance of each.

From a comparative and evolutionary standpoint, it would be interesting to know, for each species of primate, the relative contributions of each channel of communication. Rhesus monkeys have a vocal repertoire of at least seven calls in a known repertoire of fifty-nine elemental behavior patterns (Altmann 1962*a,* 1965). Emission of these elemental calls accounted for 3.0 per cent of the elemental social messages that were given; if we consider, in addition, cases in which calls were combined with some nonacoustical component to form a compound signal, 5.1 per cent of the social messages included or consisted entirely of a vocalization (based on a sample of 5,504 behavioral events [Altmann 1965]). Perhaps more revealing would be a consideration of the relative information content of each of these classes of messages. But any attempt to classify the social signals of primates into exclusive categories based on the sense organ involved in reception is misleading. Mammals in general, and primates in particular, make extensive use of multisensory constellations of information inputs. Marler (1965) concluded from his review of primate communication:

> Perhaps the most striking generalization that can be advanced from this survey of the communication signals of monkeys and apes is the overwhelming importance of composite signals. In most situations it is not a single signal that passes from one animal to another but a whole complex of them, visual, auditory, tactile, and sometimes olfactory. There can be little doubt that the structure of individual signals is very much affected by this incorporation in a whole matrix of other signals. We have seen that these composite systems are a special feature of close-range communication, transmitting information between different members of the group.

Broadcast Transmission and Directional Reception (2 in Hockett 1960*a*)

All acoustical signals have the property of broadcast transmission; yet, as a result of binaural reception, we and our simian relatives can usually tell where a vocal signal is coming from. Localizing a sound signal may depend upon several properties of vocal signals: difference in arrival time at the two ears, phase differences between the ears, the sound shadow cast by the head

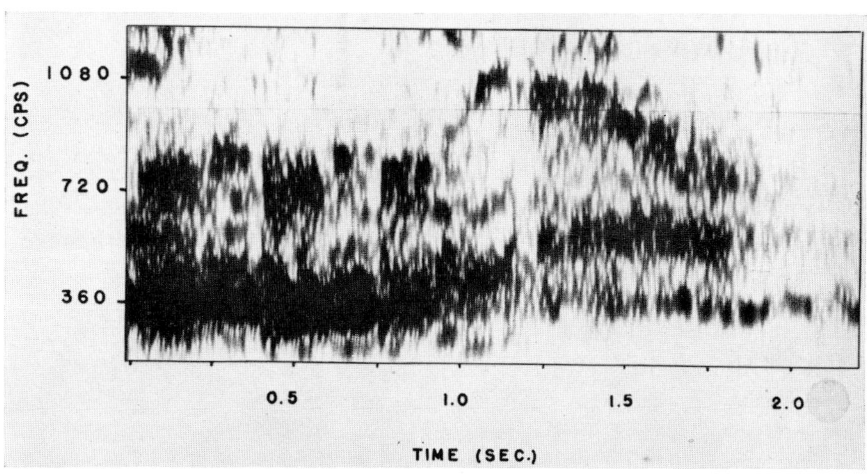

PLATE 17.1.—Spectrogram of end of a howl from an adult male howler monkey, *Alouatta palliata,* recorded by the author on Barro Colorado Island, Panama. Effective bandpass: 4.5 cps. Time scale error: $\leq$ 1 per cent.

PLATE 17.2.—Sexual presentations. In the upper figure, a young adult female (Light Tip), proestrus stage 3 (on a twelve-point scale) is presenting to a peripheral, young adult male (Kink). In the lower figure, an adult female in proestrus stage 0 is presenting closely to an adult male. Note that in each case, the female turns her head toward the male, despite the conspicuous directing components in the orientation of the female's perineum. From color transparencies of baboons in Amboseli Reserve, Kenya, by J. Altmann and the author.

PLATE 17.3.—Juvenile$_2$ male (Four) giving "yawn" as a threat. Note the poor orientation of the yawn to the relevant sense organs (eyes) of the individual toward whom the threat is being directed (adult female Light Tip). From a color transparency of baboons in Amboseli Reserve, Kenya, by the author.

onto the ear away from the sound, the ratio of direct to reverberated sounds, and familiar volume (Marler 1964).

Olfactory cues do not lend themselves readily to directional transmission or reception. From the transmitter's standpoint, the most efficient technique often is to combine olfactory messages with directional cues in other sensory modalities, as is done by presenting females (Plate 17.2). An alternative is to maneuver until upwind of the individual toward whom the message is directed.

Directional reception of olfactory signals can be accomplished by taking into account auxiliary cues, either from other sensory modalities or from wind direction (a special case of the former), or, at close range and in still air, by utilizing concentration gradients.

Addressed (Directed) Messages

Closely related to the distinctions between broadcast vs. narrow-band transmission and directional vs. nondirectional reception is the fact that some messages are directed toward particular individuals or groups, whereas others are to-whom-it-may-concern messages. Undirected, to-whom-it-may-concern messages do occur among primates, particularly in their predator-alarm calls and in their group-cohesion calls, both of which have been reported in many primates. However, most of the social signals of primates seem to consist of directed displays.

Messages may be directed in several ways. One way is to restrict the channel of transmission (for example, by using a quiet vocalization or a tactile message), so that the message is received only by certain individuals. Another is to use a private code or language known to some but not all of the recipients. A third way is to add to the message an ancillary component that indicates to whom the message is being directed.

Those components of a message that serve to direct it may be referred to as the *address* of the message. Note that addresses are, themselves, messages. They are communication about communication, that is, *metacommunication* (Bateson 1955). (Eavesdropping is the example par excellence of responding to messages that are directed toward someone else.) Itani (1963), in discussing Japanese macaque vocalizations that form his Group A, most of which are used in the coordination of group movements, writes, "few of them are directed at an individual. Some sounds are so low that they do not reach except a short distance, and others are loud and big enough to reach the whole troop. It is clear, however, judging from the manner and the line of gaze of the utterers, that generally they are not calling to a special individual."

Among primates—and doubtless among many other animals—facing and looking at the addressee is probably the most common means by which social messages are directed. (Doubtless, the efficiency of this technique is to some extent dependent upon the fact that for an animal with extensive binocular vision the position of the two eyes on the front of the face makes it fairly clear to other members of the group just who is being looked at.) Even when females present their hindquarters to a male, an act which in itself has conspicuous cues about who is the addressee, there is often visual contact or, at least, facing (Plate 17.2). Thus, one interpretation of avoiding visual contact—which has been described in rhesus (Altmann 1962*a,* Hinde and Rowell 1962), baboons (Hall 1962, DeVore 1962), bonnet macaques (Simonds 1965), gorillas (Schaller 1963)—is that it is a means of avoiding interactions. Not surprisingly, this behavior is usually given by the subordinate member of a pair, and its converse, direct staring, is usually a form of threat. Note that in such situations organs that are fundamentally receptors have come to subserve a transmitting function; according to Bateson and Jackson (1964), this is a common development in animal communication.

Particularly striking is the fact that messages may be directed toward the appropriate sense organ of the social partner. In our studies of baboons, this was not observed in vocal signals—which is not surprising, in view of crooked-line transmission of sound waves (property 7). It was, however, conspicuous in many visual and olfactory displays (Plate 17.2). Adolescent males, at an age when they first began to make use of the mouth-gape or "yawn" as a threat were very poor at directing these displays (Plate 17.3). Gradually, however, they became more skilled at directing these displays toward the face, and hence eyes, of the females whom they harassed. This gradual perfecting of the technique of directing these displays may have developed out of repeated experience with the relative communicative impact of displays that varied in the extent to which they were directed toward the relevant sense organs.

This last case points up the fact that the directing components of a message may follow a different maturational pattern than the displays themselves. Jay (1962), describing the Indian langur, *Presbytis entellus,* writes, "in the late infant-1 stage the young langur produces the earliest recognizable forms of gestures and vocalizations which will be characteristic of its adult behavior. . . . Improved coordination makes it possible to direct movements towards objects and other monkeys. . . ."

In some cases, the presence or absence of addressing components may determine whether a particular display is communicative. Jay (1962) indicates that undirected grimaces, which are given by young langurs in play, are of no communicative significance, whereas directed grimaces outside of the play situation are. In baboons, directed mouth-gaping, or "yawning," is a powerful threat; in contrast, undirected yawns are of little or no communicative significance.

The directing or addressing component of a message may have a "pointing" function. When a monkey in an agonistic situation attempts to enlist aggression from a third party, and particularly when such a third party draws near (for example, when a female comes to the aid of one of her offspring), the monkey often gives agonistic displays toward the opponent with exaggerated addressing components, thereby singling out the adversary; we have observed this in rhesus macaques and savannah baboons. Itani (1963) indicates that *Macaca fuscata* may communicate the location of a concealed invader to the troop through "triangulation" of the directing components; an alarm call and various threat gestures are given repeatedly by a monkey from 5 to 15 m from the invader, and as the monkey continues his call incessantly, using trees and rocks as a shield, he jumps about, covering a half circle in which the invader is the center.

The concept of directed messages enables one to translate into verifiable form many statements about intent. For example, "The male intended to strike the female" might be reworded "The male directed his strike at the female (but missed)." In some cases, this may clarify the empirical basis for the impression that we know the subjective state of the organism.

A peculiar form of directed behavior has been observed in rhesus, bonnet, and Japanese macaques (Altmann 1962*a,* Simonds 1965, Itani 1963) and probably occurs in other primates as well. A behavior may be directed at nobody. This is not the same as undirected behavior. For example, a rhesus female sometimes will, while presenting to a male, direct a threat toward some fixed position that is not occupied by anybody. This seems to be a particularly strong stimulus to mounting. The behavior of a female in these situations appears to be indistinguishable from that in which she presents to a male while threatening an actual third party. Itani (1963) writes of the Japanese macaque:

> An individual is attacked by a superior individual. He tries to direct the attack by uttering these sounds to a third individual nearby who has no connection with the trouble. When he finds no available third person about there, he utters these sounds toward an entirely false direction, that is, to an imaginary object, so that he can sometimes escape from his situation of being attacked.

Multiperson Communication

Another interesting aspect of the threatening-while-presenting is that it is doubly directed behavior: the female simultaneously directs two messages, one toward the male and one toward a third party, real or otherwise. Kummer (chap. 5) has described the role of such doubly directed threat-plus-present displays in the establishment of coalitions among hamadryas baboons, *Papio hamadryas.* These are cases of one-to-many communication in contrast with the more familiar, one-to-one communication for which there is one subject (sender, actor) and one object (addressee). Another form of one-many communication consists of directing a single message toward a group. Itani (1963) has classified the directed vocalizations of the Japanese macque as being either one-one or one-many and has, in addition, described a number of undirected vocalizations. In rhesus, we found many behavior patterns that are directed toward social partners in some instances, self-directed in others, and undirected in still others.

One should bear in mind the distinction between the addressees of an act and the reactors; as mentioned above, an individual may respond to messages that are not directed toward him.

According to Itani (1963), Japanese macaques may, under certain circumstances, make a rapid shift from defensive postures and vocalizations to aggressive ones. When they do, the aggressive vocalization is given with exaggerated tone. While the utterance is still directed from one individual to one other, others react because of the exaggerated tone. "The mere [unexaggerated?] sounds, however, are usually ignored by the other individuals." Cases of multiple respondents were often observed in our studies of rhesus and baboons; in particular, the behavior of the alpha males was often responded to by several members of the group, including those who had not been the addressees of his actions.

Multiple subjects, or senders, of messages introduce the possibility of many-one messages and many-many messages. For some purposes, such communication would probably be regarded as consisting of several simultaneous messages. This probably explains the lack of such categories in Itani's classification. Nevertheless, a monkey that is, for example, faced with simultaneous threats from members of a coalition may respond, not to each individual threat, but rather, to the combination.

Although few authors of reports on primate behavior have addressed themselves explicitly to the question of multiperson interactions, multiple responders seem to be more common than multiple addressees. This is not surprising, since the latter requires that an individual (the sender) simultaneously communicate with two or more individuals, whereas the former does not.

Note that multiperson interactions do not require a multiplicity of actors, addressees, or responders; a rapid succession of one-one interactions sometimes involves a series of individuals. These, then, are four sources of multi-individual, social communication. It is my impression, based on personal experience and on the literature, that in no other group of animals have directed multiperson interactions developed, either in frequency or complexity, to anything like the extent that they have in primates. The social insects, which represent the other pinnacle in the evolution of complex societies, differ markedly from primates in that most of their multiperson responses are undirected to-whom-it-may-concern messages.

Rapid Fading (3 in Hockett 1960*a*)

Sound waves are evanescent: they dissipate into white noise and thermal energy. Consequently, the spoken word does not persist. This will be true of all vocalizations, visual signals, and tactile signals as well. In fact, in the open, with any slight wind or air drift, the same is true of airborne olfactants. Indeed, except for cases of olfactory marking, all social signals of nonhuman primates exhibit this property of rapid fading.

Olfactory marking seems to be particularly well developed among prosimian primates. This is suggested by the abundance and variety of subcutaneous glands, which are particularly well developed on the face, feet, and perineum of these animals (Montagna and Ellis 1959, 1960; Montagna and Yun 1962*a, b,* 1963), and has been verified by direct observations on the behavior of various species (Ilse 1955; Jolly, chap. 1; Petter 1962).

Allaesthetic

Another consequence of the fact that language messages are acoustical is that they are *allaesthetic,* a term coined by Julian Huxley (1938) for those social signals that are perceived by the distance receptors. Such signals make possible communication between individuals that are not necessarily in contact, or even in close proximity to one another, and thereby make possible the coordination of activities among the members of a widely dispersed group. This property is shared by all acoustical messages, by visual messages except under ecological conditions in which foliage or other objects intervene (for example, in the habitat of howlers), and, to some extent, by airborne olfactants. Consequently, it is a widespread property of primate social signals.

Crooked-Line Transmission

Sound waves will travel around opaque objects, whereas light waves require a direct line between the display source and the eye. Consequently, you can hear, but not see, things that happen behind you or behind other opaque objects. (As indicated in the description of howling monkey vocalizations above, this is particularly true of low-pitched sounds, because they cast virtually no sound shadows.) Consequently, alarm calls are almost always vocalizations, and forests species are often highly vocal.

Interchangeability (4 in Hockett 1960*a;* 2 and 3 in Morris 1955; 2 and 5 in Altmann 1962*b*)

Any speaker of a human language is capable, in theory, of saying anything he can understand when someone else says it; that is, humans are interchangeable as transmitters and receivers. This property is a composite of two other properties. First, our language messages are *interpersonal* (Morris 1955); that is, all members of the group or even of the species react in a generally predictable way to each pattern of social behavior. This is not to deny the importance of individual differences and those due to age and sex. Indeed, to the extent that males and females differ in their responses to common patterns, the social signals are not interpersonal.

For the most part, primate social signals are interpersonal. A threat gesture, for example, is interpreted as such by members of both sexes and by individuals of all ages except for the youngest of infants.

Second, the signs of human language, and most primate signals, are *com-*

signs; that is, they are a part of the behavioral repertoire of all members of the society, at least at some stage in their life history. There are exceptions—primarily cases of sexual diethism. Only large adult male gorillas (*Gorilla gorilla*) roar (Schaller 1963, one exception cited), only adult male langurs whoop (Jay 1962). In many cases, these differences between males and females are clearly the result of sexual dimorphism. The female howler monkey, in which the specialized laryngeal structures are not as highly developed as in the male (Kelemen and Sade 1960), can produce only an approximation of the howl. The females of various species of Old World anthropoids have, periodically, a swollen sexual skin (Eckstein and Zuckerman 1956), which is made particularly conspicuous by their sexual presentations (Plate 17.2), although the signaling function of this swelling has been doubted, at least in rhesus and baboons (Rowell, chap. 2). For obvious reasons, intromission (but not mounting) by females will be impossible in all species.

The distinction between comsigns and interpersonal signs can be made clear by considering on the one hand the mute who understands all that he hears (and for whom our language does not consist of comsigns) and on the other, the deaf person who talks (but for whom the signs of spoken language are not interpersonal).

Multiple Coding Potential

This last illustration, concerning the deaf and the mute, holds true only so long as we restrict the domain of the discussion to the vocal-auditory channel. But people with such communication handicaps have no such restriction. They may communicate through other sensorimotor channels, for example, through writing or through sign language. According to Sebeok (1962), human language may be the only communication system that has a multiple coding potential, that is, that permits the transmutation of a set of signs from one sensory modality to the other. Although this transmutation is most easily carried out between the auditory and the visual channels, other channels can also be utilized (for example, touch, as in the case of Helen Keller).

Such a transmutation would seem to require, at the least, a synonymy among messages in the various sensory modalities. (It may also require a synonymy or at least isomorphism of generative grammars in the human case.) Synonymous messages in nonhuman primates are discussed below.

Total Self-Feedback (5 in Hockett 1960*a*)

You always hear what you, yourself, say, or at least, you hear sounds that are fairly isomorphic with what you say. Such feedback provides the basis for adjusting the social signal toward some norm, the image of which is carried in the central nervous system. Without such feedback adjustment, our speech would be much less precise, and many of the subtleties that are now communicated through fine control over vocal production would be impossible.

All vocalizations in primates have this property, although self-feedback may be masked under some conditions, for example, loud noise from wind or other animals. Visual signals will provide total self-feedback only if they are from a part of the body that the transmitting animal itself sees, or if the position and movement of the displaying structures can be determined through proprioceptors. (Note that in the latter case, the isomorphic feedback is through a different channel of communication.) It is not clear just how much of a female's sexual skin she herself can perceive. Olfactory messages may provide little self-feedback under some wind conditions. In general, virtually no actual data are now available on the amount of self-feedback that is available to primates under various ecological conditions and in various channels.

Specialization (6 in Hockett 1960*a*)

A communicative act is specialized to the extent that its direct energetic consequences are biologically irrelevant to anything but communication.

Some primate communication is unspecialized. As a group of baboons slowly moves across the savannah, each individual so adjusts the direction and rate of his movements, relative to the others, that he remains with the group. Yet, the messages—the alternate feeding on one plant, then moving on to the next, stopping, and again feeding—have a direct biological consequence (obtaining food), hence such messages are unspecialized. In contrast, most of the communicative acts in primate communication systems apparently are specialized: they seem to have no other function.

Many social signals that are now completely specialized may have evolved from basically unspecialized but communicative acts, which Darwin (1872) referred to as "serviceable associated habits."

Triggering (6 in Hockett 1960*a*)

Related to the specialized nature of many primate behavior patterns is the fact that they trigger responses in other members of the social group in the sense that the energy of the responses to these messages is not the energy in the messages themselves. This property is so nearly universal in animal communication that it has, as indicated earlier, been used as the defining property (Hockett 1958, 1960*a*). Indeed, there are only a few minor exceptions, such as the influence of one wrestler on the behavior of another, of a ballet teacher as she repositions the body of a student, or the repositioning by a grooming monkey of the body of the groomee (and not always then); that is, the exceptions are cases in which one individual physically manipulates the body of the other. (Bullock [1957], in a well-known article on the trigger concept in biology, uses this term only for discrete signaling systems.)

Semanticity (7 in Hockett 1960*a*)

For words in human languages and for some animal signals, there is a relatively fixed association between elements of the message and elements in the real world. In this sense, such semantic messages "stand for" things. The object, event, or relation for which the message stands is its *denotatum* (pl. *denotata*).

Alarm calls, which are common among nonhuman primates, and food calls are instances of semantic messages. Baboons, for example, have a loud, two-phased bark that is given when a leopard has been sighted. Struhsaker (chap. 16) has described a variety of alarm calls in vervet monkeys.

Incipient movements (Tinbergen 1952) are semantic: the redundancy in the complete movements enables one to reconstruct the wholes from the parts, and essentially the same response may be given to each. In a sense, then, these parts are icons for the wholes.

Metacommunicative status indicators, which occur in rhesus macaques (Altmann 1962*a*), Japanese macaques (Itani 1954), hamadryas baboons (Kummer 1956), and many other primates, are semantic. These status indicators are not agonistic behavior per se but are, rather, communication about such behavior; that is, they are a form of metacommunication. Thus, unlike real aggression, they cannot change the dominance relations between two individuals but can only communicate about them. In contrast to a predator call, they stand for a relationship rather than for an object.

Threats that involve a show of weapons are semantic. The gaping "yawn" of baboons, when used as a threat, appears to be a canine display (Darwin 1872, Zuckerman 1932) and probably should be interpreted as a semantic message. Yawning also occurs in baboons in two other situations: when relaxed and presumably stimulated by low oxygen tension in the blood (true yawning) and in certain conflict or anxiety-producing situations (Bolwig 1959*b*). However, the threat yawn differs from the other two in at least one respect: it is directed toward another member of the group.

The waggle dance of the domestic bee, *Apis mellifica,* justly famous as a result of the studies of Karl von Frisch and his colleagues (Frisch 1955, Lindauer 1961) is the classic example in animal communication of semantic communication (Haldane and Spurway 1954). Because this behavior has been so widely studied, is so well known, and so often helps to clarify our concepts, I will make repeated reference to it in what follows. The dance communicates to other workers the distance, direction, and quality of a new-found source of nectar. The communication of all of this information is relevant to our discussion, but for simplicity we will only consider certain aspects of the communication of direction. Direction to the nectar, relative to the sun, is communicated by the angle between the waggle portion of the dance, relative to the line of gravitational force. Thus, this component of the message stands for, or denotes, certain properties of the nectar: the dance is semantic. (A recent publication by Johnson and Wenner [1966] indicates the possibility that the bee dance is not, in fact, communicative. If true, our paradigm is ill-chosen, but the distinctions to be made here are no less real.)

At this point, it is easy to be led into a verbal trap. If a bee's dance indicates something about its new-found source of nectar, can we not say, for example, that a baboon's threat indicates his anger; that when a female chimpanzee presents and stands for mounting, this behavior expresses her receptivity; or that a monkey's grimace expresses its fear and submissiveness? And are not these too, therefore, semantic messages? Perhaps the reader has seen through the trick. The female chimp is said to be receptive because under certain circumstances she presents, and the male baboon is said to be angry because he gives this type of response, not the other way around. To say that the female chimp's behavior indicates, or communicates, her receptivity to the other chimps, where "indicates" is used in the sense of semantic communication, is to say that her messages are their own denotata!

Of course, the female chimp's behavior, like her sexual skin swelling, may indicate *to the scientist* that she is receptive, but that is another matter. There is no difficulty in using statements like "The female chimp's behavior indicates that she is receptive" as long as we do not confuse them with statements like "The bees' dance indicates nectar." To do so is to obscure an important distinction in the design of animal communication systems.

From a semiotic viewpoint, this confusion results from a failure to distinguish pragmatic from semantic aspects of communication (Morris 1955). It is a result of what Kaplan (1964) calls "functional ambiguity" of concepts and is a commonplace error of drive theory. "Receptive" and "angry" are dispositional terms (Kaplan 1964): they indicate the behavior that would be exhibited if certain conditions were fulfilled. A major problem with such dispositional terms in behavioral research is that they lend themselves too readily to reification. They thereby deceive us into believing that they have become causes, explanations, or denotata of behavior.

Arbitrary Denotation (8 in Hockett 1960*a*)

Semantic messages are either iconic or arbitrary representations to the extent that they do or do not resemble in physical contours what they denote. More precisely, a message is iconic to the extent that properties of the denotata can be mapped onto properties of the corresponding messages.

Human language, with the exception of onomatopoietic forms, consists entirely of arbitrary denotations. The predator calls of vervets and of baboons are basically arbitrary: there is no obvious resemblance between the contours of the call and the contours of the predators. However, to the extent that such calls become louder and more frequent as the danger becomes more imminent (Struhsaker, chap. 16), they are iconic, in that a one-to-one mapping is possible between certain properties of the message and certain properties of what the message denotes. With few exceptions, the semantic social signals that have been studied in primates so far are arbitrary representations.

Arbitrary Communication (8 in Hockett 1960*a*)

For semantic messages, and for those that are not, one may consider whether or not the relation between message and response is arbitrary in the same manner as was done above for the relation between message and denotatum; that is, messages that do not resemble, in physical contours, the responses that are given to them are arbitrarily related to these messages. ("Iconic' is, in this case, an inappropriate antonym for "arbitrary," because of its traditional usage in studies of symbols.) Among primates—and other animals as well—there often seems to be a rough correspondence between the intensity (generally some combination of rate, amplitude, and duration) of displays and the intensity of the responses to them. An intense threat, for example, will almost always be met with an agonistic response of approximately corresponding intensity. The same rough matching of intensities may result from those signals whose effect is to alter the distribution of responses. For example, in Amboseli we observed that baboons merely looked briefly around them in response to faint (distant) alarm barks that were repeated at wide intervals (approximately one per minute), whereas in response to loud, rapid alarm barks, their responses were markedly different and much more intense. To the extent that intensity or other properties of messages are matched in the responses to these messages, the relation between message and response is not arbitrary.

Of course, the most extreme cases of nonarbitrary communication are cases of mimicry, in which the response so closely matches the stimulus that we put them both into the same category. Some species of primates seem to be particularly prone to this. Carpenter (1940) writes: "A marked characteristic of gibbon vocalizations is that calls stimulate similar calls in associated animals."

Digital vs. Analogical Communication (9 in Hockett 1960*a*)

There are several, related pairs of terms:

count	measure
discrete	continuous
digital	analogic(al)
quantitized (stepped)	graded
all-or-none	more-or-less

Language messages are primarily quantitized, all-or-none signals, discrete rather than continuous. The result is a digital communication system. In contrast, some animal signals, such as the waggle dance of bees, contain communicative, continuous variations and are used in analogical communication.

Indeed, several authors (for example, Bateson and Jackson 1964, Haldane 1955, Sebeok 1962) have proposed that digital communication is one of the distinguishing features of human language. Sebeok (1962) wrote:

This paper has as its theme the hypothesis that whereas subhuman species communicate by signs that appear to be most often coded analogly, in speech (contrary to the opinion of certain linguists) some information is coded analogly and other information is coded digitally. The digital mechanism of speech may, therefore, be regarded as a late development in the phylogenetic series and perhaps a uniquely human faculty.

Analogical communication should not be equated with continuous gradations in messages: the demonstration that there is a continuous array of variations in a class of signals or messages is not enough to show that a particular form of communication is analogical. Morphological and physiological variations in the structures that produce the display or vocalization insure that there will be such continuous variability in signals. (These variations may in some cases facilitate individual recognition and may serve as a potential source of new messages in the evolution of social communication; consequently, there might, in fact, be selection for such variability.) We need to show that the messages have *functional continuity*. What is required is (1) a continuous array of messages, (2) a continuous array of responses to these messages, and (3) a one-to-one mapping between messages and responses.

The messages in such an analogical communication system need not be semantic, but if they are and if the semantic representation is also analogical, then two additional conditions must be fulfilled. There must be (4) a continuous array of denotata (the objects, events, or relations that are represented by the semantic message) and (5) a one-to-one mapping between denotata and message. For example, in the bees' waggle dance, variations in one component of the message, the angle of the waggle relative to gravity, correspond with variations in the direction to nectar, relative to the sun. Other workers respond with flights from the hive in corresponding directions (the response, or interpretation).

We distinguished above between *analogical representation,* which is a relation between denotata and messages, and *analogical communication,* a relation between messages and responses. Analogical communication may be nonrepresentational, and an analogical representation, if ignored by the recipient, may not be communicated. Or, if an analogical representation is communicated, the information at these two stages may be different. This information content can be measured. Haldane and Spurway (1954), in a pioneering paper in what they call "ethological cybernetics," calculated for the bee dance the amount of information received by the workers; that is, they based their calculations on the variability in directions taken by workers after the workers had observed a dance. The calculations indicated that 2.54 bits of information are received, although the actual amount, considering certain problems of experimental design, may be as high as 3 or 4 bits. By an extension of this technique, it would be possible to measure the information that is transmitted by the dancer in its analogical representation of distance. As a result of inevitable errors of reception, the information that is conveyed by the dancer must be more precise. On the basis of limited data, Haldane and Spurway estimated that the representation contains about 5 bits of information. Thus, each worker receives on the average over half, but not all, of the information that has been transmitted.

The type of analogical communication that is usually considered involves continuous variations in the form, or structure, of the response. There is, however, another analogical function of continuous variations in signals—namely, that in which a continuous gradation in signal structure corresponds to a continuous gradation in response probabilities. Suppose that a submissive behavior pattern varies in intensity, and that the greater the intensity, the less likely is an aggressive response from the partner. The aggressive act or acts themselves need not have any significant grades. Rather, it is the *likelihood* of the aggressive acts that intergrades continuously as a result of gradations in the *form* of the submissive behavior pattern. (Note that it is impossible to change the probability of just one behavior pattern. If, in our illustration, aggression becomes less likely, then the probability of other courses of action is correspondingly increased. There may be shifts in probabilities among these other patterns as well. Thus, what should be considered is not just the likelihood of aggression, but the probability distribution of all courses of action.) Although it is not possible at present to provide a single instance of this kind of communication from the literature on nonhuman primates, it is my impression, based on many observations, that this is one of the most common functions of communicative variations in displays, not only among primates, but also among other animals as well.

There are indications of graded social signals in Japanese macaques (Itani 1963), rhesus macaques (Rowell 1962), red colobus, *Procolobus verrus* (Booth 1957, Hill and Booth 1957), langurs (Jay 1962), baboons (Hall and DeVore 1965, Bolwig 1959*b*), chimpanzees, *Pan troglodytes* (Reynolds and Reynolds 1965), and gorillas (Schaller 1963). However, as indicated above, this is just what one would expect of any social signal. The question is whether such gradations have any communicative significance. Struhsaker (chap. 16) indicated the possibility of continuous semantic communication in the alarm calls of vervets in Amboseli, and the same seemed to me to be true of the alarm calls of baboons in the same area, but it is very difficult to obtain adequate samples of such rare events.

The papers by Rowell on rhesus, and by Hill and Booth on red colobus, are of particular interest in this regard because these authors have indicated the possibility that for these species, many or all of their vocalizations are part of a single graded system, which in the case of rhesus is said to be multidimensional —that is, a result of gradations in several physical parameters (volume, pitch, duration, and so forth). Marler (1965) points out that "Rowell's descriptions also demonstrate correlations with a continuously varying set of social and environmental situations. . . ." But note that this is a reference to the conditions that elicit these vocalizations and not to the responses that are given to them. There is no indication that the communication process that results from these vocalizations is of the analog type. To my knowledge, neither analogical representation nor analogical communication has ever been demonstrated in the communication system of any species of nonhuman primate. However, when one considers that studies of primate communication are in their infancy, no great significance should be attached to this lack of instances.[1]

The variability (functional or otherwise) within each class of signals should not obscure the fact that there are discontinuities between the major classes of communicative behavior patterns in primates (see Plurisituational Messages, below). The same is true for social communication in other animals, including bees. There apparently are complete discontinuities between the bee dance, the signal by which bees indicate the need for water, and the signals by means of which bees indicate the presence of the queen (Lindauer 1961). Thus for bees, and doubtless for communication systems of other animals as well, the question of whether the system is digital or analogical is a false antithesis. Such systems are digital systems in which one or more components function analogically. This is comparable to a computer that can measure and can then count the things that it has measured.

Relation between Continuity and Iconicity

Of the four possible combinations of iconicity or arbitrariness, with continuity or discreteness, there is one restriction in semantic systems: continuous messages with arbitrarily assigned meanings would be swamped in errors of communication.

In order to clarify this relation between continuity and iconicity, let us turn again to the communication of direction by the waggle dance of bees. A worker has found a source of nectar at 95° from the hive, relative to the sun. The worker returns and does a dance whose waggle component is approximately 95° from vertical (Fig. 1). (If it turns out that the waggle movement is not the actual cue, we would have to rephrase our illustration, but the underlying logic would be the same.) Deviations in the angle of these waggles from the true angle to the nectar are the first source of error in the communication; they are errors of transmission, or representation. After observing this dance, other workers depart and search for the nectar. Inaccuracies in their perception of

[1] Rowell proposes that we should not use terms such as *scream, squeal, growl, bark,* and so forth because these calls may all be part of a single continuum. However, even the demonstration of a continuum of vocalizations would not rule out the practicality of terms that stand for regions or nodal points along the continuum any more than the continuum of wavelengths of light precludes the use of discrete names of colors. Beyond that, a continuum can always be approximated, to whatever degree of accuracy is required, by an appropriate discrete system. Such approximations are often the most practical way of recording data from continua: "The female had a sexual skin swelling, stage 8," or, "The two-phased bark was given at volume 4."

the dance, as judged by the correlation between the form of the message and the dispersal in the subsequent flight of the bees, result in further errors of reception. Now, if the representation and the interpretation are both iconic, small errors in the transmission of the message or in its interpretation will result in correspondingly small errors in the over-all communication. Every point along the continuum of possible directions (top line, Fig. 1) has a corresponding point (or rather, a line segment, considering also the errors of representation) in the continuum of messages that represent these directions (middle line), and these, in turn, correspond with points (or line segments) in the continuum of flights by the other workers (bottom line).

But what would happen if either the representation or the interpretation were arbitrary rather than iconic? Suppose that a dance at a 20° angle represented nectar at 95°, and that dances that were slightly different from 20° arbitrarily represented other directions (Fig. 2). Even if the other workers made no errors in their perception of the angle of the waggle, any slight error in the message would result in very large errors in the resulting flights by other workers. And any errors in reception would only make matters worse. Note that these large errors in communication do not result from the fact that nectar at 95° was represented by a dance at 20°, but rather, from the fact that messages slightly different from 20° arbitrarily represented other directions, which were, consequently, often very different from 95°. Thus, in any continuous and semantic communication system, the use of an arbitrary representation would result in such gross errors of transmission that natural selection against any such system would be very powerful. As far as is known, all naturally occurring continuous representations are iconic.

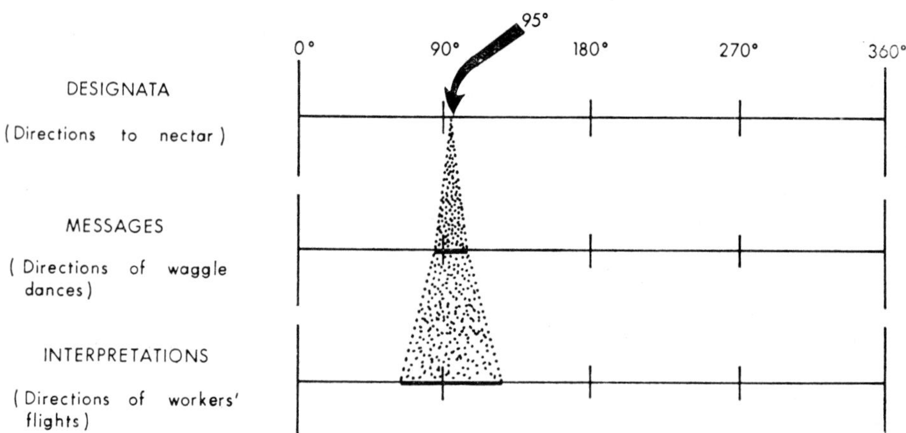

Fig. 1.—Bee dance with iconic representation. For clarity, the 360° of the circles have been represented by straight lines, and the amount of error has been exaggerated. Note that small errors in the message's representation or in its reception result in correspondingly small errors in transmission.

The result, then, is that analogical representation in nature will always be iconic. There may be a second restriction. Sebeok (1962) writes, "the most interesting point about the property of arbitrariness is this: that it is a logical consequence of the digital structuring of the code. The connection has been shown by a mathematician, Mandelbrot [1954], who has further shown that the discrete character of linguistic units necessarily follows from the continuous nature of their substratum. . . ." If Mandelbrot is correct, then there will only be two forms of semantic representation: those that are analogical and therefore iconic, and those that are digital and therefore arbitrary. Unfortunately, Mandelbrot's paper was not available to me, but the conclusion, on the face of it, does not seem reasonable. I can, for example, speak of the first step, the second step, and so forth, on a staircase; that is, I can represent the order of the stairs, in space, by the numerical order of the integers. This is an iconic representation that uses digital messages.

We have considered discrete classes of displays that elicit responses from any number of discrete classes and continuous messages that elicit a continuum of reponses. A graded continuum of signals may elicit one or another stepped responses, as anyone who has ever driven a car with wet, "grabby" brakes knows,[2] and conversely, responses that form a continuum might be elicited by a stepped series of signals. I know of no instances of either of these last two forms from the literature on social communication in primates. It should be noted that in both cases the communication process is digital. Consequently, such systems might be either iconic or arbitrary.

[2] I am indebted to Charles Hockett for this illustration.

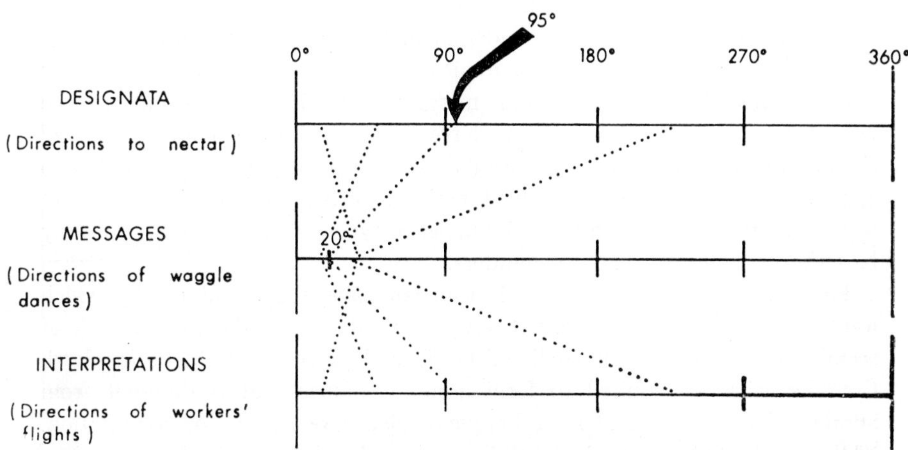

FIG. 2.—Fictitious bee dance with arbitrary representation. Dances at 20° represent nectar at 95°, but dances near 20° arbitrarily represent nectar in other directions. With such a design, small errors in transmission would result in gross errors in interpretation.

Displacement (10 in Hockett 1960*a*)

Semantic messages may be displaced in time or space from what they denote. Baboon alarm calls were given for more than an hour after an unsuccessful attack by an adult male leopard. They were given after the baboons had left the water hole basin in which the attack had taken place and had circled through the trees to the opposite side of the water hole from where the leopard, if still present, was no longer visible. Their barking was directed toward the basin of the water hole. Such barking serves to alert the baboons of every group in the area to the danger.

As another example of displacement in semantic signals, metacommunicative status indicators, such as the upright tail posture in rhesus macaques (Altmann 1962*a*) and Japanese macaques (Itani 1954), are displaced in space and time from the agonistic events that they denote.

Productivity (Openness) (11 in Hockett 1960*a*)

A human can talk about things that he has never talked about before—indeed, that may never have been talked about in the entire history of man—and be understood (sometimes!): our language is *open*, or *productive*. Beyond that, our language has *universal openness:* we can talk about anything. The universal set consists of those things that are denotable. (To prove this to yourself, try to name a counterexample.) By the same token, the universal set consists of those things that are discussable.

The bee dance, like any continuous, semantic (and hence iconic) set of messages, is necessarily open. With an infinite supply of meaningful dances at its disposal, a worker bee can indicate the presence of nectar in directions and at distances that are new to its experience and to that of all the other workers in the hive and still be understood. Yet this is a limited sort of productivity. The bee dance only communicates about the location of new nectar and hive sites. Another set of iconic messages is used by bees to indicate the need for water. But bees seem to have no method for communicating *ad libitum* about anything new. Each of these analogical processes is limited to a narrow range of communicative functions. Just so, analog computers are well adapted to carry out a limited number of types of computations, but they lack the flexibility of digital computers.

For a discrete system, such as human language, the techniques by which openness may be attained are not obvious. I will discuss them at the risk of seeming naïve to professional linguists. Several processes might be involved. Consider what is required to form a new sentence that is different from another, similar sentence. One technique involves a reordering, or permutation, of the words. ("The glass is light" is recognized as being different from "The light is glass.") Of course, not all permutations yield meaningful sentences ("Is light glass the."); there are certain grammatical or syntactic constraints on these combinations. Second, we can substitute other words ("The glass is focused."). These two techniques produce different sequences of words, and these are interpreted (responded to) in different ways.

Rhesus monkeys, like humans, interpret messages differently, depending upon the sequence of messages in which they are imbedded (Altmann 1965). Similarly, Itani (1963) indicates that in Japanese macaques there are some cases of responses to series of vocalizations that differed from the responses to any component. However, such systems are still closed. With a repertoire of r behavior patterns, there will be $(r^{\lambda+1}-1)/(r-1)$ sequences of, at most, length λ.

This immediately leads to two other possible mechanisms: limitless vocabulary and limitless sentence length. Our vocabulary is open: we can always coin new words. We have, in addition, metacommunicative "defining" mechanisms for rapidly making such news words understood (although a definition *sensu stricto* is seldom involved). However, if words differ from one another as a result of a finite and closed set of distinguishing features (Jakobson, Fant, and Halle 1952), then there will be a finite and closed set of possible words no longer than some fixed length: a limitless vocabulary requires a limitless word length.

Next, there is no limit to the length of meaningful sentences. This does not mean that our sentences are ever actually infinitely long. Sentence length is potentially infinite in the sense that we can always add one more word to the longest sentence ever used. (In this sense, life span is potentially infinite in that one cannot specify an age, x, to which it is possible to live but for which it is impossible to live to be x plus 1 second. Nevertheless, no one lives forever.)

We see then that neither permutations nor substitutions, nor coinage of new words provides us with a limitless communication system unless we are willing to let our words or sentences go on ad infinitum. But this is unrealistic: we do not say new things by using words or sentences that are longer than anyone else has ever used. Perhaps the important thing is not that the number of possible combinations of design features in words, or of words in sentences, is limitless, but that it is inexhaustible. That is, if within our lifetime or within the lifetime of our language it would be impossible to say, hear, read, or write all possible sentences up to some reasonable length, then to us as language users our language is inexhaustible and, in essence, open. Calculations of the sort indicated by Miller, Galanter, and Pribram (1960) probably would show that this is, in fact, the case.

Duality of Patterning (12 in Hockett 1960*a*)

In human languages, there are conventions in terms of (1) the shortest *meaningful* elements and (2) the minimum, *meaningless,* differentiating components; that is, language has a duality of patterning. A significant duality occurs when a system uses a relatively small stock of meaningless components to build a large number of meaningful elements. Suppose (1) that when a dog both barked and wagged his tail, this had one meaning, (2) that these two components were also combined with other patterns to give messages with different meanings, but (3) that neither pattern was communicative by itself. This would, then, be a case of duality. Such combinations might be either sequential (barks, then wags tail) or synchronic (barks while wagging tail). The former were discussed in the last section above. The latter are called "compound messages" (Altmann 1962*a*).

Of the Indian langur, Jay (1962) writes, "often a particular gesture may have one meaning when associated with a certain vocalization and a very different meaning when associated with another, or when given alone. If a vocalization signifying aggression or annoyance is given by a less dominant monkey and combined with a strong submissive gesture such as presenting, the presenting outweighs the aggressive gesture in the perception of the receiver."

Perhaps the most frequently reported cases of compound messages are those in which the communicative significance of (that is, response to) a display depends upon the identity or at least age-sex class of the emitter; that is, the transistory messages that constitute the display are combined with the more abiding messages that permit the identification of the individual or of its age-sex class. Beyond that, of course, the displays themselves may differ. Jay (1962) writes of the Indian langur, "if an infant does direct a recognizable threat to an adult, the adult does not respond by chasing, but the same gesture displayed by a large juvenile invariably results in aggressive threats by the adult."

Compound communicative acts occurred frequently in rhesus monkeys. In fact, sixty-four of one hundred twenty-three observed communicative acts consisted of such compound patterns. Beyond that, the same elemental patterns were combined in different ways to give messages that were interpreted in different ways by other monkeys, as judged by the response distribution. However, such duality differed from that found in human language in that the elemental units, like the compound units, were meaningful social signals because of the criteria used in establishing the units.

In order to study the meaningless but differentiating ingredients, one must break down these elemental messages into their components, as is being done by van Hooff for primate facial expressions. In a preliminary report (1962) he shows how each facial expression can be broken down into component positions of the eyes, eyelids, eyebrows, ears, mouth corners, and so forth. (Note that each elementary behavior pattern, in our sense, is a compound pattern to van Hooff. His elements are the meaningless but differentiating ingredients.) At this stage, one begins to wonder: Is not this duality of patterning a result of, or perhaps a restatement of, the more general fact that organisms have a hierarchical organization? Does not one always obtain meaningless but differentiating ingredients by working at the next lower level of biological organization? Molecules of estrogen and testosterone differ only in radicals and side chains on the common ring structure, yet these radicals and side chains alone do not act as hormones.

Ambiguity, Synonymity, Stereotypy, and Variability

Several properties center on the number of classes of denotata, messages, or responses. (See also, the section dealing with the number of actors and addressees.) To the extent that a semantic message has more than one denotatum, it is an *ambiguous* representation. Conversely, if a single denotatum may be represented by more than one class of messages, these messages are *synonymous* representations. The bee dance, for example, is ambiguous: it is also used to represent the distance and direction to potential new hive sites. An ancillary cue—the place where the dance is done—resolves the ambiguity (Lindauer 1961).

One can also speak of ambiguity and synonymity in the relation between message and response. Thus, to the extent that a message elicits more than one response, it is an ambiguous signal, and to the extent that two signals elicit the same responses, they are synonymous signals. Obviously, assessment of the degree of ambiguity or synonymity of signals requires a knowledge of the distribution of responses that each social signal elicits. To date, such distributions have been published only for rhesus monkeys (Altmann 1965).

An individual's choice among a set of synonymous messages may be related to efficiency of communication. In our studies of rhesus and baboons, we have noticed a striking sensitivity of group members to the dominant males. The result is that these males can communicate with greater efficiency, either by a less intense variant of a display or even a different display—one that apparently involves a smaller expenditure of energy. Similarly, Jay (1962), in describing Slate, the dominant male in a group of langurs, writes:

> In addition to the economy of gestures characteristic of Slate's interactions, possible because of his clearly defined dominance, Slate was less vocal than other males. If Slate were concerned in a dispute his usual vocal response was belching. He did not need to vocalize his emotional states to the extent that other adults did since all were aware of his presence and reacted immediately to his subtle postural cues. Slate's slight shifts of position and body tension carried more import than similar gestures or movements by any other troop member. Reliance on gestures rather than sharp or loud vocalizations was characteristic of Slate.

Thus, one of the major assets of high dominance status is an economy of communication.

To the extent that messages of a particular class elicit responses of but one class, the responses are said to be *stereotyped;* if, at the other extreme, all classes of response in the repertoire are equiprobable, then the responses to that display are maximally unstereotyped, or random. Primates are time-binding animals, however; they base their behavior not only on the immediately preceding messages, but also on information in antecedant events. These antecedant messages place additional constraints upon the behavior, resulting in increased stereotypy. If data are available on the contingencies between messages and responses, an index of mean response stereotypy (averaged over all message classes) can be calculated (Altmann 1965). To date, the necessary data are available only for rhesus monkeys (Fig. 3). Similarly, one might want to consider the extent to which there is a many-one relation between classes of messages and of responses. An obvious extension leads to an index of message stereotypy.[3]

Plurisituational Messages and the Problem of Repertoire (plurisituational messages: 3 in Altmann 1962b; 4 in Morris 1955)

[3] The term "stereotyped" is sometimes used in another sense in the behavioral literature. A social display is said to be steretotyped if it recurs in much the same form in each instance. For example, Morris (1957) has pointed out that in many cases displays tend to occur with a *typical intensity*. "Invariable," suitably modified, would probably be a better term than "stereotyped" to describe such patterns, particularly because when the requisite data are available, statistical tests of variability can be used to measure this parameter of a message (Struhsaker, chap. 16).

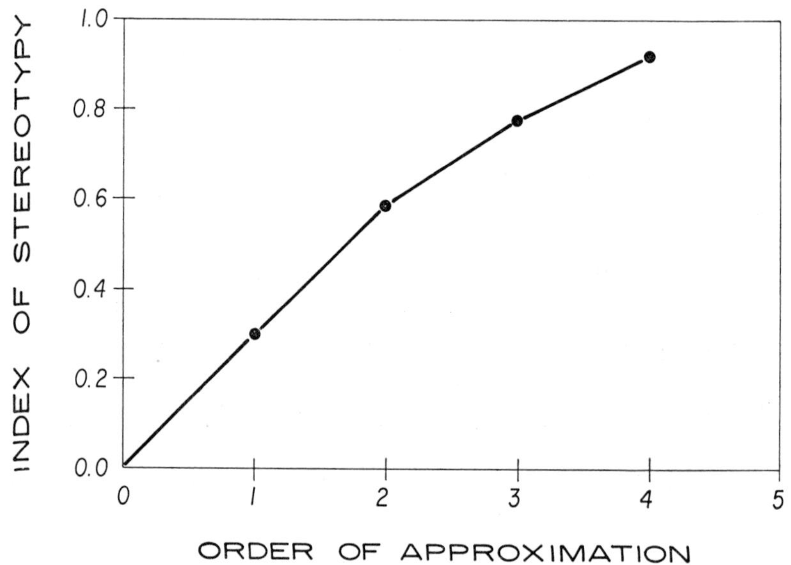

Fig. 3.—Index of sterotypy in the social communication of rhesus monkeys. Successively higher orders of approximation are obtained by taking into account the information from progressively earlier, antecedent behavioral events in the sequence of communicative interactions. The index has a range from 0 to 1, where 0 represents behavior that is completely independent of preceding events, and 1 represents completely stereotyped behavior, that is, behavior that is completely predictable from (completely determined by) the preceding behavior of the animals. The technique for computing the index is described in Altmann 1965.

A general property of animal communication systems is that they consist of plurisituational messages: each pattern of behavior recurs in many situations. No communicative act is unique to a particular event. This is equivalent to saying that there are more communicative behavior patterns than there are behavioral events and that no behavior pattern consists of only one event. Thus, communication has a hierarchical structure: each behavioral event is a member of a class of events—a pattern. Anyone who has made sound-spectrographic or motion-picture analyses of behavior will be aware that no two behavioral events are identical. The communication system works because of the ability of the individuals in the society to assign these unique behavioral events to classes.

These behavior patterns form the core of any analytic study of social behavior. Ethologists have clearly recognized this (Tinbergen 1951); the first task in studying the behavior of a species is, we are told, to draw up an *ethogram*—essentially a catalogue of behavior. Unfortunately, ethologists have never, to my knowledge, made clear the criteria for establishing the categories or patterns in an ethogram.

Actually, considerable agreement has been obtained by various people working independently on the same species, for example, the catalogues of rhesus behavior that were developed independently by Altmann (1962*a*, 1965), Chance (1956), Hinde and Rowell (1962; also Rowell and Hinde 1962) and Reynolds (1966). But when differences develop that cannot be explained on the basis of differences in empirical data, an explication of the criteria ceases to be a game in logical reconstruction (to be left for philosophers of science) and becomes an important issue in the gathering of data and the construction of theories.

To my knowledge, the only extensive discussion of the criteria of cataloguing natural units of social behavior grew out of our studies of social behavior of rhesus monkeys (Altmann 1962*a*, 1965; see also Hebb and Thompson 1954; Rodriguez Delgado and Delgado 1962; Nissen 1958). They may be summarized as follows (Altmann 1965):

> Like other problems in classification, categorizing the units of social behaviour involves two major problems: when to split and when to lump. If one's goal is to draw up an exclusive and exhaustive classification of the animals' repertoire of socially significant behaviour patterns, then these units of behaviour are not arbitrarily chosen. To the contrary, they can be empirically determined. One divides up the continuum of action wherever the animals do. If the resulting recombination units are themselves communicative, that is, if they affect the behaviour of other members of the social group, then they are social messages. Thus, the splitting and lumping that one does is, ideally, a reflection of the splitting and lumping that the animals do. In this sense, then, there are natural units of social behaviour.

As Struhsaker (chap. 16) has realized, these criteria are necessary but not sufficient. They are *criteria of sequential demarcation;* they indicate how we break up the temporal continuum of each individual's behavior to obtain behavioral events. But they fail to indicate how we decide when any two events are members of the same class or category. We need criteria that indicate when two behaviors are, in some sense, the same and when they are different. That is, we need *criteria of membership* for establishing and distinguishing between the categories that constitute the behavior patterns.

Our choice of criteria of membership must take into account several properties of messages that have been discussed—in particular, sequential stereotypy, continuous variations in classes of messages, and analogical communication. The relevance of a lack of total response stereotypy for any signal, or of total signal stereotypy for any response, is that the responses cannot be used to demarcate the classes of signal categories (although it may be suggestive of how to do this). To do so would be to commit the errors of calling two patterns the same pattern because they elicited the same responses and of dividing one pattern into several because it elicited a number of responses. Of course, the responses are crucial in deciding which behavior patterns are communicative, but that is another matter.

For continuously variable messages in which the variations are without significance, there seems to be no problem; the variants are considered to be members of the same class or pattern. As a consequence, the boundaries between classes in the repertoire correspond with discontinuities in the array of signals.[4]

It is less obvious how one is to demarcate elemental messages for messages in which the continuous array of variants are functional, either in semantic representation or in communication. Struhsaker (chap. 16) proposes that each such signal variant be regarded as a different element in the catalogue. For example, each bee dance that differed in the angle of the waggle would be regarded as a distinct item in the catalogue. His reason is that to do otherwise is to obscure the fact that analogical systems have an infinite source of meaningful messages and are thus open systems. But this argument can be turned around: to call every different message a different pattern in the repertoire is to obscure the fact that real and important discontinuities do occur in the repertoire. Beyond that, the repertoire of every species that includes even one such set of analogical messages would be infinite. Hence all such repertoires would be indistinguishably large.

[4] The physical parameters that one measures on signals (for example, duration, frequency, amplitude, volume, color, length, breadth, angle) are almost all continuous variables. Consequently, each includes an infinity of possible values. Infinite sets cannot be defined by listing all their members in the way that is sometimes done with finite sets. They can only be defined by giving rules of membership. In general, the values of each physical parameter of a signal or display can be mapped into the points along a line segment and thus into a contiguous subset of the real numbers. That is, a set of such measurements is defined by first establishing an ordering rule, so that the measurements form a linear array (usually, an array of magnitudes), and by then indicating the limits along the array that bound the set.

The problem, of course, is to bring out the important fact that such analogical communication systems, unlike digital systems, are invariably open, yet to prevent obscuring the fact that in any species there is a finite set of such systems, each of which is sharply demarcated from the others. Anything as simple as a list of behavior patterns, or the number of items in such a list, cannot convey that much information. The solution would seem to be to recognize that a repertoire of socially significant behavior patterns consists of a finite set of message classes, each sharply demarcated from the others yet each consisting of a continuous array of behaviors, and that in some cases the variants of a pattern may be components of a functional, analogical communication system.

Prevarication

Hockett (1963, sec. 2.14) has recently considered the properties that underlie the fact that linguistic messages can be false and they can be meaningless in the logician's sense of unverifiable through observation. According to Hockett, prevarication depends upon semanticity, displacement, and openness:

> Without semanticity, a message cannot be tested for meaningfulness and validity. Without displacement the situation referred to by a message must always be the immediate context, so that a lie is instantly given away. Without openness, meaningless messages can hardly be generated. False ones can: a gibbon could, in theory, emit a food call when no food had been discovered.

The ability to say that which may not be so (lies, fictions, errors, superstitions, religions, hypotheses, science, delusions, and ambitions) lays the foundations for some of man's greatest achievements, as well as for some of his basest acts, and much of his mental illness.

To these properties that underlie prevarication, we must add a crucial one—discreteness, or discontinuity. Without this, it would be impossible to say "no," and hence to be able to cope with false statements. The reason that discontinuity, or discreteness, is necessary is much the same as the reason that a continuous semantic message will not be arbitrary: Only by introducing a discontinuity can one keep the system from being swamped by gross errors in transmission. Thus the introduction of discontinuity to such a system enables it to assign truth values to statements—that is, to make both statements and their negations.

To illustrate this argument, let us design a fictitious bee that prevaricates with a continuous message, the waggle dance as a representation of direction. Our bee will use dances whose waggles point within the right semicircle to represent the presence of nectar in various directions and dances within the left semicircle to represent the absence of nectar in various directions (see Fig. 4). The communication system of real bees, which was described above, permits bees to tell lies (a worker could indicate a source of nectar where there was none) but provides no mechanism for contradicting another message: if a worker flies out to the location indicated by another bee and does not find a source of nectar there, he cannot convey this fact. But our fictitious bees can. They can make statements and the negations of statements. They can contradict each other. (Such a system might, in fact, be useful to bees: it might enable them to forage more efficiently by avoiding duplication of effort.)

Our fictitious prevarication system has, however, one serious flaw. Suppose, for example, that a thorough search for nectar between 0° and 3° (distance, too, being specified) had been fruitless—or rather, nectarless. The worker returns and dances. But note that a small negative (counterclockwise) error in the message or in its reception would result in the interpretation that nectar was present in that direction. Such gross errors in communication would produce such inefficiency that there would be strong selection against any system of this kind.

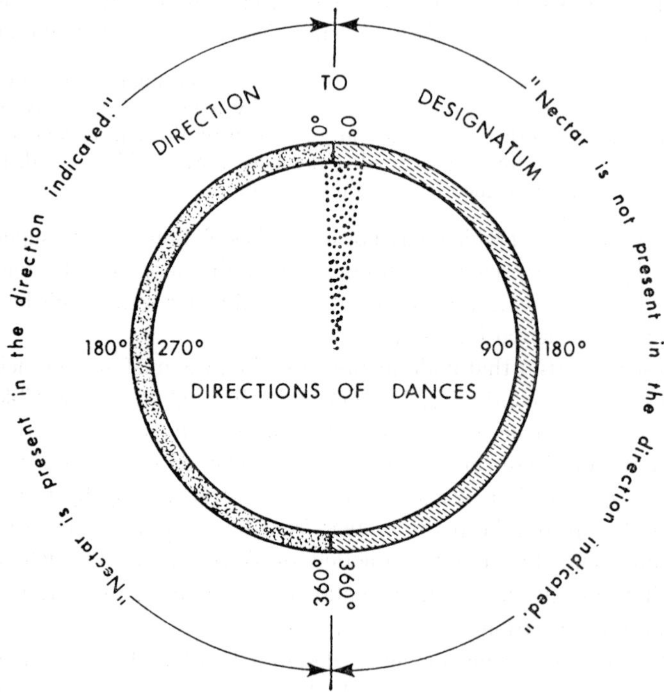

FIG. 4.—Model for dances of prevaricating bees, using continuous messages. The presence of nectar is indicated by any waggle directed within the left hemisphere; the absence, in the right hemisphere. Note that the presence or absence of nectar in any direction may be indicated. The dotted segment indicates the directions of the waggle components in a dance that represents the absence of nectar at 2°, but that has an error in transmission. Note that this slight error might lead to the interpretation that there is nectar at 1°.

The solution to this design problem is to introduce a discontinuity between the class of affirmative messages and the class of negative messages (Fig. 5). Now, any slight error in a dance that indicates the absence of nectar near 0° would, at worst, be meaningless.

A comparable argument can be made for any such prevaricating system. In order to avoid gross errors, there must be a discontinuity between the set of affirmative messages and the corresponding set of negative messages.

The idea that negations are only possible in digital communication has been developed independently by Bateson and Jackson (1964) in a discussion of the relations between digital and analogical communication.

Avoidance of ambiguity probably has been a major factor selecting for discontinuities between the physical structure of each display pattern in the repertoires of animals.

Traditional Transmission and Learnability (13 in Hockett 1960*a*, and sec. 2.16 in Hockett 1963)

As Hockett points out, human genes are not specific to the idiosyncrasies of any one language but are permissive of all. But the genes are not passively

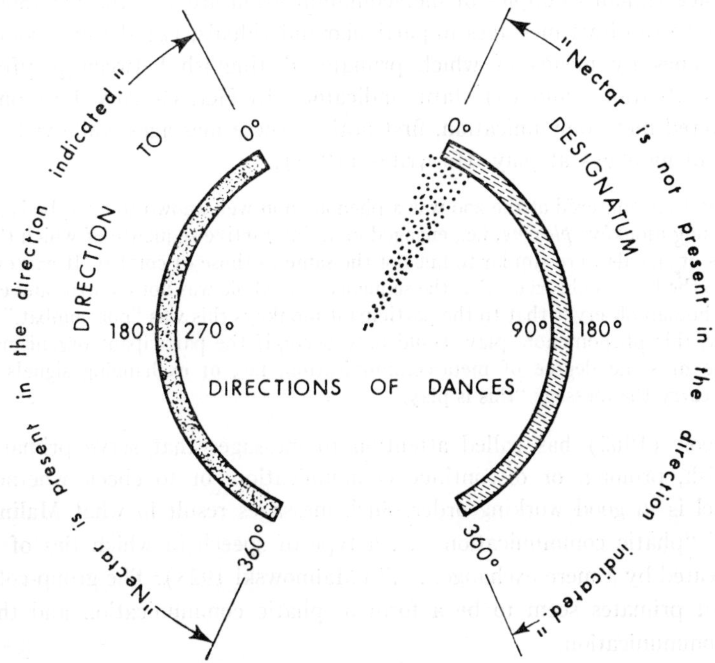

FIG. 5.—Model for dances of prevaricating bees, using discontinuity between the negative and affirmative messages. This discontinuity helps prevent the type of errors that are inherent in the system shown in Figure 4.

"permissive." The human phenotype includes a strong positive drive toward participation in the communicative interactions of the society, a drive that can be frustrated only by the most radical isolation. Furthermore, a speaker of one language can learn another.

Since these two design features are not really structural properties, and since so little is known about the genetics and cultural inheritance of communication in primates, nothing more will be said here about these features of language.

METACOMMUNICATION

Humans have a marvelous ability to communicate about their own communication, what Hockett (1963, sec. 2.15) refers to as the "reflexiveness" in language. In humans, this reflexiveness results from the universality of our language (see section on productivity), but it is not restricted to linguistic communication; in general, communication about communication is referred to as "metacommunication." Something as subtle as the twinkle of an eye may tell you whether a statement is to be interpreted seriously or as a joke.

Nonhuman primates, too, can metacommunicate. Perhaps the most commonplace simian examples of metacommunication are (1) the metamessages that serve to direct messages to particular individuals, (2) the metacommunicative cues by means of which primates distinguish between playful and serious situations, and (3) status indicators. In fact, Gregory Bateson, who discovered metacommunication, first noticed these messages while watching a group of monkeys at play. He writes (1955):

> What I encountered at the zoo was a phenomenon well known to everybody: I saw two young monkeys *playing*, i.e., engaged in an interactive sequence of which the unit actions or signals were similar to but not the same as those of combat. It was evident, even to the human observer, that the sequence as a whole was not combat, and evident to the human observer that to the participant monkeys this was "not combat."
>
> Now, this phenomenon, play, could only occur if the participant organisms were capable of some degree of meta-communication, i.e., of exchanging signals which would carry the message "this is play."

Sebeok (1962) has called attention to messages that serve primarily to establish, prolong, or discontinue communication, or to check whether the channel is in good working order. Such messages result in what Malinowski named "phatic communication . . . a type of speech in which ties of union are created by a mere exchange. . . ." (Malinowski 1923). The group-cohesion calls of primates seem to be a form of phatic communication and thus of metacommunication.

One difficulty in studying metacommunication has been that there has been no formal theory of the basic nature of metacommunication in sufficiently operational terms that it was possible to know just when you had an example of it. Indeed, many people have remained skeptical that metacommunication is basically different from other forms of communication.

I would like to propose a theory of the structure of metacommunication. It is, in brief, that metacommunication is a stochastic process, each of whose components is a stochastic process. That is, when we say that metamessages affect the interpretation of other messages, we mean that the sequential contingency between message and response is, in turn, contingent upon a metamessage. Such sequential metacontingencies between messages and the responses that they elicit are the essence of metacommunication. Thus, the metamessages that convey the information that a particular threat is a play threat are affecting the contingency between the threat by one individual and the response to that threat on the part of another.

NATURAL UNITS OF BEHAVIOR

The question has often been raised whether there are natural units of behavior or whether all units of observation on behavior are arbitrary. A classification of communicative behavior is natural to the extent that it fulfils either of two criteria, which we will call the "logical" and the "biological."

The logical criterion has been described by Kaplan (1964):

> What makes a concept significant is that the classification it institutes is one into which things fall, as it were, by themselves. It carves at the joints, Plato said. Less metaphorically, a significant concept so groups or divides its subject-matter that it can enter into many and important true propositions about the subject-matter other than those which state the classification itself. Traditionally, such a concept was said to identify a "natural" class rather than "artificial" one. Its naturalness consists in this, that the attributes it chooses as the basis of classification are significantly related to the attributes conceptualized elsewhere in our thinking. Things are grouped together because they resemble one another. A natural grouping is one which allows the discovery of many more, and more important, resemblances than those originally recognized.

The biological criterion of natural classification is that the fundamental classifications be based upon the demarcations, both sequential and categorical, that occur in the natural habitat of the species. This is not to say that evidence for the natural classification cannot be obtained from animals in laboratories, outdoor enclosures, zoos, island colonies, and so forth. On the contrary, for certain purposes such artificial colonies often have definite advantages over natural habitats and are highly recommended in preparation for any field study. But the naturalness of a classification obtained under such conditions is to be judged by the extent to which it reflects what happens in the animals natural habitat. This criterion is, implicitly, an evolutionary one: it is an attempt to provide information on how the animals cope with the problems that they face in the environments that have selected for and that maintain the characteristics of each population and species.

I am convinced that in the study of animal communication the most natural classifications, in the logical sense, will be those that are most natural in the biological sense. This conviction could not be proved, but only illustrated, from the existing corpus of research—and from Plato's metaphor. Indeed, its proof can only come from the whole future development of research in the behavioral sciences.

SUMMARY

In this chapter, I have attempted to clarify the basic nature of social communication, discuss structural properties of social communication systems and review relevant data on the structure of primate behavior, discuss criteria for establishing natural classifications of behavior, and present an outline of a mathematical model of metacommunication.

Many structural properties that are universal in human language are known to occur in various species of nonhuman primates, some of which combine several of these properties. Inadequacies in the available data on social communication among nonhuman primates make it impossible to say whether any species of primate other than man combines all of these properties. Consequently, it is not yet possible to test Charles Darwin's contention (1874) that the behavior of man differs from the behavior of other animals in degree, not in kind.

Perhaps more important in the long run will be the contribution of such structural analysis of communication to an understanding of how the social behavior of man and other primates functions and how it has evolved.

ACKNOWLEDGMENTS

This chapter was written while the author was the recipient of research grants (GB 683 and GB 2879) from the U.S. National Science Foundation for "Field Studies of Primate Behavior" and "Analysis of Data on Baboon Behavior," and a grant (MH 07336–01) from the National Institute of Mental Health, U.S. Public Health Service, for the study of "Mammalian Social Behavior."

REFERENCES

Altmann, Stuart A. 1959. Field observations on a howling monkey society. *J. Mammal.* 40:317–30.

———. 1962*a*. A field study of the sociobiology of rhesus monkeys, *Macaca mulatta*. *Ann. N.Y. Acad. Sci.* 102 (Art. 2): 338–435.

———. 1962*b*. Social behavior of anthropoid primates: analysis of recent concepts. In *Roots of behavior*, ed. Eugene L. Bliss. New York: Harper.

———. 1965. Sociobiology of rhesus monkeys. II. Stochastics of social communication. *J. Theoret. Biol.* 8:490–522.

———. 1966. Vocal communication in howling monkeys. (7.5 ips tape) Library of Natural History Sounds, Laboratory of Ornithology, Cornell University.

Andrew, R. J. 1963. The origin and evolution of the calls and facial expressions of the primates. *Behaviour* 20(1–2):1–109.

———. 1964. The displays of the primates. In *Evolutionary and genetic biology of primates,* ed. John Buettner-Janusch. 2 vols. Vol. 2, pp. 277–309. New York: Academic Press.

Bastian, Jarvis. 1965. Primate signalling systems and human language. In *Primate behavior: field studies of monkeys and apes,* ed. I. DeVore, pp. 585–606. New York: Holt, Rinehart and Winston.

Bateson, G. 1955. A theory of play and fantasy. *Psychiat. Res. Rept. A.* 2:39–51.

Bateson, Gregory, and Jackson, Don D. 1964. Some varieties of pathogenic organization. *Res. Publ. Assoc. Nerv. Ment. Dis.* 42:270–83.

Birch, Herbert G. 1952. Communication between animals. Cybernetics. . . . Trans. Eighth Conf., pp. 134–172, Josiah Macy, Jr. Foundation.

Bolwig, Niels. 1959*a*. Observations and thoughts on the evolution of facial mimic. *Koedoe* 2:60–69.

———. 1959*b*. A study of the behaviour of the Chacma baboon, *Papio ursinus. Behaviour* 14:136–63.

———. 1964. Facial expression in primates with remarks on a parallel development in certain carnivores (a preliminary report on work in progress). *Ibid.* 22:167–92.

Booth, A. H. 1957. Observations on the natural history of the Olive Colobus monkey, *Procolobus verus* (van Beneden). *Proc. Zool. Soc. Lond.* 129(3):421–30.

Bullock, Theodore H. 1957. The trigger concept in biology. In *Physiological triggers,* ed. Theodore H. Bullock, pp. 1–16. Washington, D.C.: American Physiological Society.

Carpenter, C. R. 1934. A field study of the behavior and social relations of howling monkeys (*Alouatta palliata*). *Comp. Psychol. Monogr.* 10(2):168 pp.

———. 1940. A field study in Siam of the behavior and social relations of the gibbon (*Hylobates lar*). *Ibid.* 16(5):212 pp.

Chance, M. R. A. 1956. Social structure of a colony of *Macaca mulatta. Brit. J. Anim. Behav.* 4(1):1–13.

Cherry, Colin. 1961. *On human communication: a review, a survey, and a criticism.* New York: Science Editions.

Collias, Nicholas, and Southwick, Charles. 1952. A field study of population density and social organization in howling monkeys. *Proc. Am. Phil. Soc.* 96:143–56.

Crawford, Meredith P. 1939. The social psychology of the vertebrates. *Psychol. Bull.* 36:407–46.

Darwin, Charles. 1872. *The expression of the emotions in man and animals.* London: Murray.

———. 1874. *The descent of man and selection in relation to sex.* 2d ed. New York: P. F. Collier.

DeVore, Irven. 1962. The social behavior and organization of baboon troops. Unpublished dissertation, University of Chicago.

Eckstein, P., and Zuckerman, S. 1956. Changes in the accessory reproductive organs of the non-pregnant female. In *Marshall's physiology of reproduction,* 3d ed., ed. A. S. Parkes, 2 vols. London: Longmans Green, vol. 1: part 2, pp. 543–654.

Frisch, Karl von. 1955. *The dancing bees,* trans. Dora Ilse. New York: Harcourt, Brace.

Haldane, J. B. S. 1953. Animal ritual and human language. *Diogenes* 4:61–73.

———. 1955. Animal communication and the origin of human language. *Sci. Progr.* 43:385–401.

Haldane, J. B. S., and Spurway, H. 1954. A statistical analysis of communication in *Apis mellifera* and a comparison with communication in other animals. *Insectes Soc.* 1(3):247–83.

Hall, K. R. L. 1962. The sexual, agonistic and derived social behaviour patterns of the wild Chacma baboon, *Papio ursinus. Proc. Zool. Soc. Lond.* 139:283–327.

Hall, K. R. L., Boelkins, R. C., and Goswell, M. J. 1965. Behaviour of patas monkeys, *Erythrocebus patas,* in captivity, with notes on the natural habitat. *Folio Primatol.* 3:22–49.

Hall, K. R. L., and DeVore, Irven. 1965. Baboon social behavior. In *Primate behavior: field studies of monkeys and apes,* ed. I. DeVore, pp. 53–110. New York: Holt, Rinehart and Winston.

Hebb, D. O., and Thompson, W. R. 1954. The social significance of animal studies. In *Handbook of social psychology,* ed. Gardner Lindzey. 2 vols. Cambridge, Mass.: Addison-Wesley, vol. 1, pp. 532–561.

Hill, W. C. Osman, and Booth, A. H. 1957. Voice and larynx in African and Asiatic Colobidae. *J. Bombay Nat. Hist. Soc.* 54(2):309–21.

Hinde, R. A., and Rowell, T. E. 1962. Communication by postures and facial expressions in the rhesus monkey (*Macaca mulatta*). *Proc. Zool. Soc. Lond.* 138:1–21.

Hockett, Charles F. 1958. *A course in modern linguistics.* New York: Macmillan.

———. 1959. Animal "languages" and human language. In *The evolution of man's capacity for culture,* ed. J. N. Spuhler, pp. 32–39. Detroit: Wayne State University.

———. 1960*a*. Logical considerations in the study of animal communication. In *Animal sounds and communication,* ed. W. E. Lanyon and W. N. Tavolga, pp. 392–430. Washington: American Institute of Biological Sciences.

———. 1960*b*. The origin of speech. *Sci. Am.* 203:89–96.

———. 1963. The problem of universals in language. In *Universals of language,* ed. Joseph H. Greenberg, pp. 1–22. Cambridge, Mass.: M.I.T. Press.

Hockett, Charles F., and Ascher, Robert. 1964. The human revolution. *Current Anthropol.* 5:135–68.

Hooff, J. A. R. A. M. van 1962. Facial expressions in higher primates. *Symp. Zool. Soc. Lond.* 8:97–125.

Huxley, Julian S. 1938. Threat and warning coloration in birds, with a general discussion of the biological functions of colour. Proc. 8th Int. Orn. Cong., pp. 430–55. Oxford (1934).

Ilse, D. R. 1955. Olfactory marking of territory in two young male Loris, *Loris tardigradus lydekkerianus,* kept in captivity in Poona. *Br. J. Anim. Behav.* 3:118–20.

Itani, Junichiro. 1954. *Takasakiyama No Saru.* Tokyo: Kobunsha.

———. 1963. Vocal communication of the wild Japanese monkey. *Primates* 4(2): 11–66.

Jakobson, R., Fant, C. G. M., and Halle, M. 1952. *Preliminaries to speech analysis.* Cambridge: M. I. T. Press.

Jay, Phyllis C. 1962. The social behavior of the langur monkey. University of Chicago doctoral dissertation.

Johnson, Dennis, and Wenner, Adrian M. 1966. A relationship between conditioning and communication in honey bees. *Anim. Behav.* 14:261–65.

Kaplan, Abraham. 1964. *The conduct of inquiry.* San Francisco: Chandler.

Kelemen, G., and Sade, J. 1960. The vocal organ of the howling monkey (*Alouatta palliata*). *J. Morphol.* 107:123–40.

Kummer, H. 1956. Rang-Kriterien bei Mantelpavianen. Der Rang adulter Weibchen im Sozialverhalten, den Individualdistanzen und im Schlaf. *Rev. Suisse Zool.*, 63(16):288–297.

Lindauer, Martin. 1961. *Communication among social bees.* Cambridge, Mass.: Harvard University.

Malinowski, Bronislaw. 1923. The problem of meaning in primitive languages. In *The meaning of meaning,* ed. C. K. Ogden and I. A. Richards, Suppl. 1, pp. 451–510. New York: Harcourt, Brace.

Mandelbrot, Benoit. 1954. Structure formelle des textes et communication. *Word* 10:1–27.

Marler, Peter. 1964. Developments in the study of animal communication. In *Darwin's biological work: some aspects reconsidered,* ed. P. R. Bell, pp. 150–206. New York: Wiley.

———. 1965. Communication in monkeys and apes. In *Primate behavior: field studies of monkeys and apes,* ed. I. DeVore, pp. 544–84. New York: Holt, Rinehart and Winston.

Miller, George A., Galanter, Eugene, and Pribram, Karl H. 1960. *Plans and the structure of behavior.* New York: Holt.

Montagna, William, and Ellis, Richard A. 1959. The skin of primates. I. The skin of the potto (*Perodicticus potto*). *Am. J. Phys. Anthropol.* 17:137–62.

———. 1960. The skin of primates. II. The skin of the slender loris (*Loris tardigradus*). *Ibid.* 18:19–44.

Montagna, William, and Yun, Jeung Soon. 1962*a*. The skin of primates. VII. The skin of the great bushbaby (*Galago crassicaudatus*). *Am. J. Phys. Anthropol.* 20: 149–66.

———. 1962*b*. The skin of primates. X. The skin of the ring-tailed lemur (*Lemur catta*). *Ibid.* pp. 95–118.

———. 1963. The skin of primates. XVI. The skin of *Lemur mongoz. Ibid.* 21:371–82.

Morris, Charles. 1955. *Signs, language, and behavior.* New York: Braziller.

Morris, Desmond. 1957. "Typical intensity" and its relation to the problem of ritualisation. *Behaviour* 11:1–12.

Nissen, Henry W. 1958. Axes of behavioral comparison. In *Behavior and evolution,* ed. Anne Roe and George Gaylord Simpson, pp. 183–205. New Haven: Yale University Press.

Petter, J. J. 1962. Recherches sur l'écologie et l'éthologie des Lémuriens malgaches. *Mém. Mus. Natl. Hist. Nat.*, Ser. A, Vol. 27.

Révész, G. 1944. The language of animals. *J. Gen. Psychol.* 30:117–47.

Reynolds, Vernon. 1966. The social behaviour repertoire of rhesus monkeys (*Macaca mulatta*) in a captive colony. In preparation.

Reynolds, Vernon, and Reynolds, Francis. 1965. Chimpanzees in the Budongo Forest. In *Primate behavior: field studies of monkeys and apes,* ed. I. DeVore, pp. 368–424. New York: Holt, Rinehart and Winston.

Rodriguez Delgado, R., and Delgado, J. M. R. 1962. An objective approach to measurement of behavior. *Phil. Sci.* 29:253–68.

Rowell, T. E. 1962. Agonistic noises of the rhesus monkey (*Macaca mulatta*). *Symp. Zool. Soc. Lond.* 8:91–96.

Rowell, T. E., and Hinde, R. A. 1962. Vocal communication by the rhesus monkey (*Macaca mulatta*). *Proc. Zool. Soc. Lond.* 138:279–94.

Schaller, George B. 1963. *The mountain gorilla: ecology and behavior.* Chicago: University of Chicago Press.

Sebeok, Thomas A. 1962. Coding in the evolution of signalling behavior. *Behav. Sci.* 7:430–42.

———. 1965. Animal communication. *Science* 147:1006–14.

Sebeok, Thomas A., Hayes, Alfred S., and Bateson, Mary Catherine. 1964. *Approaches to semiotics.* The Hague: Mouton.

Shannon, C. E. 1948. A mathematical theory of communication. *Bell Syst. Tech. J.* 27:379–423, 623–56.

Simonds, Paul E. 1965. The bonnet macaques in south India. In *Primate behavior: field studies of monkeys and apes,* ed. I. DeVore, pp. 175–96. New York: Holt, Rinehart and Winston.

Southwick, Charles H. 1962. Patterns of intergroup social behavior in primates, with special reference to rhesus and howling monkeys. *Ann. N.Y. Acad. Sci.* 102 (Art. 2):436–54.

Stark, D., and Schneider, R. 1960. Larynx. In *Primatologia: handbook of primatology,* ed. H. Hofer, A. H. Schultz, and D. Stark, 3(2):423–587. Basel and New York: S. Karger.

Tinbergen, N. 1951. *The study of instinct.* Oxford: Clarendon Press.

———. 1952. "Derived" activities: their causation, biological significance, origin, and emancipation during evolution. *Quart. Rev. Biol.* 27(1):1–32.

Ullrich, Wolfgang. 1961. Zur Biologie und Soziologie der Colobusaffen (*Colobus guereza caudatus* Thomas 1885). *Zool. Gart.* (*NF*), 25:305–68.

Wiener, Norbert. 1948. Time, communication, and the nervous system. *Ann. N.Y. Acad. Sci.* 50:197–220.

Zhinkin, N. I. 1963. An application of the theory of algorithms to the study of animal speech: methods of vocal intercommunication between monkeys. In *Acoustic Behaviour of Animals,* ed. R.-G. Busnel, pp. 132–80. Amsterdam: Elsevier.

DISCUSSION OF COMMUNICATION PROCESSES

THOMAS A. SEBEOK

My role as discussant here is insidious. I came as the representative of another discipline—a linguist among primatologists—and what I experience is this: I hear everything through a kind of coarse linguistic filter into which a substantial array of important data and some subtle ideas are fed; but my filter transmits only the frequencies to which it is attuned, and even these, as you will appreciate, come through distorted and with significant loss of detail.

When I define myself as a linguist, this is too narrow a limitation, because linguistics is only one of the anthropological disciplines—the one which is devoted to certain semiotic processes—"semiotic" not in the medical sense of the word, but in Charles Sanders Peirce's use of this term, implying the analysis and comparison of various systems of signs which, in this context, involve, broadly, the coding of information in cybernetic control processes and the consequences that are imposed by this categorization where living animals function as input/output linking devices in a biological version of the traditional information-theory circuit with a transcoder added (Sebeok 1965). In the systems under consideration here, consisting of a transmitter–channel–transcoder–channel–receiver, the relaying mechanisms are members of the order Primates, such as Hall's patas monkeys or Struhsaker's vervet monkeys. Ultimately, it will indeed be necessary to evaluate the structure of primate communication fully against the background of other sign systems used by man (semiotics; Sebeok, Hayes, and Bateson 1964) and elsewhere in the animal kingdom (zoosemiotics).

Anthropology is, of course, literally a branch of primatology. It is devoted to four interdependent topics, each of which may be said, in its specialized

Thomas A. Sebeok, Research Center in Anthropology, Folklore, and Linguistics, Indiana University, Bloomington, Indiana, U.S.A.

way, to focus on tools—those instruments which man, severally and uniquely, employs to overcome the limitations of nature. It was Benjamin Franklin who is often reported to have first defined man as inventor, *Homo faber,* the tool-making animal; although he was right, we must guard against interpreting man's inventions too simplistically.

Basic to anthropology is a concern with the highly specialized structures which characterize *H. erectus,* especially his long, mobile arms and hands, free for holding, for carrying, and above all, for making and using tools (Napier 1963). Physical anthropology studies the primary with the best grip and best eye-hand coordination, and those parts of the brain related to such control.

It has long been known that primates (as well as the sea otters acutely observed by Hall and Schaller 1964) are mammals that habitually use tools as a means of obtaining certain types of food. Jane Goodall has revealed and vividly recounted that some chimpanzees in the wild even modify natural objects to make tools. Man, however, is the only animal that makes tools for making tools—what we now call "machine tools." I would label the academic study of this capability, in its numerous ramifications, cultural anthropology.

Man, furthermore, is the only animal that practices incest prohibition, and we may regard this taboo as his fundamental tool for the weaving of social networks; in its multifarious manifestations, the study of man's sense of order form the subject matter of social anthropology.

Fourth and last, man is both uniquely and globally endowed with a set of tools which workers in my profession, linguistics, consider the universal building blocks of language: these are called "distinctive features." Distinctive features are the ultimate units of language, the atomic particles, if you will, of linguistic structure, organized in accordance with a binary code of utmost efficiency, so that whatever phase of the speech event is approached, the elicitation of its correlates must yield a distinct, unambiguous, "yes" or "no" response. What is basic and general in the structure of the expression in this form of human communication is, first, the minimal system of oppositions grounded on maximal distinctions; and secondly, the rank order according to which this elementary phonological component pervades the more complicated syntactic and other constructions. This is another way of saying that the theory of distinctive features and phonetic universals constitutes a part of general linguistic theory (the part which, so far, has been most concretely and substantively realized).

In light of the threefold division of human tools suggested here, supplemented by Altmann's insightful reference to those properties of social communication that he deems necessary for us to be able to say "not," it may be instructive to cite Kenneth Burke's four-clause definition of man. This great critic and polymath, in an essay "On Words and The Word,"

(Burke 1961) tells us that "Man is (1) The symbol-using animal (2) Inventor of the negative (3) Separated from his natural condition by instruments of his own making (4) And goaded by the spirit of hierarchy." Stripped of its rhetorical flourish, the first category seems to me to comprehend semiotics, notably linguistics; the third one, cultural anthropology; and the fourth one, social anthropology. The second category is a dramatistic statement of Altmann's position which, in turn, leads me to wonder about the precise nature of the differences between prevarication and a diversionary display of the kind Hall describes in the patas, and the phenomenon of mimicry in animals, especially of the Batesian type. Thus it would be useful to have a detailed explanation—in Altmann's and Hockett's terms—of the behavior of certain vespine audio-mimics—for example, the fly *Spilomyia hamifera* Lw., which displays a wingbeat rate of 147 strokes per second while hovering near the wasp *Dolichovespula arenaria* F., which it also closely resembles in color pattern (Gaul 1952). Since this wasp flies with 150 wing strokes per second, presumably the two flight sounds are indistinguishable to predators, and fly-catching birds are deceived—or, on the contrary, we would need to assume that this visual and auditory convergence is due to chance.

Altmann—as Peter Marler (1961) did before—finds certain distinctions introduced by the philosopher Charles Morris (1946) useful to him in his logical analysis of animal communication. Thus an investigator of zoosemiotic behavior may choose to focus on the zoosyntactic or zoosemantic or zoopragmatic aspects of the total event, with differing degrees of probable illumination; Marler, for instance, went on to explore a pragmatic approach and to perform a very interesting analysis of the song of the chaffinch into signals which function as identifiors, designators, appraisors, and prescriptors. If some such framework could be consistently adopted by students of primate signaling behavior, this might serve as a basis for comparative studies; as in linguistics, not inventory but system is the base of typology, and to comprehend a system a mere listing of its components is insufficient. The principle of ordered division must take ever deeper roots in the study of social interactions in primates.

Crosscutting the tripartite formulation mentioned, Morris also stakes out three possible semiotic procedures: pure, descriptive, and applied. Pure zoosemiotics is concerned with the elaboration of theoretical models, that is—in the broadest sense, as illustrated by Altmann's chapter or by the parallel but wholly independent work of the Russian scholar N. I. Zhinkin (1963) on vocal intercommunication of chimpanzees—with the development of a language designed to deal scientifically with animal signaling behavior. Descriptive zoosemiotics comprehends the study of animal communication as a natural and as a behavioral science, as exemplified by the chapters in this section providing partial ethograms (in Tinbergen's sense) of patas monkeys and vervet monkeys. Applied zoosemiotics deals with the exploitation of the valuable and diverse data so obtained for the benefit of man, but utilitarian applications in primatology—with scattered exceptions, such as the employment of a female baboon reported to have served as a goatherd (Hoesch 1961)—clearly remain tasks for the future.

What is the linguist's share in this complex effort? Those of us who are interested in your endeavors seek chiefly to disclose the biological and anthropological origins of human communication as we search for the answers to particular questions such as these: what are the anatomical and physiological correlates of verbal behavior and what sensory and cognitive specializations are required for language perception; what motivates the onset and selective accomplishment of language learning in the development of human infants, as in other behavior programs certain releaser mechanisms begin to function at predetermined stages of maturation; why do subhuman forms (as did Viki "Hayes") lack the capacity to acquire even the beginnings of language; how can present evolutionary theory account for the uniqueness of both form and behavior of language specialization in man; and what is the genetic basis for language propensity, man's species-specific biological endowment? As the English zoologist R. J. Pumphrey (1951) has pointed out, there are two schools of thought about the implications of inquiries of this sort: "All are agreed that human speech differs in material particulars from the speech of other animals. There are those who, like Darwin, believe in a gradual evolution, but there have been others who have believed that speech is specifically a human attribute, a function *de novo,* different in kind from anything of which other animals are capable." Altmann does not commit himself to Darwin's contention, but reconstruction of the phylogeny of distinctive features—what Hockett and Ascher (1964) have treated as the inception of duality of patterning—has clearly not yet progressed beyond mere speculation. While, by the way, I share Itani's uneasiness at the unwarranted promotion of *Hylobates* in the context of their discussion of the "opening of the call system," his own claim (made at the Symposium but not included in this book) for the alleged survival of the calling sounds of Japanese monkeys as interjections in language seems to me quite illusory. In Edward Sapir's apt analogy (Sapir 1939), drawn from the varying national modes of pictorial representation, "A Japanese picture of a hill both differs from and resembles a typical modern European painting of the same kind of hill. Both are suggested by and both 'imitate' the same natural feature. Neither the one nor the other is the same thing as, or, in any intelligible sense, a direct outgrowth of the natural feature. . . . The interjections of Japanese and English are, just so, suggested by a common natural prototype, the instinctive cries, and are thus unavoidably suggestive of each other."

Moynihan's expert description, at the Symposium, of communication in *C. moloch* was organized around the properties of the four alternative and complementary channels this animal employs: olfactory, tactile, visual, and acoustic. His observations (although they are not published here) invite reflections on the nature of osmic and optical signs, and of those involving pressure changes—tactile and auditory. Roman Jakobson (1965) has asked "Why is audible speech the only universal, autonomous and fundamental vehicle of communication" in man? He pointed out the preponderantly auditory character of nonrepresentational signs, in contradistinction to the mostly visual character of primarily representative signs which display a factual similarity or contiguity with their objects. The psychologist Frank A. Geldard anticipated Jakobson when he remarked (Geldard 1960) that "Vision is the great spatial sense, just as audition is the great temporal one." Simultaneity is the principal structuring device used in visual coding, where relational comparisons have to be made, and successivity in auditory coding, where data presented in rapid serial order are to be resolved. It is the second semiotic type which "implies a compulsory hierarchical arrangement and discrete components"—in brief, the distinctive features of human discourse.

A typology of the sign systems used by animals in the different sensory modalities at their disposal remains to be worked out. At this stage, one can merely compare in summary fashion some obvious advantages or disadvantages of the four systems mentioned by Moynihan (and also alluded to by Altmann in his verbal presentation):

1. In *chemical systems,* which may involve the distal organs of smell and the proximal organs of taste, pheromones tend to function as yes/no signals; that is, a particular scent is either produced or it is not. Whether any animal can modulate the intensity or pulse frequency of pheromone emission to formulate new messages is unknown, but we do know that, once emitted, the odor is very likely to persist and thus to convey a message after the departure of its source from the site. The one great advantage of chemical signals is, therefore, their capacity to serve as vehicles of communication into the future. This capacity, whereby an individual can send messages to another in his absence and, by a delayed loop, even to himself, is analogous to the human use of script.

2. In *optical systems,* which presuppose reflected daylight in diurnal species such as most monkeys and bioluminescence in some other animals, patterns of activity are highly variable as to shapes and colors, and in time and range of intensity. Visual signs are thus both flexible and transient: they can be rapidly switched on or off. They are useful when fine discrimination is needed; when unfamiliar material is to be comprehended and may be available for repeated scanning; when, as Geldard (1960) emphasizes, reference data have to be immediately available "or where simultaneous (or nearly simultaneous) relational comparisons have to be made"; when the receiver must promptly select limited data from a large repertoire of information; and when unfavorable environmental conditions hamper other forms of reception.

3. *Tactile systems,* being subject to wide fluctuations in intensity and time, are particularly useful for the transmission of quantitative information (concerning, for example, distance) and when the visual and auditory channels are overloaded.

4. *Acoustic systems* involve a minute output of energy, and their transient character makes accurate timing possible. With the proper receiving organs, sounds—especially those in the higher frequencies—can be more or less precisely directional when it is advantageous that information about the sender's location be broadcast, while, on the other hand, with the use of lower frequencies, the sender's whereabouts may be kept concealed (cf. Hall's notion of "adaptive silence"). Sound fills the entire space around the source and thus does not require a straight line of connection with the receiver: the signals can travel around corners and are not usually interrupted by obstacles (unlike visual displays which, for instance in the howlers, are "often impossible because of intervening foliage," as Altmann reminds us). The flexibility in frequencies, intensities, and patterns is also important in that this allows for considerable specific variations as well as, within a species, for individual variation with many shadings and emphases.

An interactionist model of the social event directs attention to an intricate problem: the message that is received is rarely—not even when the channel is maximally efficient and noise-free or when several mutually redundant channels are simultaneously used—identical with the message that was sent. How a primate extracts from the message received the message "intended" by the sender and how this animal supplies the data needed to fill in the inevitable gaps incident to communication are the sorts of questions that must have led Altmann to the notion of a stochastic process (which he then imaginatively applied in his studies of the rhesus macaque). However, it is well to remember that a communication system presents two considerably different facets when regarded from the two ends of any channel, and that the probabilistic aspect takes precedence only at the receiving end of the chain: the many ambiguities which bedevil the recipient of a message are doubtless unequivocal for the sender. Looked at in this way, the two variables that need to be disentangled, in each case, are the signals generated by the sender and the contribution of the receiver, the match between them being likely to improve depending on the density of successful feedback transactions. Finally, we must ask how the human observer—the cryptanalyst perforce located outside of the system—goes about supplying this missing link and the extent to which the process depends upon his image of the "beast in view."

REFERENCES

Burke, K. 1961. *The rhetoric of religion.* Boston: Beacon Press.

Gaul, A. T. 1952. Audio mimicry: an adjunct to color mimicry. *Psyche,* 59:82 f.

Geldard, F. A. 1960. Some neglected possibilities of communication. *Science,* 131: 1583 ff.

Hall, K. R. L., and Schaller, G. B. 1964. Tool-using behavior of the California sea otter. *J. Mammal.,* 45:287 ff.

Hockett, C. F., and Ascher, R. 1964. The human revolution, *Current Anthropol.,* 5: 135 ff.

Hoesch, W. 1961. Über ziegenhütende Bärenpaviane (*Papio ursinus ruacana* Shortridge). *Z. Tierpsychol.,* 18:297 ff.

Jakobson, R. 1964. On visual and auditory signs. *Phonetica,* 11:216 ff.

Marler, P. 1961. The logical analysis of animal communication. *J. Theoret. Biol.,* 1: 295 ff.

Morris, C. 1946. *Signs, language and behavior.* New York: Prentice-Hall.

Napier, J. 1963. The locomotor functions of hominids. In *Classification and human evolution,* ed. S. L. Washburn, pp. 178 f. Chicago: Aldine.

Pumphrey, R. J. 1951. *The origin of language.* Liverpool: The University Press.

Sapir, E. 1939. *Language.* New York: Harcourt, Brace.

Sebeok, T. A. 1965. Animal communication. *Science,* 147:1006 ff.

Sebeok, T. A., Hayes, A. S., and Bateson, M. C. 1964. (eds.) *Approaches to semiotics.* The Hague: Mouton & Co.

Zhinkin, N. An application of the theory of algorithms to the study of animal speech: methods of vocal intercommunication between monkeys. In *Acoustic behaviour of animals,* ed. R.-G. Busnel, pp. 132 ff. Amsterdam: Elsevier.

EDITOR'S COMMENTS

STUART A. ALTMANN

This symposium has raised a large number of research problems about primate behavior. Indeed, if this symposium has not provided the reader with more questions than answers, then we have failed in one of our basic tasks. It would be presumptuous to try to spell out all of these problems. It may, however, be worthwhile to try to single out several key problems that may markedly influence research. The comments will be restricted to field studies, since this is the area with which I am most familiar, but many of these comments will doubtless apply, *mutatis mutandis,* to laboratory studies.

Substantive Problems

Doubtless, our greatest need is for more and better data. To date, almost all field studies have been species-oriented, in contrast with most laboratory studies (Altmann 1965*a*). This fact, combined with the desire on the part of many primate fieldworkers to be identified with a particular species, has led to studies of a fairly wide variety of species. However, the sampling among primate groups has been uneven.

1. Tupaiidae (tree "shrews"). Tree "shrews" have not yet received the attention that they deserve. One extensive field study has recently been completed, and several colonies have been established.

2. Lemuridae, Indriidae, Daubentoniidae (lemuriform lemurs). Recent work has done much to enlighten us about the ways of lemurs. With many of these animals, however, conservation problems are now acute. Several forms may never be thoroughly studied in their native habitats.

3. Lorisidae (lorisiform lemurs) and Tarsiidae (tarsier). We have yet to see an extensive field study of any member of these groups.

4. Cebidae, Callithricidae (New World monkeys). One species of howlers has, so far, received much more attention from fieldworkers than any other New World monkey. One species each of the spider monkey, the night monkey

(dauroucoulis), and, most recently, the titi are the only other New World monkeys for which more than casual observations have been made. This large group of monkeys deserves much more attention than it has received so far, particularly because of the rapidly growing interest in them as laboratory animals.

5. Cercopithecidae (Old World monkeys). Old World monkeys have received far more attention than any other group of primates. This is not surprising in view of the excellent observational conditions that are available for some of them. Several species in this group have had sufficient preliminary, descriptive studies so that now intensive, analytic field and laboratory studies can be carried out with great profit. This valuable background information, combined with good observation conditions and the highly social disposition of these animals, insures that they will continue to be our best-known group.

Yet, even in this group, there are some striking gaps. To date, we have only the most limited data on such interesting forms as drills, mandrills, geladas, mangabeys, and the Celebes ape. The same is true for many species of macaques, despite good headway with rhesus, bonnet, and Japanese macaques. Despite the promise of several fascinating evolutionary problems, only two of the many species of *Cercopithecus* have received any serious attention from field students of behavior.

6. Pongidae (apes). The most conspicuous gaps here are the paucity of data on the pigmy chimpanzee, siamang, and orang-utan.

7. Hominidae. The one extant species is a highly aberrant form. It has been moderately well studied in several habitats.

The basic need, with every species, is for more and better data. In the area of communication, we particularly need detailed descriptions of the signals (displays, vocalizations, and so forth), of the responses to these messages, and of the correlations between the two. Other relevant contingencies, such as the age, sex, and family relations of the participants are also needed. Indeed, when we have, for each species, some of the data that are necessary to describe the structural properties of the communication process (chap. 17), we will have gone a long way toward an understanding of primate social behavior.

Practical Field Problems

The planning and execution of a primate field study involve a number of seemingly mundane details—everything from estimating costs in a grant application and selection of suitable types and reliable makes of field vehicles, binoculars, microphones, and so forth to techniques for recording complex social interactions under field conditions. For example, many students may be unaware (1) that one compact dictation recorder, the Minifon P55, has proved unreliable under field conditions and could not be adequately repaired by the manufacturer, and (2) that one of the lowest priced 16 mm movie cameras, the Beaulieu R16, is also one of the best for general field use. Such problems of logistics should not be belittled. When field recorders were used, we got about five times more data than we did using paper and pencil; increases in accuracy and richness of detail were comparably great. Any factor that will so greatly increase the productivity of fieldworkers is hardly mundane.

Ray Carpenter has repeatedly advocated that much more attention be paid to logistical problems and has recommended that one or more central bases of operation be established. Considering the present activity in primate field studies, such planning might pay off both scientifically and financially. For example, no less than nine field vehicles were purchased or rented in East Africa at various times during 1963 and 1964 for primate field studies. Only one of these vehicles was transferred, at the end of that study, to another primate field project.

The fieldworker can still get much useful information from an early guide to naturalistic studies (Scott 1950), and a brief discussion of primate field techniques has been compiled by Schaller (1965), but there is a need for a much more comprehensive, practical manual.

Problems of Data Analysis

The actual analysis that one does will, of course, depend upon one's species, data, interests, and so forth. Each type of analysis brings its own problems. There is, however, a recurrent practical problem in working with data from prolonged, intensive studies of highly social animals: that of bulk. Consider, for example, the plight of someone who comes back from an extensive primate field study with, say, 3,000 pages of field notes. One of his first tasks will be to index these data. Yet 3,000 pages with an average of about thirty index entries per page means alphabetizing 90,000 entries! It is a waste of time and effort to undertake such a task by hand. Several alphabetizing programs are available for computers. Or consider a slightly more complex task. A study of the communication process involves a study of sequences of events, fundamentally message-response sequences. A species like the rhesus has approximately one hundred twenty-three message patterns, and for each of these there is a like number of possible responses. Thus, one's basic sequential data on message-response contingencies would be tabulated in 123^2 (=15,129) categories. The problem is compounded when longer sequences are considered. I once estimated that to do this tabulation by hand for our rhesus data would have taken approximately 20 years. It was accomplished in far less time with the aid of a computer.

Electronic data processing may be useful in problems involving either great complexity or great bulk. Very often, the primate field student's problem will be of the latter kind. For many purposes, an ordinary punch-card sorter, which anyone can learn to use in an hour, or a desk calculator is all that is needed; other problems may require a computer. A rule of thumb in our computing center is that if a task will take you more than 3 or 4 days of steady work on a desk calculator, it pays to look into the use of computers.

To those who come into primate field studies from disciplines with a long tradition of analytic research, the uses and abuses of electronic data processing will be familiar. But for those from other fields, the whole idea may seem an anathema. Yet, to try to compete with a machine in a job that the machine can do better is sheer foolishness.

Problems in Constructing and Testing Theories

As more and more adequate data on primate social behavior are becoming available, the task of relating and organizing them is growing in magnitude. Fundamentally, most of the social phenomena that we study, as well as the dynamics of the concomitant evolutionary processes, are quantitative (Altmann 1962). This is often indicated by the quasiquantitative statements that we make: male langurs are more aggressive than females; rhesus vocalizations form a multidimensional continuum; sexual selection is insignificant compared with selection for optimal biomass in the evolution of sexual dimorphism in baboons; the socioeconomic sex ratio is more variable in *Macaca fuscata* than in *Alouatta palliata;* dominance hierarchies among females are linear; subordinate adult males always lead baboon progressions; and so forth.

Present field and laboratory studies are producing a surfeit of such generalizations—too often unaccompanied by any supporting evidence, or perhaps by a few well-chosen illustrations quoted directly from field notebooks. Part of the task is to single out those hypotheses whose substantiation and testing are of particular importance. In a research area such as this, in which generalizations sometimes seem to outnumber published data, this is a particularly difficult task.

One of the major contrasts between research in much of physiology and that in behavioral research on highly social animals is the accessability of data. The experience of a fieldworker, watching the complex of interactions in a society of primates would be comparable, say, to a neurophysiologist sitting atop the thalamus, his toes dangling in the cerebrospinal fluid as he watches the interplay of impulses shunting to-and-fro. With many primates, the problem is not how to get data, but which data to select and what to do with them. After many months of exposure to an active, readily observed population of primates, there is a feeling that one knows fairly well how the social system works. And indeed, one often does. But discovery should not be confused with justification. It is all too easy for us to say that the grimace serves to suppress aggression, that tail postures of macaques are status indicators, or that gamboling is a metamessage and never provide the evidence upon which such hypotheses are based.

This leads immediately to another problem. Judging by current literature from field studies, it has not always been obvious how such evidence can be supplied or just how our hypotheses can be tested. Indeed, in many cases, it is not obvious just what the null hypothesis is. For example, in studying dominance relations for which data are obtained by unbiased sampling and can be summarized in the form of a square matrix with undefined main diagonal, one can usually make a strong case against the weak null hypothesis that agonism can be directed with equal probability in either direction for all possible pairs. In such a case, each defined cell of the matrix would be equiprobable. But a stronger null hypothesis is that the results are just what one would expect, considering the differences with which each individual monkey participates in agonistic situations (see Goodman 1963 for the construction of this null hypothesis from the matrix marginals).

As John Kemeny (1959) has pointed out, the social sciences will probably serve in the future, as the physical sciences have in the past, to stimulate the development of mathematics. While the behavioral sciences have yet to trigger anything comparable to the development of the calculus, primate field studies have already turned up one unsolved mathematical problem, that of estimating repertoire size (Altmann 1965*b*). Another problem, that of accounting for the distribution of group sizes in a population, has existed in implicit form since Carpenter's first census of howler monkey groups on Barro Colorado Island (Carpenter 1934). But work toward a solution has taken place only in the last few years, stimulated, however, by work on group sizes in another primate, *Homo sapiens* (Goodman 1964). To date, these models have not been applied to populations of nonhuman primates or, to my knowledge, of any nonhuman animal.

I do not mean to imply that theory construction in this area is at present being held up pending work in pure mathematics but only to emphasize that work will be needed at every level. No, on the contrary, in many cases the necessary mathematical work has already been carried out. Much work in linear algebra, for example, may be directly relevant to the description and analysis of agonistic relations (Glanzer and Glaser 1959, Rapoport 1950). The basic concept of the stochastic analysis of communication took approximately 8 years from its formulation by Shannon (1948) for sequences of messages until it was applied to message-response sequences in a primate field study, in our 1956 study of rhesus. It has also been applied, in at least rudimentary form, to the behavior of several nonprimate animals: bitterlings (Wiepkema 1961), mice (Grant 1963), guppies (Baerends, Brouwer, and Waterbolk 1955), ducks (Dane and van der Kloot 1964), fiddler crabs (Hazlett and Bossert 1965), glandulocaudine fishes (Nelson 1964) and pigeons (Fabricius and Jansson 1963). Nevertheless, stochastic analysis is not a panacea for the study of animals' communication. There are many communicative relations that simply cannot be represented in this manner (Altmann 1965 and chap. 17 of this volume).

There is a certain danger, in the present rash of primate field work, for empirical work to progress unreasonably faster than the theoretical. In an area such as this, in which the empirical work is so difficult, expensive, and time-consuming, and may, for some vanishing species, soon be impossible to replicate, this could be most unfortunate. The construction of adequate theories will generally depend upon a corpus of data that fulfil certain minimum requirements, for example, sample size, unbiased sampling, and so forth. The testing of these theories will often require specified observations.

What could be more discouraging than to spend much time, money, and effort in an arduous and prolonged study only to discover, subsequently, that some readily available and crucial type of data had not been gathered, or that some simple, systematic sampling technique might have enabled you to test some of your hypotheses? Of course, these are risks that will always be present, but they can be minimized by some cognizance of what we intend to do with our data. The moral is obvious: the empirical and theoretical work should proceed in parallel. This does not imply that everyone must be equally adept at both. Doubtless, a certain natural division of labor will take place. But if the empirical scientist expects models that are relevant to his situation, he must, in turn, provide adequate data for the construction and testing of these models.

Field and Laboratory

The traditional view of the distinction between field and laboratory studies is that field studies, by their very nature, must always be descriptive, inexact, nonanalytic—at best a source of ideas—and that only within the context of a laboratory setting can one make the kind of controlled, precise, reliable, quantitative studies that are required to test hypotheses.

If Carpenter's pioneering research (Carpenter 1942) did not dispel this myth, then it certainly should be by several recent field studies. The misunderstanding stems, in part, from the fact that the initial phase in the study of a new species is inevitably descriptive, and that for many species of primates we are still very much in this initial phase. Beyond that, however, there seems to be a basic misunderstanding both about the role of controlled conditions in behavioral research and about what can be accomplished in the field. All of the cherished criteria of scientific research—representative and adequate sampling, reliability, replicability, and so forth—can be met under appropriate field conditions.

I do not believe that there is any fundamental difference between field and laboratory studies. The differences that there are lie not in their logic but in their logistics. In particular, very large phenomena at higher levels of biological and social organization can usually be best studied under field conditions, whereas certain very small processes, and those requiring manipulations or variations that one cannot readily produce, or with profit wait for nature to produce under field conditions, may best be pursued in a laboratory context. For example, one cannot study home range utilization, intergroup relations, or responses to predators in a laboratory unless one's laboratory is, in fact, a field. Likewise, evaluation of the relative role of various selective forces must be carried out in the natural habitat. On the other hand, it is very difficult under field conditions to be able consistently to observe the details of, say, each mother-infant relation or to observe all the details that one would like of certain visual and olfactory displays. Surrogate mothers would have to be much more mobile to keep up with a free-ranging group, and they would have to be made much sexier before the males of such a group would accept them.

Even this distinction according to levels of organization shows signs of breaking down. The type of remote telemetering that Robinson (chap. 9) describes may soon enable us to study the neurological correlates of the behavior of free-living primates in their natural habitats. And telestimulation has a great potential for testing a wide variety of theories about social behavior by enabling us to produce contrafactual conditionals: we would be able to ask questions of the form "How would they behave if such-and-such were not the case?" Beyond that, there is no reason why an extension of this technique could not enable us to measure, say, the tension of various facial muscles during facial expressions or the resonance in various air sacs during vocalizations.

Such combinations of techniques that until now have been restricted either to the laboratory or to the field may often be of great value. As another example, there is now enough knowledge of serum genetics and of population dynamics in some species of Old World monkeys that one could undertake a highly illuminating study of the relation between social organization and population genetics. Although the gene loci that are now used in serum genetics may have only limited linkage with behavioral processes, their geographic distribution, within and between social groups, might reveal much about the breeding structure of the population.

Doubtless the most productive relation between field and laboratory studies of primates will be one of mutualism, involving a repeated reciprocation of ideas, techniques, and evidence. According to the traditional view, the laboratory is the place where hypotheses from field work are tested, but the converse may also be true. What is required in either case is that the hypothesis to be tested be made sufficiently explicit. That this can be done without in any way compromising the fieldworker's traditional inclination to gather a wide variety of data within a broad, ecological and evolutionary framework has now been demonstrated.

REFERENCES

Altmann, Stuart A. 1962. Primate social behavior. *Current Anthropol.* 3:100.

———. 1965*a*. Primate behavior in review. *Science* 150:1440–42.

———. 1965*b*. Sociobiology of rhesus monkyes. II. Stochastics of social communication. *J. Theoret. Biol.* 8:490–522.

Baerends, G. P., Brouwer, R., and Waterbolk, H. Tj. 1955. Ethological studies on

Lebistes reticulatus (Peters): An analysis of the male courtship pattern. *Behaviour* 8:249–334.

Carpenter, C. R. 1934. A field study of the behavior and social relations of howling monkeys (*Alouatta palliata*). *Comp. Psychol. Monogr.* 10(2):1–168.

———. 1942. Sexual behavior of free ranging rhesus monkeys (*Macaca mulatta*). I. Specimens, procedures and behavioral characteristics of estrus. II. Periodicity of estrus, homosexual, autoerotic and non-conformist behavior. *J. Comp. Psychol.* 33:113–42, 143–62.

Dane, B. and van der Kloot, W. G. 1964. An analysis of the display of the Goldeneye duck (*Bucephala clangula* [*L.*]). *Behaviour* 22:282–328.

Fabricius, Eric and Jansson, Ann-Mari. 1963. Laboratory observations on the reproductive behaviour of the pigeon. . . . *Animal Behav.* 11:534–47.

Glanzer, Murray and Glaser, Robert. 1959. Techniques for the study of group structure and behavior: I. Analysis of structure. *Psychol. Bull.* 56:317–32.

Goodman, Leo A. 1963. Statistical methods for the preliminary analysis of transaction flow. *Econometrica* 31:197–208.

———. 1964. Mathematical methods for the study of systems of groups. *Am. J. Sociol.* 52:170–92.

Grant, E. C. 1963. An analysis of the social behaviour of the male laboratory rat. *Behaviour* 21:260–81.

Hazlett, Brian A. and Bossert, Wm. H. 1965. A statistical analysis of the aggressive communications systems of some hermit crabs. *Animal Behav.* 13:357–73.

Kemeny, John G. 1959. Mathematics without numbers. *Daedalus* 88:577–91.

Nelson, Keith. 1964. The temporal patterning of courtship behaviour in the glandulocaudine fishes (Ostariophysi, Characidae). *Behaviour* 24:90–146.

Rapoport, A. 1950. Outline of a mathematical theory of peck right. *Biometrics* 6:330–41.

Schaller, George B. 1965. Field procedures. In *Primate behavior: field studies of monkeys and apes,* ed. I. DeVore, pp. 623–29. New York: Holt, Rinehart and Winston.

Scott, J. P. (ed.) 1950. Methodology and techniques for the study of animal societies. *Ann. N.Y. Acad. Sci.* 51 (Art. 6): 1001–1122.

Shannon, C. E. 1948. A mathematical theory of communication. *Bell Syst. Tech. J.* 27:379–423, 623–56.

Wiepkema, P. R. 1961. An ethological analysis of the reproductive behaviour of the bitterling (*Rhodeus amarus* Bloch). *Arch. Neerl. Zool.* 14:103–99.

Index